Electric Utility Resource Planning

In 2012, using easy-to-understand text and examples, the first edition of this book explained how electric utilities "work," and how they plan (or should plan) for the future, by:

- "Creating" a hypothetical electric utility
- Explaining how (and why) this utility will operate its system of generating units
- Guiding readers through a planning analysis for the utility, examining various resource options (solar, new gas-fueled generation, and conservation)
- Introducing four *Fundamental Principles of Resource Planning* that should guide utilities as they plan for the future

The first edition material, with significant updates, now appears as Part I of the second edition of this book. Part II of this book then presents six all-new chapters that address the challenges (and opportunities) of moving toward a zero-carbon future. Using the same hypothetical utility, with its new goal to utilize solar and batteries to serve 100% of its customers' energy with zero carbon emissions by a future "target" year, Part II of this book addresses many subjects, including:

- The enormous amount of MW of new solar and batteries the utility will need to add
- Why certain characteristics of new solar and battery additions change as increasing amounts of these resources are added
- In the years prior to achieving its zero-carbon goal, how the hourly operation of the utility's existing fossil-fueled generators, plus the new solar, will change (and why the stability of the transmission grid will be challenged)

With this second edition, author Dr. Steven Sim again applies the experience and insights he gained from more than 30 years of resource planning for Florida Power & Light (FPL). As one of the largest electric utilities in the United States, FPL has faced a multitude of resource planning challenges, including how to get to zero carbon. During this time period, Dr. Sim performed and directed thousands of analyses designed to address these challenges. He also served as an expert witness in dozens of regulatory hearings, addressing both the economics of resource options and the non-economic impacts (air emissions, system reliability, fuel usage, etc.) associated with these options.

Electric Utility Resource Planning

Economics, Reliability, and Decision-Making

Second Edition

Steven Sim

CRC Press
Taylor & Francis Group
Boca Raton London New York

CRC Press is an imprint of the
Taylor & Francis Group, an **informa** business

Second edition published [2024]
by CRC Press
2385 Executive Center Drive, Suite 320, Boca Raton, FL 33431

and by CRC Press
4 Park Square, Milton Park, Abingdon, Oxon, OX14 4RN

CRC Press is an imprint of Taylor & Francis Group, LLC

© 2024 Steven Sim

First edition published by CRC Press 2012

ISBN: 9781032294193 (hbk)
ISBN: 9781032294223 (pbk)
ISBN: 9781003301509 (ebk)

DOI: 10.1201/9781003301509

Typeset in Times New Roman
by KnowledgeWorks Global Ltd.

Contents

Part II   Moving toward Zero Carbon

Acknowledgments

I view the 40-plus years I have spent working for Florida Power & Light (FPL) as an ongoing, and very enjoyable, learning process that never stopped. The last 30 of those years have been spent performing and directing resource planning analyses for FPL and FPL's parent company, NextEra Energy.

This learning process has been aided, both directly and indirectly, by many people. Each of the individuals that are mentioned below assisted in these analyses in various ways. Some directly shared their knowledge and experience. Others suggested that they tended to look at a specific issue or analysis from a different perspective than I had taken and encouraged me to examine that different perspective as well. Other people asked probing questions that forced a deeper look into analyses that had been performed and the conclusions that had been drawn from those analyses. Still others led me to consider how the results of the analyses might be better explained and presented. All of this input was very helpful.

Regardless of the manner in which their contributions were provided, I am very grateful for the contributions all of the folks (whose names are about to be mentioned) have made to my ongoing education regarding electric utility planning and operation. These individuals have certainly helped me become a more knowledgeable resource planner than would otherwise have been the case. And, unbeknownst to them (and me) at the time, they have helped make this a better book. The names of these folks are presented in no particular order.

I shall start by thanking some outstanding folks I have had the privilege to work directly for at FPL. These include Rene Silva, Sam Waters, Nelson Hawk, Enrique Hugues, Rosemary Morley, Bob Barrett, and Scott Bores. This is a group of exceedingly bright and talented people.

The list of folks I have had the privilege to work with at FPL in resource planning evaluations is a lengthy one. This list includes (in no particular order): Dennis Brandt, Daisy Iglesias, Juan Enjamio, Rick Hevia, Jeff Benson, Severino Lopez, John Scott, Richard Brown, Andrew Whitley, Alex Demory, Remy Cain, Sydney Shivers, Kelly Bradley, Sklyer Shatkin, Raul Montenegro, Tim Wehnes, Leo Green, Anita Sharma, Dan Haywood, Richard Feldman, Jun Park, John Gnecco, Wendell Payne, Hector Sanchez, Bob Schoneck, Kiko Barredo, Frank Prieto, Gerry Yupp, Heather Stubblefield, John Hampp, Ray Butts, Gene Ungar, and Ian Nichols ("on loan" from the Land Down Under). I am sure I learned a lot more from each of you than vice versa.

A very special "thank you" goes to two other co-workers, Sharon Fischer, and Laura Black, who assisted greatly in the preparation of certain figures that appear in this book whenever the author had (yet another) "stupid author" question.

In the course of working on a large number of regulatory dockets, most involving the Florida Public Service Commission (FPSC), certain FPSC Commissioners and Staff members stand out as having repeatedly asked important and interesting questions, thus leading me over time to improve the quality of both the analyses I presented on behalf of FPL and how the analysis results were presented. Those individuals that most readily come to mind in this regard are Commissioners Susan

Clark, Terry Deason, and Joe Cresse. Among the FPSC Staff members, Jim Dean and Tom Ballinger have been particularly impressive regarding both the breadth and depth of the questions they have asked in numerous FPSC dockets. Needless to say, these questions by the Commissioners and FPSC Staff members represented valuable input in my ongoing education in resource planning. (However, I can remember a time or two when I was on the witness stand when this input, delivered in the form of difficult questions, was not always immediately "treasured.")

I have also been privileged to work with some outstanding attorneys who helped me prepare for numerous regulatory filings and hearings. Their questions and suggestions regarding many testimonies and analyses have also led to significant improvements in both the analyses themselves and in how the analysis results are presented. These improvements have also led to making this a better book. These outstanding barristers include Charles Guyton, Matt Childs, Bonnie Davis, Bryan Anderson, Jessica Cano, John Butler, Bryan Anderson, Will Cox, Kevin Donaldson, Chris Wright, Ken Rubin, John Burnett, Ken Hoffman, and Wade Litchfield. Thanks for all of the help and suggestions. (And, just to make sure that I state—in writing—the answer to the question you asked me on the first day we worked together on draft testimony: "Yes, Bonnie Davis, I did go to college.")

I have also benefited greatly from discussions over the years with resource planners from other electric utilities. Some of these individuals include the following: Glen Snider (Duke Energy Carolinas), Leon Howell and Zac Hagler (Oklahoma Gas & Electric), Ben Borsch (Duke Energy Florida), John Torpey (American Electric Power), Scott Park (Duke Energy Midwest), Joe Lynch (SCANA), Jose Aponte (Tampa Electric Company), Jack Barrar (PEPCO), and last, but not least, John Kelley, Howard Smith, Chris Habib, Jeff Grub, Theron Furr, David Schmidt, and Joel Dison (who have worked for various companies within the Southern Company system).

Although their names have been mentioned above, an extra special "thank you" to Charles Guyton, Jim Dean, Will Cox, and Hector Sanchez. Messrs. Guyton and Dean agreed to forego their pressing retirement activities to review drafts of the new Part II chapters. Will, attorney extraordinaire for FPL and NextEra Energy, donated his valuable working time to also review the Part II chapters. Hector, the chief system operator for the FPL system (as well as for the grid for most, if not all of the rest of Florida), also donated his time to review the Chapter 13 material. The comments, questions, and suggestions from all of these individuals were extremely helpful and very much appreciated. And I am very privileged to call them friends.

Finally, I owe a large debt of gratitude to my lovely and talented wife, Sue (a.k.a. "the Goddess"), for her patience and understanding as I spent time in the evenings and on weekends trying to get my thoughts for both editions of this book down on paper. (And, "yes, dear," it is now time to take you on that vacation I promised.)

To the aforementioned list of folks, and to the numerous other folks whose names I have unintentionally omitted, one more set of thanks is called for. Not only have I found the ever-evolving field of electric utility resource planning to be constantly challenging, but also I found the work to be a *lot* of fun because of these challenges and the great people I have been able to work with. I have been very fortunate in both of these respects. One could not ask for more.*

---

* OK. One *could* ask for more, but the conclusion of a book writing effort is no time to get greedy.

Author

Dr. Steven Sim has worked in the field of energy analysis since the mid-1970s. He graduated from the University of Miami (Florida) with bachelor's and master's degrees in mathematics in 1973 and 1975, respectively. He then earned a doctorate in environmental science and engineering from the University of California at Los Angeles (UCLA) in 1979 with an emphasis on energy.

During his doctoral work, Dr. Sim also completed an internship of approximately a year and a half at the Florida Solar Energy Center (FSEC), a research arm of Florida's state university system located in Merritt Island, Florida, near Cape Canaveral. His work at FSEC involved an examination of consumers' experience with solar water heaters and projections of the potential for renewable energy in the Southeastern United States.

In 1979, Dr. Sim joined Florida Power & Light Company (FPL), a subsidiary of NextEra Energy, Inc. FPL is one of the largest electric utilities in the United States. In the more than 40 years from 1979 to the time the second edition of this book is being written, FPL has experienced tremendous population growth in the geographic area it serves. Annual average growth has been roughly 74,000 net new customer accounts per year during that period. With each customer account representing approximately 2.5 people, this represents a growth of approximately 185,000 more people per year that FPL needed to serve.

Therefore, the growth that FPL had to keep up with was roughly equivalent to 1 million new people every 5.4 years. Few, if any, other electric utilities have had to face the challenges inherent in meeting continued growth of this magnitude over such an extended time. This extraordinary growth, combined with the planning issues that all utilities face, including changing environmental regulations, fuel decisions, modification/retirement of existing generating units, and changes in generating technologies, ensured that FPL's planning efforts have addressed a wide variety of resource planning challenges.

During that time, Dr. Sim has had several roles in regard to FPL's resource planning efforts. He spent approximately the first 10 years of his career at FPL designing a number of FPL's demand side management (DSM) programs. One of these, the Passive Home Program, earned a U.S. Department of Energy (DOE) award for innovation. Among the numerous other DSM programs he either designed or co-designed were the Conservation Water Heating Program (that featured heat pumps and solar water heaters) and one of the nation's most successful load management programs (FPL's residential load control program, known as the On Call Program, that has approximately 800,000 participating customers).

In the course of designing these DSM programs, he became very interested in gaining a better understanding of how DSM programs affect FPL's entire utility system. Of particular interest were how FPL's DSM programs impacted the actual operation of FPL's generation, transmission, and distribution systems, including impacts on air emissions for the FPL system as a whole, and fuel usage by all of the power plants in FPL's system.

This interest led to him joining FPL's System Planning Department in 1991. This department (since renamed the Integrated Resource Planning Department) is charged with determining when new resources should be added to the FPL system, the magnitude of the needed new resources, and what the best resources are for FPL to add to meet the continued growth in customers (and their demand for electricity), changing environmental laws, and other regulatory requirements.

Dr. Sim has found this work to be challenging, continually evolving, and always interesting. Over this time, Dr. Sim has performed, directed, and/or collaborated on many thousands of analyses designed to examine how FPL can best serve its customers given these changing circumstances, laws, and regulations. In addition to these analyses, Dr. Sim has also served as an FPL witness in dozens of hearings before the Florida Public Service Commission (FPSC) and in hearings/meetings with other Florida governmental agencies and organizations. A partial list of the subjects addressed in these hearings included the following: the economic and non-economic impacts on the FPL system of proposed new resource additions, including nuclear, coal, natural gas, solar, battery storage, and conversion of existing fossil-fueled generating units to hydrogen-fueled units; the economic and non-economic impacts on the FPL system of FPL's DSM programs that are alternatives to new power plants; analyses of generation additions versus transmission additions; the air emission and fuel use impacts on the FPL system from each of these resource options; the reliability of the utility system; and the planning processes that can be used by a utility to plan for new resources.

In these numerous regulatory hearings, Dr. Sim has provided both written and oral testimony. In so doing, he has had the opportunity to respond to a wide variety of inquiries from regulators, administrative law judges, environmental organizations, fuel suppliers, etc., regarding a host of issues. Dr. Sim has also participated in various collaborative efforts with other electric utility organizations and individual utilities, including advisory groups for the international Electric Power Research Institute (EPRI). In addition, he has served on the Southeast Electric Exchange's (SEE) Integrated Resource Planning Task Force and the resource planning group of the Florida Reliability Coordinating Council (formerly the Florida Coordinating Group). He has also served as chair for the latter two groups.

These activities have further broadened his perspective of the challenges facing other electric utilities across the country and how these challenges are being addressed. Over the years, he has spoken at a number of electric utility conferences, both national and international. At one of those conferences, the *6th National Demand-Side Management Conference*, Dr. Sim was awarded the Outstanding Speaker Award. He has also enjoyed speaking with, and to, a variety of individuals and groups that were interested in energy issues, particularly those issues that directly involve electric utilities. He has seen this interest grow in recent years as issues such as energy prices, climate change, renewable energy, and federal/state energy policy are regularly in the news.

In light of this growing public interest, Dr. Sim wrote the first edition of this book in 2012 to share the insight he had gained in over 30 years of directly analyzing energy-related issues for one of the biggest electric utilities in the United States, and in collaborating with others in the electric utility industry regarding the challenges they have faced.

In the decade since 2012 and the publishing of the first edition of this book, there have been many changes in the electric utility industry that FPL and other utilities have faced in their resource planning work. As an example, one of Dr. Sim's major projects was to direct the resource planning analyses to determine what it would take for FPL to get to zero-carbon operation in various target years. Based in large part on the results of these analyses, NextEra Energy made an announcement in June 2022 of a "Real Zero" goal of getting to zero-carbon emissions by 2045.

The utility industry changes that have occurred since the first edition of this book was published, and especially the analyses of what it takes to achieve a zero-carbon goal, led to the writing of the second edition of this book.

Dr. Sim retired from FPL in October 2022. He continues to teach a resource planning course for FPL and other NextEra Energy companies. His hope is that, through the second edition of this book (and/or the resource planning course sessions), he will be able to convey insight he has gained in over 40 years of various aspects of resource planning work for such a large, rapidly growing electric utility. And, as a result, that this insight will help to facilitate discussions on a variety of energy-related issues, leading to better informed decisions by utilities, regulators, and legislators. Finally, he hopes that you will not only find this book informative, but also enjoyable to read as well.

Part I

The Fundamentals of Electric Utility Resource Planning

1 Introduction

WHY WRITE THIS BOOK?

In my 40-plus year career at Florida Power & Light Company (FPL), I have been fortunate to have many opportunities to interact with numerous individuals, both on a one-to-one basis and in group situations, in discussions that address a variety of issues pertaining to electric utilities. These individuals generally have two things in common. First, they are not electric utility resource planners. Second, despite this unfortunate choice in their life's work,[1] they have a genuine interest in energy issues, particularly those relating to electric utilities.

These discussions frequently focus on the issue of what "resource options" FPL was considering, or had already chosen, in order to continue to provide electricity reliably and at a reasonable cost. These resource options could include any (or all) of the following:

- New electric generators fueled by natural gas, coal, uranium, or solar energy;
- Purchases of power and/or energy from new or existing electric generators owned by a neighboring utility or another non-utility party;
- Lowering (or increasing) customers' demand for, and usage of, electricity by implementing utility demand side management (DSM) programs; and
- Other alternatives to the resource options listed above including, but not limited to, the following: battery/energy storage, new transmission facilities, modernizing existing generators to make them more efficient and capable of producing more power, and converting existing fossil-fueled generators to run on hydrogen.

In regard to the discussions I have had regarding these resource options, and the decisions that were being made about them, most of the discussions were characterized by what I will term a sincere intellectual curiosity. In such cases, the individuals who were discussing the topic with me seemed truly interested in learning what specific resource options had been considered and why a particular decision was made.

However, in some cases, the discussions could be characterized as outright disbelief, and/or open hostility, being shown by the other participant who was certain that he/she already knew what the "Correct Decision" was. More than a few of these conversations were similar to the following hypothetical exchange:

Questioner: "Did the utility really decide to go with Option A (*insert appropriate name of new generator, new energy conservation program, etc.*)

---

[1] ☺

DOI: 10.1201/9781003301509-2

> instead of Option B *(insert the name of an alternative favored by the Questioner)?"*

Me: "Yes."

Questioner: "What were you _____[2] thinking of? It is obvious that the decision should have been Option B!" *(Appropriate facial expressions, tone of voice, and/or hand gestures may also be supplied by the reader's imagination.)*

However, contrary to this hypothetical exchange, most of the discussions I have had have been very pleasant. One such exchange took place early in my utility career at a League of Women Voters meeting in South Florida that was sponsoring a panel discussion regarding energy-related issues. I was invited to represent FPL as a panelist. I singled out this particular discussion because it influenced this book in two ways.

Both of these ways are related to a question I was asked at one point in the proceedings. During the discussions, one issue that emerged was whether an individual could generate one's own electricity rather than purchase it from the utility.[3] The questioning up to that point had actually been a bit intense when one lady in the audience raised her hand and then asked: *"What do I have to do to become an electric utility?"*

Blindsided by this unexpected (and somewhat odd) question, and attempting to buy time until I could think of how best to respond to her question, I blurted out: "Well, first you've got to get a smokestack."

Fortunately for me, the response resulted in a few giggles that seemed to lighten the mood considerably. The discussion then continued in a much more relaxed atmosphere and actually proved to be quite productive. Grateful for that fact, I have remembered that exchange and actually used my stumbling response to her question as the initial working title for this book (until sage advice from reviewers convinced me to use a more "proper" title).

The second, and much more important, item I took away from this particular meeting was a perception of *why* a number of questions were being asked. During the discussion, I realized, perhaps for the first time, that many (if not most) people have a number of misconceptions regarding: (i) how electric utilities actually operate, and (ii) how utilities plan for the future. Since that time, this perception has been reinforced in virtually every subsequent meeting I have participated in that discussed utility-related energy issues.

It was apparent that a basic understanding of how a utility actually operates, particularly in regard to the utility's existing generating units, would be very useful if one is to understand how utilities plan for the future. With that in mind, Chapter 2 of this book is designed to provide this basic understanding of how a utility actually operates its many generating units.

---

[2] Please fill in the blank by inserting an appropriate term such as "people," "doofuses," or "buffalo heads."

[3] I seem to recall that the phrase "Evil Empires" came up in reference to electric utilities in the course of discussing this particular issue.

This basic understanding of utility operation of its generating units in Chapter 2 sets the stage for the remaining chapters of the book. In those chapters, the other subject of the book is discussed: how do electric utilities plan for the future? We will refer to this issue as "resource planning" for a utility.

WHO IS THIS BOOK WRITTEN FOR?

The information presented in this book is primarily intended for two distinct audiences which I view as equally important. The first audience comprises a wide variety of individuals who do not work for electric utilities, but who are interested in energy issues, especially issues related to energy, electric utilities, and/or environmental issues. From my experience, a number of different types of individuals would be included in this audience. These include, but are not limited to, the following:

- Individuals working in municipal, state, and federal governmental organizations, including those who regulate various aspects of electric utility operations and planning;
- State and federal legislators who make laws that impact electric utilities and their customers.
- Journalists who report on utility and energy issues;
- Environmentalists, especially representatives of numerous environmental organizations that frequently intervene in utility regulatory hearings;
- Educators and students involved in energy and environmental education; and
- Others who simply wish to know more about energy issues.

A better understanding of how electric utilities actually operate and plan their systems, and just as important, why they do so, should help to eliminate—or at least minimize—numerous misconceptions that get in the way of meaningful discussions about electric utilities and their plans for the future. In turn, this understanding should result in more informed discussions that lead to better decisions regarding energy issues.

The second audience this book is intended for is a variety of individuals who work for electric utilities in areas other than resource planning. I believe the book's information will prove useful to people who have other jobs in utilities, including, but not limited to: executives, attorneys, designers of utility DSM programs, individuals working in the power generation, transmission, and distribution business units, individuals who operate a utility's fleet of generating units, individuals working in the power purchase business units, environmental personnel, individuals working in the regulatory affairs business units, and communications personnel.

This utility audience also includes individuals in various utilities around the United States (as discussed a bit more in Chapter 2) that have turned their attention back to the resource planning concepts discussed in this book after being away from them for a decade or more due to decisions in their respective states to change the fundamental structure of the electric utility business. Beginning in the year 2000, I began receiving a number of calls from individuals in such utilities.

The conversations often proceed with my cell phone ringing at an inconvenient time with the following type of exchange:

Caller: "Help! We used to do utility resource planning like you guys do, but we have been away from it so long that we don't remember how to do it. Can you help us?"

Me: "Yes. I will be glad to help, but right now I'm on the golf course lining up my putt for a score of 8 on this par 3 hole. I will call you back when this athletic crisis has passed."

If this book helps minimize such tragic distractions to my golf game, then I may benefit as much from this book as I hope the two types of intended audiences will.

AN OVERVIEW OF THE BOOK'S FIRST EDITION (*i.e.*, PART I OF THIS BOOK)

The first edition of this book was published in 2012. That edition consisted of an introductory Chapter 1, followed by eight additional chapters and seven appendices. Most of that material has been retained (with a number of updates) and appears as Part I of the second edition of the book. As before, Part I begins with this introductory Chapter 1.

Chapter 2 lays out the foundation for much of the rest of the book by explaining certain fundamental concepts of electric utilities. The difference between "regulated" and "unregulated" utilities is discussed, along with an explanation of why this book focuses on regulated utilities. Next, the major operating areas of an electric utility, including those areas that are the primary focus of this book, are presented. The different types of electric generating units (*i.e.*, power plants) that a utility typically utilizes are then presented along with an explanation of how a utility decides which generating units to operate at a given time. The chapter concludes by "creating" a hypothetical utility system. This hypothetical utility system will then be used throughout the remainder of the book to illustrate a variety of topics using resource planning analyses for "our" utility. The first of four "Fundamental Principles" of resource planning is introduced in this chapter (and the remaining three Fundamental Principles are introduced later in Part I of this book).

Chapter 3 introduces three key questions that utility resource planning must continually answer: (i) when (what year) does a utility need to add new resources, (ii) what is the magnitude of the needed new resources, and (iii) what is the best resource option(s) for the utility's customers. It also introduced different types of analytical approaches that are used to answer these three key questions.

Chapter 4 then utilizes the hypothetical utility system created in Chapter 2 to illustrate how analytical approaches designed to answer the first and second of the three key questions are actually implemented. The answers to these two questions are then used in subsequent chapters as we develop answers to the more complex (and, for most people, more interesting) third question.

Chapters 5 through 7 illustrate how the third key question is answered using our hypothetical utility system and two basic types of resource options. These resource options are representative of the choices a utility has to meet its future resource needs and ensure that it can maintain a reliable and economic supply of electricity.

Chapter 5 focuses on one of the two basic types of resource options. This type of resource option is a new electric generating unit (or "Supply" option). The list of Supply options that could potentially be selected by the hypothetical utility to meet its resource needs at the time this book is written includes new generating units that could be fueled by natural gas, renewable energy, and, to a lesser degree, nuclear energy. Battery/energy storage is also a viable option. In this chapter, we select a few generating unit examples and perform the economic analyses to see which of these would be the best economic choice for our hypothetical utility's customers if it chose to meet its resource needs with a new Supply option.

The focus of Chapter 6 is on the second basic type of resource option. This type of resource option does not supply additional electricity but instead changes the demand that a utility's customers have for electricity at the utility's peak hour of electrical demand and the amount of energy that customers use over the course of an entire year. These options are traditionally referred to as DSM options and the options most frequently reduce the demand for electricity. Similar to the approach used in Chapter 5, we select a few DSM examples and perform economic analyses to see which of these selected DSM options would be the best economic choice for our hypothetical utility's customers if it chose to meet its resource needs with a DSM option.

In Chapter 7, we bring together the information developed in the previous two chapters that separately focused on Supply options and DSM options. We now look at all of the Supply and DSM options examined in the previous two chapters to determine which resource option is really the best economic choice for our hypothetical utility. In addition, we then took a look at these resource options from non-economic perspectives for our hypothetical utility system (such as the types of fuel used and total system air emissions). Then, armed with the results from the economic and non-economic evaluations, our hypothetical utility then makes its selection of the best resource option with which to meet its resource needs.

Once we completed Chapters 3 through 7, the three key questions for the hypothetical utility have been answered. The analyses we performed to get to these answers for the hypothetical utility are intended to be both illustrative as to the importance of the four fundamental principles of resource planning that were introduced along the way, as well as being useful in showing how these fundamental principles are applied. However, we were not quite finished yet.

Chapter 8 discusses why the results of the preceding analyses could easily have been much different. The focus of this chapter is on a variety of "constraints" that may apply to a utility in its resource planning work. We discuss how these constraints may significantly change the resource options that a utility selects. Furthermore, these changes may be for the better, or for the worse, from the perspective of the utility's customers.

Chapters 2 through 8 comprised the bulk of the first edition of this book.[4] The next section describes what information is conveyed in the second edition of this book.

AN OVERVIEW OF THE BOOK'S SECOND EDITION (*i.e.*, PART II OF THIS BOOK)

As previously mentioned, the first edition of this book was published in 2012. As I looked back on the book in 2022/2023, I was pleased, but not surprised, to see that the book's information stood up well over the ensuing decade. That is because the first edition of the book was based on fundamental principles of electric utility resource planning, and true fundamental principles in any field of endeavor seldom change over time.

However, my look back at the 2012 book showed that there have been a number of significant changes in the electric utility industry itself. A short list of those changes includes the following:

- Costs for one resource option evaluated in the book's first edition, solar photovoltaics (PV), have significantly decreased (as have the costs for other renewable energy-based resource options and battery storage).
- Electric vehicle (EV) adoption has significantly increased, and forecasts show this growth in EV adoption continuing. The greater projected EV use results in projected increases in both electrical demand at utility peak hours and annual electricity usage.
- Social and political pressures, plus other motivations, have resulted in many, if not most, electric utilities announcing plans/ambitions to get to net zero (or zero) carbon emissions by a certain target year.
- In 2022, the Inflation Reduction Act (IRA) established a set of significant new federal tax credits for certain resource options (such as solar, wind, and battery storage, for example) to further encourage electric utilities to select zero-carbon-emitting resource options.

As a result of these and other changes, I believed it was time to update the first edition of this book by applying the fundamental principles of resource planning, which were explained in the book's first edition, to the new "conditions" brought about by the changes mentioned above. This is done in Part II of the second edition of the book which consists of five new chapters (Chapters 9–13), a "carry-over" chapter (Chapter 14) that has been extensively updated, plus the previously mentioned appendices.

---

[4] The first edition of the book also contained a chapter that provided a summary as well as the author's opinions on a variety of resource planning-related topics. The first edition also contained a number of appendices which contained a listing of four fundamental principles of resource planning, a glossary of terms, and a handful of "mini-lessons" that provide additional information about certain topics. That material has now been moved into Part II in the second edition of the book. This material will be mentioned again later in this chapter.

Here is an overview of Part II of this book:

- Chapter 9 explains that our hypothetical utility has set a goal of getting to zero carbon by a certain target year and plans to do so with a combination of PV and battery resources. This chapter also provides an introduction to what will follow in each subsequent chapter of Part II.
- Chapter 10 looks at how many new megawatts (MW) of solar photovoltaics (PV) and battery storage our hypothetical utility will have to add by the target year to achieve its zero-carbon goal. Several estimation approaches for determining this are utilized in this discussion.
- Now that our utility has projected how many MW of new PV and batteries are needed to reach its zero-carbon goal, Chapter 11 focuses on PV resources. Certain characteristics of PV that resource planners must consider are discussed. The declining cost of PV over the last decade is examined first. Then the 2022 IRA's new federal tax credits for PV are discussed. As part of this discussion, we return to the resource analyses performed in Chapter 5 of Part I to show how the results for the PV resource option would change after accounting for the 2022 IRA's federal tax credits. The firm capacity values of PV for both summer and winter are discussed next. We see that the firm capacity value characteristics of PV are different in several ways than the firm capacity values for conventional generating units, and that this poses challenges for resource planning.
- Similar to how Chapter 11 focused on PV resources, Chapter 12 focuses on battery storage resources. The discussion starts by looking at projected installed costs for batteries. A characteristic that is unique to battery/energy storage is that these resources are alternately an electric load and a generation resource (and that cycle is then repeated many times). The chapter then focuses on the firm capacity value of batteries and its relationship to battery duration. A simple method with which to determine the needed duration of batteries used to meet a utility's reserve margin criterion is presented. We see that the needed duration for batteries also has cost implications.
- Chapter 13 broadens the discussion to examine more than just a resource planning perspective. This chapter looks at the ramifications that our utility's move toward a zero-carbon goal will have on the utility's system operations and transmission planning functions. A few selected days in future years are examined to see how our utility would likely operate its generation fleet with ever-increasing amounts of PV being added. This examination addresses both conventional generating units and the new PV facilities. Next, a discussion of how inverter-based resources (IBRs), such as PV, pose challenges regarding maintaining system stability that transmission planners, system operators, and resource planners will need to address.
- Chapter 14 is a "carry-over", but extended, chapter from the first edition of the book.[5] This chapter provides a summary of what we have learned in the preceding 13 chapters. It also presents opinions of mine regarding

---

[5] This appeared originally as Chapter 9 in the first edition of the book.

almost two dozen resource planning-related topics. Both the summary and the opinions have been extensively updated, in large part to address various facets associated with a movement toward zero-carbon.

The book then closes with 7 appendices. The first appendix, Appendix A, contains a list of what I call my four "Fundamental Principles" of electric utility resource planning. Each of these four principles has been previously introduced earlier at an appropriate point in the book. Appendix A contains an easy-to-reference listing of these principles. Appendix B contains a glossary of terms used in the book that are not commonly used by people who are outside of the electric utility and/or energy industries.

The remaining appendices, Appendices C through G, present five "mini-lessons" about various subjects addressed in the book. Rather than including these mini-lessons in the main body of the book, this approach is designed to accomplish a couple of objectives. First, this approach should allow a reader who may already understand the concept in question in a particular chapter to continue reading straight through the chapter without having to wade through an explanation he/she doesn't need.

Second, this approach provides an option for readers who may be unfamiliar with the concept in question. Such a reader can opt to continue reading through the chapter to understand the overall message and then go back to the appropriate appendix to delve into the concept. Alternatively, the reader could stop when the concept is introduced, go to the appropriate appendix, and then return to the main text. In either case, the appendices will (hopefully) provide a quick and easy-to-understand explanation of that concept.

ARE WE KEEPING IT SIMPLE?

To many (if not most) folks who have read to this point, the entire subject matter of electric utility operation and planning may seem complex and a bit daunting. A reader may be questioning whether the subject can be discussed in simple enough terms to make it understandable. The answer is, in a word, "yes." In addition, I hope to make the discussion interesting.

As mentioned earlier, one of the book's two intended audiences is a variety of individuals who do not work in the electric utility industry. The objective is to inform these individuals, but to do so without bogging down the discussion with unnecessary utility industry jargon or going into an unwarranted level of detail.

Therefore, the book purposefully attempts to simplify a number of complex subjects while maintaining the basic correctness of the concepts that are discussed. The simplifications used are those that have proven useful in my past discussions with various individuals and groups outside of the electric utility industry.

Individuals who are well schooled in the principles discussed in the book, particularly those folks working for electric utilities who are intimately familiar with resource planning work, will certainly recognize where these simplifications have been made. It is hoped that these folks will understand why the simplifications have been made.[6]

---

[6] If not, it is hoped that they will "get over it" with the passage of an appropriate amount of time and/or counseling.

A FEW WORDS REGARDING ASSUMPTIONS USED IN THE BOOK

When discussing how electric utilities operate, and how they plan for the future, it is most helpful to do so by means of examples. For that reason, the book is full of examples, especially example calculations in both Parts I and II of this book that involve various utility costs and other values. In order to perform those calculations, a number of assumptions needed to be made regarding the cost and operating characteristics of existing generating units and new resource options, fuel costs, environmental compliance costs, etc. Therefore, in our discussions, we shall first make a number of assumptions, and then we shall use the assumptions in a variety of calculations.

A few words about these assumptions are in order. First, the values for the assumptions used in the book are generally representative of actual and/or forecasted values that have been used by various utilities in the years immediately preceding (and during) the writing of the second edition of this book. The different individual utility forecasts varied considerably, but when the forecasts are viewed as a group, the assumption values we are using typically fall within the range of values encompassed by those forecasts. Moreover, the assumed values are typically not necessarily "tied to" any particular utility or particular year but are more general, representative values.

Second, the reality for an electric utility is that the assumptions it is using at any point in time regarding costs and operating characteristics are not only subject to change but are also usually continually changing. In real life, this necessitates that analyses are typically reworked when new assumption values become known and are accepted as improved assumption values. As a consequence of these new assumptions, the results from the new analysis may or may not agree with the previous analysis results.

Therefore, in regard to this book, virtually any assumption that has been made and used in our calculations may or may not be "accurate" at the time the book is read if one were to attempt to compare those values to current values. In fact, it is safe to say that the more years between the time this book is written and the time it is read, the more likely it is that a given assumption value used in the book will no longer reflect the then-current value.

However, this fact should not concern the reader. This is because the goal of the book is *not* to definitively demonstrate that a specific Resource Option A is better than a specific Resource Option B. Instead, the objective of the book is to explain and demonstrate *the process of how resource planning in electric utilities is (or should be) performed* in order to determine whether one resource option is better than another one. With this in mind, the assumptions used throughout the book should be viewed as basically "placeholder" values that allow us to discuss resource planning concepts and analytical approaches using quantitative examples. These concepts and analytical approaches will remain valid regardless of any subsequent change in assumption values in the future.

In other words, the focus should be on the resource planning concepts and analytical approaches that will be discussed. The assumed values are simply a necessary tool with which to discuss these concepts and analytical approaches.

A COUPLE OF DISCLAIMERS

This book can be thought of as having two types of information. The first type of information is factual information. An example of this type of information is the fact that utilities perform economic analyses. The second type of information can be considered "the author's opinion." My opinions are largely confined to Chapter 14. In regard to the second type of information, my opinions, it is prudent to offer a disclaimer or two.

The first of these is that the thoughts and opinions expressed in this book are solely mine. As such, they do not necessarily reflect the opinions of my long-term former employer, FPL, or of its parent company NextEra Energy. In addition, any errors or omissions that appear in this book are solely my responsibility.[7]

Now that we have dispensed with the obligatory disclaimers, it is time to return to the objective of understanding how electric utilities operate and plan for the future. So we will now dive into the electric utility pool, shallow end first, and paddle over to Chapter 2, where we will learn some basic facts about how electric utilities actually work.

---

[7] Unless, of course, a suitable scapegoat can be found.

2 How Does an Electric Utility Actually "Work"?

TWO "TYPES" OF ELECTRIC UTILITIES

Generally speaking, virtually all electric utilities were once structured to operate in a similar way. This utility operational structure is characterized by having all three main functions associated with electricity production and delivery all "under the same roof" of a single electric utility company. These three main functions are:

i. Generation: the production and/or procurement of electricity (the operation of the utility's own electric generating units/power plants and/or the purchase of power from other entities' power plants).
ii. Transmission: the movement of bulk quantities of electricity from these power plants throughout the utility's service area (utilizing the large transmission towers and electric lines that cross long distances); and,
iii. Distribution: the subsequent movement of electricity from these bulk transmission lines to individual customers' homes and businesses (utilizing lower voltage electrical lines that typically run along streets and into customers' homes and businesses).

In such a structure, one utility company is the customer's sole choice for providing all aspects of electric service. In this structure, the utility has a monopoly position. In exchange for this monopoly position, virtually all aspects of the utility's retail business (*i.e.*, sales to the ultimate users of electricity) are overseen by a state or local regulatory agency or other authority. (The Federal Energy Regulatory Commission, FERC, has authority over various other aspects of utility business such as transactions between utilities.)

For example, the retail business aspects of investor-owned electric utilities (*i.e.*, utilities for which you can purchase shares of its stock) are generally regulated by a state's Public Service Commission, Public Utility Commission, or an equivalent state organization. Municipal-owned utilities are often regulated by an arm of the municipal government. Certain other organizations that provide electric service, such as the Tennessee Valley Authority (TVA) and Bonneville Power Authority (BPA), are more directly tied to federal government oversight.

However, in the last decades of the twentieth century, a number of states modified the regulatory structure of investor-owned utilities in an attempt to encourage greater competition in the electric utility industry and, hopefully, lower electric rates for customers. The modifications frequently resulted in the utility breaking up, or "spinning off," one or more of the three functional areas (generation, transmission, and/or distribution) into separate, independent companies that are now not all

DOI: 10.1201/9781003301509-3

owned/controlled by the original utility company. For reasons of practicality (for example, how many sets of wires and poles do you want running alongside your street?), customers have little or no choice in regard to the distribution company whose wires connect the customer's home or business.

However, the customer, either directly or indirectly, will have a choice at least in regard to the source(s) of electricity generation that he/she is served from. In such a case, the transmission function is also regulated in a way that attempts to ensure that no one generation company has an unfair advantage over another in regard to delivering its power to the distribution company for subsequent delivery to customers' homes or businesses.

This first (or original) utility structure is one I will refer to as being a "traditional" (or "regulated") environment for utilities, while the second structure is one I will refer to as a "deregulated" environment. Although the "deregulated" term is commonly used, it is a bit of a misnomer because there is still considerable regulation in such an environment, although the regulation takes different forms.

For this book, we will be discussing utility operation and planning under the "traditional" utility structure. There are several reasons for this. First, the vast majority of the generation, transmission, and distribution facilities in operation at the time this book is written came into existence under a traditional utility structure. At the very least, a focus on the traditional utility structure will aid in understanding how decisions were made that shaped the development of electric utility systems. Second, many states in the United States continue to operate under a traditional utility regulatory structure.

Third, the basic principles discussed in this book are not only true from the perspective of a traditional utility structure, but many of these principles are also relevant in an unregulated utility structure.[1] Finally, the track record to-date of the unregulated utility structure "experiment" is, in my opinion, a mixed one at the time this book is written. After an initial move into a deregulated utility structure, a number of states and utilities have moved back—in various ways and to varying degrees—toward a regulated utility structure, and toward the type of integrated resource planning issues and analyses that are discussed in this book. (Which led to the previously mentioned cell phone call I unfortunately received on the golf course while I was lining up my putt for an 8.)

WHOSE PERSPECTIVE WILL BE TAKEN?

The discussions in this book will take the perspective of an electric utility's customer, not an electric utility's shareholder. Although the perspective of an electric utility's customer is important in both regulated and unregulated utilities, it is arguably more important in a regulated utility. In fact, resource planning in a regulated utility will likely predominately, if not exclusively, focus on the customer perspective.

---

[1] For example, the discussions regarding utility system reliability and system air emissions remain relevant to a significant degree for unregulated utility systems although this analysis is generally performed not by an individual utility alone, but by another party. Conversely, the discussion regarding the economic analyses of Supply options would be significantly different if discussing unregulated utility systems.

WHAT ASPECTS OF AN ELECTRIC UTILITY WILL WE FOCUS ON?

We just briefly introduced what is generally thought of as the three main functional areas of an electric utility in a regulated structure: generation, transmission, and distribution. In addition, there are other separate functional areas of a utility system that resource planning personnel regularly coordinate with in the course of the planning work. Examples of these other functional areas include load forecasting, fuel cost forecasting, system operations, transmission planning, and distribution planning. Each of these functional areas is a complex discipline with its own set of fundamental principles, issues, regulations, and analytical approaches.

Recognizing that resource planners typically interact with subject matter experts in each of these functional areas as part of a utility's resource planning process, this book will focus on two areas. The first of these is the electric generation area. This area encompasses all of a utility's existing electric generating units. It also encompasses new electric generating unit options that may include, at a minimum, the following: natural gas-fueled units, nuclear units, renewable energy options (such as solar and/or wind), and battery/energy storage. Furthermore, the generation area also encompasses purchases of power from these types of facilities, regardless of whether such facilities are owned by other utilities or by non-utility organizations.

The second area we will focus on is the Demand Side Management (DSM) area. The DSM resource options we will primarily address are those DSM options that lower the demand for, and the usage of, electrical energy. These DSM options are often referred to as either energy conservation (or energy efficiency) programs or as load management (or demand response) programs. Collectively, we will refer to these simply as DSM programs/options because they can affect the amount of electricity that customers demand and use.

Because the scope is on generation and DSM resources, it can be said that the book's focus is on all resource options with which a utility may meet its current and future electrical load.[2] And, as described above, these resource options can be categorized in two types of resource options: (i) Supply options (new power plants/generating units) that provide more electricity to meet the utility customers' demand for electricity, and (ii) DSM options that lower, or otherwise change, the customers' demand for electricity and their usage of electricity.

One has to start somewhere in this discussion of Supply and DSM options. Therefore, we will start with Supply options. Accordingly, only generating units will be discussed in the remainder of this chapter. The choice to focus first on generating units is beneficial because it not only allows one to understand how a utility system is operated but also, as we shall see, it is helpful in understanding how both Supply and DSM options are evaluated. Once we have laid the foundation for how Supply options are operated on a utility system, we will turn to the topic of DSM options in subsequent chapters.

---

[2] In Chapter 13, we will broaden the discussion to take a look at certain aspects of system operation and transmission planning.

TYPES OF GENERATING UNITS A UTILITY MAY HAVE

The number of generating units a utility will have may vary considerably depending upon the "size" of the utility, *i.e.*, according to how many customers the utility serves. For example, at the time the writing of the second edition of this book began, Florida Power & Light (FPL) served approximately 5 million customer accounts (*i.e.*, electric meters), or about 11 million people. It served these customer accounts and people with 80 generating units that ranged widely in size, *i.e.*, in megawatt (MW) capacity. Smaller utilities will typically have a smaller number of generating units.

Although the actual number of generating units a utility has may vary significantly from one utility to another, the *types* of existing generating units a utility has typically varies relatively little from one utility to another. For purposes of discussion, we will group the majority of existing generating units into five basic types of conventional (or non-renewable energy-driven) generating units that are in existence at the time this book is written. The five basic types of conventional generating units are:

1. *Steam-generating units fueled by natural gas (with oil as a secondary or backup fuel)*: These units are typically units that were placed in-service many years ago. The design of these units is relatively simple. Natural gas (or oil) is burned as fuel in a boiler and the heat from the combustion of the fuel is used to convert water into steam. The steam then enters a steam turbine that, in turn, drives an electric generator to produce electricity.
2. *Steam generating units fueled by coal*: These units are similar in basic principle to the previously mentioned steam generating units that burn natural gas (or oil), but these steam units burn pulverized (or fluidized) coal instead. Like their natural gas-fueled (or oil-fueled) counterparts, most of these coal-fueled steam units are older units.
3. *Gas/combustion turbine units*: These generating units also burn natural gas (or oil as a backup fuel), but the basic design of this type of generating unit is different from that of the steam units. Fuel is burned in a combustor, and the resulting hot gases rotate a turbine that, in turn, drives an electric generator. This type of generating unit is typically relatively small in size, and the generating unit can be started much more quickly than either of the steam units listed above. Older units with poor fuel efficiency are commonly referred to as gas turbines, while newer units with design improvements that result in greater fuel efficiency are commonly referred to as combustion turbines.
4. *Combined cycle units*: This type of generating unit is basically a combination of steam and combustion turbine technologies. These combined cycle units not only operate primarily on natural gas but are also capable of burning oil as a secondary or backup fuel. (These units are also capable of burning coal that has been converted into gaseous form by a separate "gasifier" facility. However, these "integrated gasification-combined cycle units" are rare.) A combined cycle unit uses one or more combustion turbines to produce electricity as described earlier. Then the heat from the gases that exhaust from the turbine is captured in what is termed a "heat recovery steam generator." This heat recovery steam generator operates similarly to

steam generating units by using the captured heat to turn water into steam. The steam is then used to produce additional amounts of electricity. As a result, combined cycle units are significantly more fuel-efficient than steam, gas turbine, or combustion turbine generating units.

5. *Nuclear units*: These units are somewhat similar to steam-generating units, but only in the sense that heat is used to produce steam by heating water. However, in nuclear units, the heat is not derived from burning fuel in a boiler. Instead, it is derived from heat that emanates from the nuclear fuel (uranium) itself.

At the time the first edition of this book was published in 2012, these five types of conventional generating units comprised the vast majority of generation in the United States. Since that time, a number of other types of generating units have emerged as mature technologies suitable for large-scale utility use. As a result, these technologies have now been incorporated by many utilities. These other types of generating units, which are primarily renewable energy-based, include solar photovoltaic (PV) facilities, solar thermal technologies that produce steam, wind turbines, and battery/ energy storage facilities. In addition, there are other types of generating units, such as hydroelectric facilities, which have been in existence for many years in specific geographic areas where such water storage/electric generation facilities are feasible.

At least three (and often four) of the five types of conventional generating units listed above are likely to be found in any given utility system, either as a generating unit owned and operated by the utility or as the originating source of electrical power that is purchased by a utility from another entity. At the time the second edition of this book is written, these five types of conventional generating units, in aggregate, also comprise the majority of the total MW of generating capacity in operation across the United States. For these reasons, our discussion in the early portions of this book will focus on these five types of conventional generating units.[3] And, as we shall see in Part II of this book, the addition of renewable energy-based generators impact how the conventional generating units must be operated. Therefore, we will start by understanding how these conventional generating units have typically been operated.

These five types of conventional generating units vary greatly in regard to the cost of constructing them, and the cost of operating and maintaining them. Both of these cost considerations are important when a utility decides which type of generating unit to build. However, once a generating unit has been built and is now part of a utility's fleet of existing generating units, the operating cost of the unit becomes of primary importance in regard to daily operation of this fleet. The cost of constructing a generating unit has now become a moot point and plays no part in deciding whether or when to operate that generating unit.

The primary "operating cost" is the cost of burning fuel to generate electricity from a particular generating unit. This primary operating cost is determined by two factors: (i) the cost of the fuel, and (ii) the efficiency with which the generating unit

---

[3] Two of the "other" types of generating units, PV and battery storage, are discussed later in both Parts I and II of this book.

burns that fuel to produce electricity. The next section of this chapter focuses on this operating cost. (Later in the book, when the subject of how utilities select new resource options, such as new conventional and/or renewable generating units, is discussed, we will return to consider the construction and maintenance costs of these generating units and how these costs are captured in the analyses.)

HOW DOES A UTILITY DECIDE WHICH GENERATING UNITS TO USE?

Because one of the basic assumptions in this book is that we are discussing electric utilities in a regulated structure, it is correct to say that utilities typically operate their generating units to meet two objectives: (i) to safely provide reliable electric service to all customers at all times (within acceptable electrical voltage limits); and (ii) to keep electric rates low by generating electricity at the lowest operating cost.[4]

I realize that the last statement is one that some readers may question, but it is accurate. Utilities in a regulated structure are constantly trying to reliably operate their generating units in a manner that minimizes the total operating costs of the units. But how is the operating cost of an individual generating unit determined? I'm glad you asked.

The operating cost of any of the previously mentioned five general types of generating units is primarily determined by a simple multiplication of two factors previously mentioned: (i) the cost of the fuel, and (ii) the efficiency with which the generating unit burns the fuel. These two factors determine the operating cost of each generating unit on a utility system. Then, once the operating cost for each generating unit on a utility system is known, that information is used to determine when each generating unit is actually used by the utility as it uses (or "dispatches" in utility parlance) its fleet of generating units in the order of lowest operating cost (these are typically dispatched first) to highest operating cost (these are typically dispatched last).

The first factor, the cost of the fuel, may vary considerably from one year/month/day to the next. This is particularly true for natural gas and oil. This fluctuation in fuel price is often referred to as volatility in the price of the fuel, and we see it in everyday life. We all recognize how the price of gasoline we buy at the pump may vary significantly over a short time frame, and how the price may move in either direction—higher or lower.

Gasoline prices are directly driven in large part by the price of oil. Therefore, the volatility in gasoline prices we regularly see at the pumps is caused in large part by volatility in oil prices that utilities see from day to day. Natural gas prices have historically also been relatively volatile over longer time periods. However, natural gas prices can also be volatile in the short-term as evidenced by the significant increase in natural gas prices experienced in 2021 and 2022. (However, natural gas prices returned to prior lower cost levels in 2023.)

In contrast, the cost of coal and nuclear fuel (that are used, respectively, in coal steam units and nuclear units) typically varies very little from one year to the next.

---

[4] In striving to meet these two objectives, the utility will also have to adhere to any regulations (such as for air emissions and water usage) that may be applicable.

In utility/energy parlance, these two types of fuel are said to have relatively little cost "volatility" over time compared to natural gas and oil.

At the time the second edition of this book is written, oil is not commonly used as a primary fuel by electric utilities due to its cost and the fact that higher levels of air emissions result from burning oil compared to using natural gas. For these reasons, oil is almost exclusively a secondary or "backup" fuel for electric utilities in the United States that is used only when a sufficient natural gas supply is temporarily unavailable. Therefore, in the remainder of this book, the assumption is made that natural gas will be the primary fuel for steam units that could burn either natural gas or oil. Any reference to oil will be as a backup fuel.

All of these fuels typically are discussed in terms of different "units" of measurement. For example, natural gas is typically spoken about in terms of cubic feet of gas, coal in terms of tons of coal, and oil in terms of barrels of oil. In order to be able to meaningfully compare the relative cost of these fuels, a common unit of measurement is needed. This common unit of measurement is the British Thermal Unit (BTU), which represents the amount of heat required to raise the temperature of one pound of water by 1°F at sea level.

The BTU unit of measurement allows the costs of different fuels to be compared in regard to the price per a set amount of heat content (*i.e.*, the price per a set amount of BTUs) of these fuels. This is commonly done by stating the price for each fuel in terms of the cost in dollars to supply one million BTUs, abbreviated as $/mmBTU.

Although these costs, particularly those for natural gas and oil, may vary significantly over time, it is necessary in this book to assume cost values for these fuels in order to illustrate the concepts that will be discussed throughout the remainder of this book. Therefore, costs for each of the four types of fuel have been assumed and are presented in Table 2.1 below. These cost values are representative of actual costs for each of these four types of fuel that occurred or were forecasted at some point during the time this book was written.

The perceptive reader (you, I'm sure) will likely notice that, at the time you read this book, the actual costs for one or more of these fuels will be different, and perhaps significantly different, than the values shown in Table 2.1 below. Do not despair. As explained in Chapter 1, it is recognized that many assumption values, including those for fuel costs, will continually change over time. However, the resource planning concepts and analytical approaches that are discussed in the book remain valid regardless of assumption values used in this book to demonstrate those concepts and

TABLE 2.1
Representative Fuel Costs

Type of Fuel	Representative Fuel Cost ($/mmBTU)
Natural Gas	$ 6.00
Oil	$10.00
Coal	$ 2.00
Nuclear Fuel	$ 0.60

analytical approaches. Moreover, the assumed fuel cost values we are using are not supposed to represent fuel costs for any particular year.[5]

If one were to assume fuel cost values that are different than those that appear in the table, then the specific numerical results that illustrate the discussions that follow in the remainder of this book would change. However, and much more importantly, the fundamental utility resource planning concepts, principles, and processes that are discussed throughout the book will remain valid regardless of the cost assumptions used.

Furthermore, electric utilities recognize that there is great uncertainty regarding the projected costs of future fuel costs. Utilities recognize that forecasts that attempt to address a number of years in the future are likely to be inaccurate, particularly the more years in the future the forecast attempts to address. This is true of virtually any type of forecast and forecasts of future fuel costs are no exception. As we shall see in later chapters, forecasted fuel prices are an important input in numerous analyses that electric utilities undertake.

For this reason, utilities typically address the uncertainty inherent in any single fuel cost forecast by using multiple fuel cost forecasts in many of their analyses. For example, a utility may first develop a "medium cost" fuel cost forecast. It then may develop at least two other fuel cost forecasts; one with lower projected costs (a "low" fuel cost forecast), and one with higher projected costs (a "high" fuel cost forecast), than in the "medium" fuel cost forecast.

Then, as appropriate for the type of analysis that will be performed, these multiple fuel cost forecasts may be utilized in the utility's analyses. In practice, the use of multiple fuel cost forecasts increases the number of analyses that the utility will perform. However, this allows a number of perspectives regarding how the analysis results may differ depending upon how actual fuel costs in the future may turn out.

For purposes of this book, the use of a single fuel cost forecast is sufficient to demonstrate the principles, concepts, and processes of utility resource planning that will be discussed. Therefore, only one set of fuel cost values will be used throughout the book.

We can now turn our attention to the second factor that determines the operating cost of a generating unit: the efficiency at which a generating unit burns fuel to produce electricity. This efficiency is termed the "heat rate" of the generating unit, and it is expressed in terms of the number of BTUs of fuel that are required to produce 1 kilowatt-hour (kWh) of electricity. (A kilowatt-hour, or kWh, of electricity is 1,000 watts (W) of electricity delivered or used for 1 hour. For example, a 100 W light bulb burning for 1 hour consumes 100 watt-hours (Wh) of electricity. Ten such light bulbs burning for 1 hour would consume 1,000 Wh, or 1 kWh, of electricity.)[6]

---

[5] In fact, as will be seen in later chapters, we have conspicuously avoided any mention of a specific year by using terms like "Current Year" and referring to later years as one, two, etc. years from the "Current Year."

[6] Two concepts should be introduced at this point: electrical demand (or power) and energy. The term "demand" (or "power") refers to the amount of electricity being demanded (or produced) at any one instant in time. Demand is often referred to in terms of watts, kilowatts (kW), and megawatts (MW). Energy refers to the amount of electricity being used or produced over a specific period of time (which is typically measured in terms of hours). Energy is often referred to in terms of kilowatt-hours (kWh), megawatt-hours (MWh), and gigawatt-hours (GWh). Please refer to Appendix B for further explanation/ definition of each of these terms.

The efficiency of a generating unit is analogous to the miles-per-gallon (mpg) efficiency rating of an automobile, except that the "direction" of the reference is reversed. For an automobile, the "direction" is how much "product" (miles of travel) you get for a standard unit of fuel (a gallon of gas). The higher the mpg number is, the more fuel-efficient the car is. This direction is reversed for the heat rate efficiency value for electrical generating units. Here the direction is how much fuel (BTUs) is needed to produce the product (1 kWh of electricity). Therefore, the *lower* the BTU per kWh (BTU/kWh) heat rate value is, the more fuel-efficient the generating unit is because it takes less fuel to produce a kWh of electricity.

In contrast to fuel prices that change, often significantly and quickly, over time, a heat rate remains relatively constant over the life of a generating unit. The heat rate is typically at its lowest in the first few years a generating unit operates when the plant is "new and clean." The heat rate typically increases a bit over time (*i.e.*, becomes less fuel-efficient) until the unit undergoes a planned maintenance overhaul designed to return the generating unit to optimum efficiency. (This planned maintenance overhaul is similar in basic concept to the need years ago with older automobiles to take them in for service for a "tune-up" once the automobile is driven a pre-determined number of miles.)

After this planned maintenance overhaul, the generating unit's heat rate is lowered so that it is closer to the original heat rate value of the generating unit at the time the generating unit was new. This cycle of increasing, then decreasing heat rates will be repeated numerous times over the 25-year (or longer) life of a generating unit. However, these fluctuations in heat rate values over the life of the unit are generally in a relatively narrow range, and an average heat rate value for a generating unit over its operating life is often assumed for each type of generating unit for purposes of utility resource planning (and will be assumed in this book).

The assumptions that will be used as representative heat rates for the five different types of conventional generating units a utility is likely to have currently as part of its generation fleet are presented in Table 2.2 below. (In order to simplify the discussion at this point, we will ignore—for now—the newer combustion turbine units and list only the older, less efficient gas turbine units.)

TABLE 2.2

Representative Heat Rates for Existing Generating Units

Type of Existing Generating Unit	Representative Heat Rate (BTU/kWh)
Combined cycle	7,000
Steam-coal	10,000
Steam-gas (oil)	10,000
Nuclear	11,000
Combustion/gas turbines	14,000

As previously mentioned, the fuel portion of the operating cost of individual generating units is determined by multiplying the cost of the type of fuel the generating unit uses or burns ($/mmBTU) by the efficiency at which the generating unit uses this fuel to generate electricity (BTU/kWh). The operating costs for the five basic types of generating units using the previously presented representative costs for the four fuel types, and the representative heat rates for the five types of existing generating units a utility is likely to currently have are presented in Table 2.3.

TABLE 2.3
Representative Operating Costs for Existing Generating Units

(1)	(2)	(3)	(4)	(5)
Type of Existing Generating Unit	Primary Type of Fuel	Representative Heat Rate (BTU/kWh)	Representative Fuel Prices ($/mmBTU)	Representative Operating Cost ($/MWh)
Combined cycle	Natural gas	7,000	$6.00	$42.00
Steam-coal	Coal	10,000	$2.00	$20.00
Steam-gas(oil)	Natural gas	10,000	$6.00	$60.00
Nuclear	Uranium	11,000	$0.60	$ 6.60
Combustion/gas turbines	Natural gas	14,000	$6.00	$84.00

In Table 2.3, Column (1) lists the five types of conventional generating units a utility may have. Column (2) lists the primary fuel the generating unit is assumed to burn. Columns (3) and (4) show, respectively, the representative heat rate and fuel cost assumptions that were presented earlier.

Column (5) uses the information in Columns (3) and (4) to calculate a representative operating cost for these types of generating units. The formula used to derive the values in Column (5) is: Column (3) × Column (4) × (1,000 kWh/MWh) × (1 mmBTU/1,000,000 BTU). The resulting value is presented in terms of dollars per megawatt-hours ($/MWh). A MWh is 1,000 kWh (i.e., 1 MWh = 1,000 kWh). Operating costs of generating units are often expressed in terms of $/MWh.[7]

As shown in Table 2.3, the combination of the highest heat rate (i.e., lowest efficiency) and the assumed cost for natural gas, results in gas turbines being the most expensive type of generating unit to operate on a $/MWh basis. Closely following gas turbines in regard to operating cost is a steam unit that burns natural gas as its primary fuel. Considerably less expensive is the combined cycle unit. It has the lowest heat rate (i.e., the highest efficiency) and provides a much lower operating cost than either gas turbines or steam units that burn natural gas.

The converse is true of steam units that burn coal and of nuclear units. Although neither of these units has a low heat rate, the costs of coal and nuclear fuel are much lower than the assumed costs of natural gas. Consequently, the cost of generating a

---

[7] Note that the operating cost values in Column (5) account for fuel costs only. Variable operating and maintenance (O&M) costs of the generator are not included in this calculation in order to keep this discussion simple. Variable O&M costs are typically in the range of approximately $2.00/MWh or less, and often in the $0.10/MWh to $0.20/MWh range, for fossil-fueled units. Nuclear units are often assumed to have near-zero or zero variable O&M costs. Therefore, the omission of variable O&M costs in Column (5) does not significantly change the relative operating costs shown in this table.

MWh of electricity with coal-fired steam units and nuclear units is much lower than with units that burn natural gas (given the assumptions we are using).

Based on this information, one might be tempted to ask: *"Why would a utility operate anything other than coal or nuclear generating units?"* A similar, and directly related, question may also come to mind: *"Why would a utility build any type of generating unit other than a coal or nuclear unit?"*

But if we attempt to answer these questions now, we are getting ahead of ourselves. In order to meaningfully answer questions such as these, one needs to know more about utility systems, especially a utility's fleet of generating units. Operating on the premise that anything you create yourself is something you will understand better than something else that is just handed to you, we will take that approach to learn more about utility systems by "building" a utility system.

LET'S CREATE A HYPOTHETICAL UTILITY SYSTEM

We will begin the process of "creating" a hypothetical utility system by first creating a profile of customers' electricity usage for a hypothetical utility. After all, utility systems are designed to serve their customers in regard to how these customers actually use electricity. Therefore, it is the logical place to begin.

There are many ways to look at customers' electricity usage. One perspective is to look at electricity usage during one 24-hour day. Another perspective is to look at the total electricity usage during one calendar year. A typical year (*i.e.*, a non-leap year) has 365 days of 24 hours each. Therefore, there are 365 days/year × 24 hours/day = 8,760 hours during a typical year. During those 8,760 hours, there will be 1 hour in which the highest amount of electricity is used, and 1 hour in which the lowest amount of electricity will be used. There will also be 8,758 other hours in which the electrical demand falls between these highest and lowest electrical demand values.

The hourly "demand" values for electricity are typically discussed in terms of a megawatt (MW) of electricity and the highest level of electricity usage during a calendar year's 8,760 hours is termed the "peak" demand for electricity or the peak load. Note that even within that hour, the actual electrical demand will vary from minute to minute. However, for analysis purposes, the load over this peak hour is averaged to provide a more convenient hourly unit of measurement.

The demand for electricity for most utilities is driven in large part by the outside temperature. Consequently, the highest or peak demand for electricity during the year is likely to occur on a summer afternoon on a really hot day or on a really cold winter day. In general, higher electrical loads typically occur in the summer and winter months, with lower loads occurring in the spring and fall months. There may be significant differences in the peak load from one month to another, especially as the calendar moves from one season to the next.

There are also significant differences in the electrical loads from one hour to another during a daily 24-hour cycle. The hourly electrical loads over the course of a day are generally driven by two factors: the outside temperature during each hour and what customers are doing at a given hour (*i.e.*, whether they are active at home or at work during the day and early evening, or whether they are asleep at night and in the early morning). For these reasons, electrical loads typically increase during the day and then decrease during the night.

We begin to create our hypothetical electric utility system by assuming that our utility system is a summer peaking utility.[8] We then construct a representative electrical load for the summer day during which our utility's highest/peak load for the year is reached. A peak load of 10,000 MW is assumed for our utility. A graphic depiction of hourly loads for a given period of time is referred to as a "load curve" (or "load profile."). Figure 2.1 presents a daily load curve of our utility system's electrical load during the 24 hours of its summer peak day.

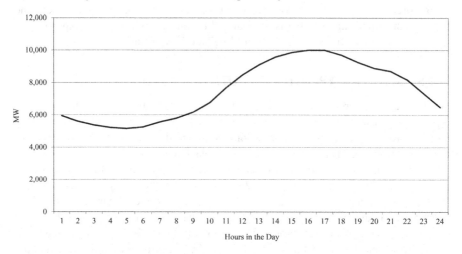

FIGURE 2.1 Representative peak day load curve for a hypothetical utility with a summer peak load of 10,000 MW.

The first value on the left side of the graph represents the hour from midnight to 1 a.m. (This is commonly referred to as "hour-ending 1" or simply as hour 1.) The loads for the remaining 23 hours of the day are then shown as one moves from left to right across the figure.[9]

When viewing this figure starting from the hour 1 "starting point" on the left-hand side, we see that electrical loads decrease in the early morning hours when outside temperatures are cooler (or at least less hot) and people are asleep, increase during the day as the outside temperature rises and people are active, then decrease again as darkness falls and people return to sleep.

---

[8] The term "utility" is occasionally used instead of "utility system" in places throughout the book. This is done merely for convenience. However, the reader should always keep in mind that an electric utility is truly a complex system of numerous generating units (as well as transmission and distribution lines, etc.). As we shall see in later chapters, as new resources are added to this system, there are numerous impacts on the operation of the existing generating units that are parts of the system.

[9] To ensure that the assumed loads for our hypothetical utility are representative of actual utility loads, the hourly electrical load values presented in Figure 2.1 are loosely based on a particular summer-peak day load of FPL's. The FPL data was used as a starting point from which to create the values for our hypothetical utility. FPL's highest load that day was slightly greater than 20,000 MW. For our hypothetical utility, this value has been reduced in the graph to a highest load of 10,000 MW and all other hourly loads have been reduced proportionally.

The key item to notice in Figure 2.1 is how much variation in electrical load there can be in even a single 24-hour day. For this day on which the highest load for the year (10,000 MW) is reached at hour 17 (or the 4-to-5 p.m. hour), the electrical load still drops significantly in the night-time hours to approximately 5,150 MW at hour 5 (or 4-to-5 a.m.). Therefore, the highest load of the day is almost double the lowest load of the day. (These hourly variations in electrical load have significant implications when a utility considers solar and/or battery storage options as future resource additions. This will be discussed in Part II of this book.)

Expanding this look to all 8,760 hours in a year gives another perspective of how electrical loads vary over the course of the year for our utility system. Figure 2.2 presents a graphic representation of these electrical loads for each of the 8,760 hours in a typical calendar year in a special version of a load curve that is typically referred to as "an annual load duration curve."

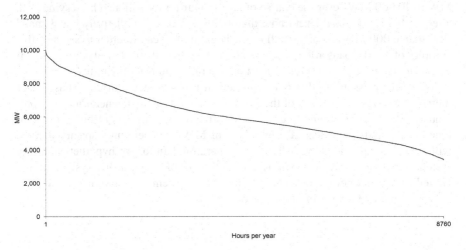

FIGURE 2.2 Representative annual load duration curve for a hypothetical utility with a peak load of 10,000 MW.

In an annual load duration curve, the electrical loads are not shown in chronological order, *i.e.*, the loads do not start with the 1 a.m. load on January 1, followed by the 2 a.m. load on January 1, etc. Instead, the highest load experienced during the year, regardless of what month, day, or hour of the day the highest load occurred on, is represented by the left-most value on the graph. This is the 10,000 MW value we just saw on the summer day graph in Figure 2.1.

The second highest load experienced during the year is then presented immediately to the right of the 10,000 MW value. This sequence is followed until the lowest load during the year (approximately 3,400 MW) is presented as the last value on the far right of the graph. Therefore, our utility system obviously has loads that are even lower than the 5,150 MW lowest daily load for our summer peak day that was shown previously in Figure 2.1.

As can be seen from comparing the previous load curve for the summer peak day with the annual load duration curve, the shape of the graph of the annual load duration curve is significantly different than the shape of the graph for the summer peak day.[10] From this curve, one can see there is even more variation in electrical loads over the course of a year (a range of 3,400 to 10,000 MW) than was for the summer peak day.

In addition, there are many hours that can be labeled as relatively high-load hours and many other hours that can be labeled as relatively low-load hours. If one looks across the graphed line to where the approximate mid-point of the annual load curve is, we see that this is roughly where the graphed line crosses the 6,000 MW level. What this means is that, for our hypothetical utility system, approximately half of the hours in the year will have an electrical demand of greater than 6,000 MW and approximately half of the hours in the year will have an electrical demand of less than 6,000 MW.

Therefore, in regard to the operation of our hypothetical utility's fleet of generating units, the utility will be serving a very wide range of electrical demand levels (3,400 MW to 10,000 MW) over the course of a year. Our utility will also be serving both a relatively high demand level (more than 6,000 MW), and a relatively low demand (less than 6,000 MW), in about half of the hours in the year. In order to see how this "profile" of hourly loads affects the operation of the hypothetical utility system's generating units, we will next "create" a set of generating units for the utility.

We start by assuming that our hypothetical utility system currently has one or more generating units of each of the five types of conventional generating units previously discussed: (i) steam-natural gas, (ii) steam-coal, (iii) gas turbine, (iv) combined cycle, and (v) nuclear. The numbers of MW of generating capacity that we will assume are provided by each type of generating unit for our hypothetical utility system is based on a rough composite of a number of actual utility systems in the United States and not on any one specific utility system. The assumed number of MW supplied by the type of unit is shown in Table 2.4.

TABLE 2.4

Assumed Capacity (MW) by Type of Existing Generating Unit for the Hypothetical Utility System

Type of Existing Generating Unit	MW	% of Total MW
Combined cycle	3,000	25.0%
Steam-coal	3,500	29.2%
Steam-gas (oil)	3,500	29.2%
Nuclear	1,000	8.3%
Combustion/gas turbines	1,000	8.3%
Total =	12,000	100.0%

---

[10] Figure 2.2 also uses FPL data as a starting point, or "seed," from which to create the annual load curve for our hypothetical utility. The 8,760 hourly load data from FPL was again reduced so that the highest hourly load was 10,000 MW and all other hourly loads were reduced proportionally.

The astute reader (you again) will ask the obvious question: *"Why does the utility have 12,000 MW of generating units if its highest load is only 10,000 MW?"* The answer to the question has two parts. First, recall that regulated utilities are under an obligation from their respective regulatory agencies to provide electrical service to all customers at all times. Second, generating units, like automobiles and many other complex machines, may break unexpectedly (thus requiring time in order to repair the generator) and they also need periodic planned maintenance. If a specific generating unit either unexpectedly breaks, or if it has been taken out of service for planned maintenance, it cannot operate to supply electricity during that time period.

Therefore, a utility system must have more generating units, or electric generating capacity (MW), than its highest load alone would seem to dictate. This additional, or "reserve," generating capacity will be further discussed in Chapter 3.

NOW LET'S OPERATE OUR HYPOTHETICAL UTILITY SYSTEM[11]

UTILITY SYSTEM OPERATION ON THE SUMMER PEAK DAY

Now that we have made an assumption regarding the amount of capacity that each type of conventional generating unit can supply to our utility system, we can gain some insight into how our utility will actually operate those generating units. First, we recall that the representative fuel-based operating costs of the different types of generating units are previously presented in Table 2.3. This table shows that nuclear generating units were the least expensive units to operate at $6.60/MWh. Therefore, the utility will operate nuclear units as often as possible. In utility parlance, the nuclear units will be "dispatched" first to meet the utility's electrical load before any other type of generating units are dispatched.

As previously shown in Table 2.4, we have assumed there is 1,000 MW of nuclear capacity on the utility system. By superimposing this amount of capacity on the summer peak day load duration curve previously presented in Figure 2.1, we get a visual picture of how much of the utility's load during the 24-hour period of the summer peak day can be supplied by nuclear generation. This picture is shown in Figure 2.3 that is presented on the next page.

As shown in this figure, the shaded area at the bottom of the figure represents our utility system's nuclear generation capacity. It is clear that the nuclear capacity does not "reach up" to the actual load level that is projected for any hour. Thus, nuclear capacity can supply only a relatively small portion of the electrical load even during the early morning hours (hours 1 through 8, or 1 a.m. to 8 a.m.) when the electrical load is at its lowest. (Recall that our utility system's lowest load on this peak summer day was approximately 5,150 MW and that our utility system's nuclear capacity can supply only 1,000 MW.)

Nevertheless, assuming that the nuclear generating capacity is available to operate (*i.e.*, the nuclear units are not broken nor have the units been taken out of service

---

[11] I realize that you may not have operated one of these babies before, but relax, you can handle it.

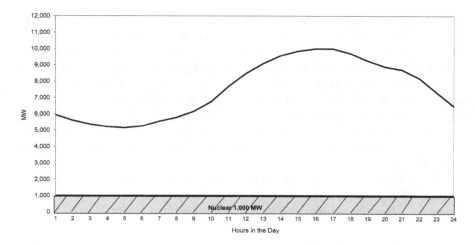

FIGURE 2.3   The potential contribution from nuclear generation during the summer peak day.

for planned maintenance or refueling), the utility will operate the nuclear generation during all 24 hours because of the very low operating cost of nuclear units.

Our utility must now dispatch at least one more type of generating unit in order to meet the demand for electricity. Returning to Table 2.3, it is apparent that coal-fueled steam units are the next most economical generation to operate at $20.00/MWh. From Table 2.4, we see that our utility is assumed to have 3,500 MW of coal-fueled generation. Figure 2.4 below, now shows how much of the utility's load during the 24-hour period is supplied by the combination of nuclear and coal.

As shown in this figure, the combination of our utility system's nuclear and coal generation (a combined total of 4,500 MW) represented by the total cross-hatched

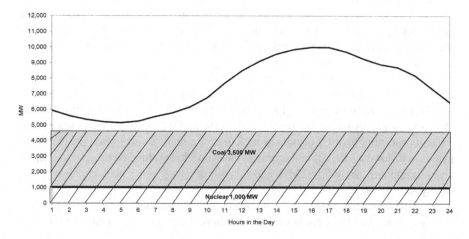

FIGURE 2.4   The potential contribution from nuclear and coal generation during the summer peak day.

area in the figure still doesn't meet even the lowest electrical load (5,150 MW) experienced during this peak day.

We will now "cut to the chase" and superimpose all of our utility system's capacity that can be supplied by each of the types of generation on this summer peak day load duration curve. This is presented in Figure 2.5.

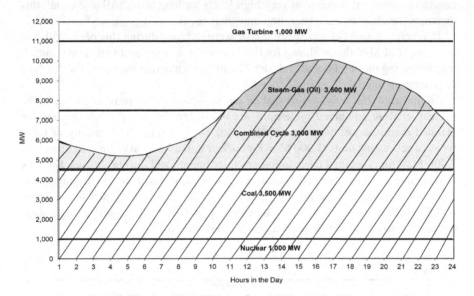

FIGURE 2.5 The potential contribution of all types of generation during the summer peak day.

As shown by the third (and lightly shaded) area in this figure, the potential contribution from the next most economical type of generating units, combined cycle, can potentially be used in combination with nuclear and coal units to meet all of the load on the summer peak day from approximately midnight through 11 a.m. and then again during the hours from about 11 p.m. to midnight. Note that all of the combined cycle capacity (3,000 MW) will be used on this peak day.

As shown by the fourth (and darker shaded) area, steam-natural gas units would also be needed in order for our utility to meet the remaining load during approximately hours 11 through 23 (11 a.m. to 11 p.m.). However, not all of the 3,500 MW capacity of steam-natural gas units are in operation.

The figure also seems to indicate that there is a "cushion" of exactly 1,000 MW from the top of the load curve (10,000 MW) to the bottom of the gas turbine band (11,000 MW). The logical question to ask is: *"Why aren't the gas turbine units used at all?"*

The answer to the question is that Figure 2.5 presents an idealized picture. What is shown on the graph for the MW of each type of generating unit represents the maximum possible contribution that each type of generating unit can make.

The values shown in the figure assume that (i) no generating units of any type have broken down and are, therefore, not in service on this summer peak load day, and (ii) no generating units have been taken out of service for planned maintenance on that day.[12] But having generating units out of service because the units have unexpectedly broken can definitely happen on peak load days. Although a generating unit could break on any day, it is somewhat more likely to happen on a peak or other high load day. This is because a peak or high load day is frequently preceded by a number of high load days, weeks, or even months during which many of the utility system's generating units are operated at very high levels for long hours. All else equal, this increases the chances that generating units may break.

If generating units of any specific type were to break during this peak load day, the amount of MW that is shown for this type of generating unit would be reduced, thus shrinking the height of the relevant "band" shown across the figure for that type of generating unit.

To illustrate this, let's assume that on our utility's summer peak day, none of the nuclear, coal, or combined cycle units are out of service (*i.e.*, they are not out for planned maintenance, nor have they broken). Therefore, the full capacity of these types of generating units (1,000 MW for nuclear units, 3,500 MW for coal units, and 3,000 MW for combined cycle units) are fully operational during the day. But let's also assume that a number of the steam-natural gas units break during the day (or the day before), reducing the available capacity from these units from their full capacity of 3,500 MW down to 2,000 MW. Therefore, 1,500 MW of steam-oil/gas capacity is unavailable that day. This case is depicted in Figure 2.6.

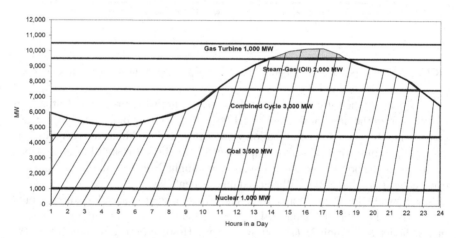

FIGURE 2.6   The potential contribution of all types of generation during the summer peak day (assuming a reduction of 1,500 MW of steam-gas (oil) capacity.

In this case, the former height of 3,500 MW for the steam-oil/gas units now shrinks to a height of 2,000 MW. In turn, the unchanged height of 1,000 MW for gas turbines now "drops down" in the figure so that some, but not all, of the gas turbine capacity is used to meet our utility's very high loads during the mid-to-late afternoon

---

[12] Utilities typically strive, when possible, to schedule planned maintenance so that this work does not occur on peak load days. Therefore, one would not expect planned maintenance to diminish the generating capability of any type of generating unit on the summer peak day or the winter peak day.

hours. (Gas turbines might also be used if the actual peak load on this summer day exceeded the projected peak of 10,000 MW.)

This example in which our utility system needs to operate gas turbines due to the unavailability (whether due to breakage or planned maintenance) of units in the other types of generating units that are more economical to operate than gas turbines, is actually quite representative of when and why utilities acquire and operate these expensive-to-operate gas turbine units.

UTILITY SYSTEM OPERATION OVER THE COURSE OF A YEAR

Based on the insight we have gained from the discussion just completed regarding how our utility's different types of generating units will be operated on the summer peak day, we can now develop a conceptual look of how these types of generating units will be operated over the course of a year. Returning to the annual load duration curve presented earlier in Figure 2.2, we now make two changes to that figure.

First, we superimpose the maximum capacity of each type of generating unit (as we just did for the summer peak day load curve). Second, we cross-hatch the area that is actually under the annual load duration curve line in each of the bands representing the amount of capacity offered by each type of generating unit. The cross-hatched area for each type of generating unit that is under the annual load duration curve line shows how much of each generating unit type is being operated at any given hour and helps demonstrate how many hours each type of generation is projected to operate over the course of a year for our utility system. The result is presented in Figure 2.7.

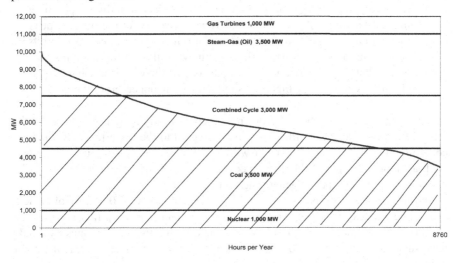

FIGURE 2.7 The potential contribution from all types of generation during the course of a year.

This picture gives us the advantage of a much broader perspective (8,760 hours instead of 24 hours) from which to view how the generating units of our utility will be operated. Reminding ourselves that this picture represents an idealized scenario (no unit breakage or planned maintenance is represented), we can use this broader perspective to draw certain general conclusions about how our utility system's different types of generating units will be operated:

1. Due to their lowest operating costs, all of the nuclear generating capacity will be used all of the time these units are available to operate. (This is shown by the entire nuclear generating capacity band being shaded for all hours.)
2. The next most economical-to-operate type of generation, steam-coal units, will have a very high percentage of their capacity being used in all hours of the year. In addition, all of the coal-fueled capacity will be used except for the very lowest load hours of the year. (This shows up as a small unshaded area at the far-right side of the coal generation band.)
3. Combined cycle unit generation will be operated for many of the hours of the year. The higher the load gets, the greater the amount of the combined cycle generation that will be used. As seen by the relatively large unshaded area to the right of the load curve that is within the combined cycle band, only a portion of the combined cycle generation will be operated during many of the hours in the year. There are also some hours in the year in which no combined cycle generation is used.
4. The steam-natural gas capacity will play a much smaller role in the utility's operation over the course of a year. The small, shaded area for steam-natural gas generation on the upper left-hand side of the graph shows that these steam-natural gas units will be used only in those hours when the electrical loads become quite high.
5. Finally, although the gas turbine capacity is not shaded in this graph, these units have a potential role to play as discussed earlier for the summer peak day load curve. As significant amounts of other types of generating units become unavailable at various times of the year, the gas turbine units may be used, but only in a relatively few hours during the year. These hours typically represent the very high loads in the extreme upper left of the graph.

These general observations for our hypothetical utility system will be valid for many, if not most, electric utility systems. In regard to the importance of this low-cost-to-high-cost "dispatch order" of when the types of generating units are operated, a simple example of the differences in fuel costs that can occur between the operation of certain types of generating units is helpful.

In this example, let's assume that our hypothetical utility's 1,000 MW of nuclear generating capacity is out of service for 24 hours and that our utility supplies the 24,000 MWh of energy (= 1,000 MW × 24 hours) that the nuclear capacity would have supplied entirely with some of its steam-natural gas units. What is the additional fuel cost from such an occurrence?

Using the assumed heat rate (11,000 BTU/kWh) and fuel cost ($0.60/mmBTU) for the nuclear capacity, we see that the cost of supplying 24,000 MWh, or 24,000,000 kWh, for this one day from the nuclear capacity would have cost $158,400 (= 24,000,000 kWh × 11,000 BTU/kWh × 1 mmBTU/1,000,000 BTU × $0.60/mmBTU).

However, because we have assumed the nuclear capacity is out of service for this 24-hour period, the 24,000 MWh that will now be supplied by some of the steam units burning natural gas will cost $1,440,000 (= 24,000,000 kWh × 10,000 BTU/kWh × 1 mmBTU/1,000,000 BTU × $6.00/mmBTU).

Therefore, our utility system will incur an additional $1,281,600 of additional fuel cost (= $1,440,000–$158,400) for this one 24-hour period. This simple example points out the reason for, and the importance of, operating a utility's generating units in a dispatch order of low-cost-to-high-cost because the utility's customers will typically be charged the full fuel costs of supplying electricity.[13]

Utilities also refer to the dispatch order of generating units in terms other than "low cost" and "high cost." A common naming convention is used instead that inherently recognizes the operating cost of the generating units by referring to how much of the time during the year a generating unit will operate. For example, nuclear units are typically referred to as "baseload" units because they continually operate to serve load that is always present for the utility system; i.e., "base" loads, due to the low operating cost of these units. For most utilities, coal-fired units are also considered baseload units. At the other end of the spectrum, the gas turbines are referred to as "peaking" units because they typically operate only when electrical loads are at or near the utility's peak load due to their high operating costs.

In between these two types of generating units are the combined cycle units and the steam-natural gas units. These types of units are often referred to as "intermediate" units because they typically operate at loads that fall between base loads and peak loads. (On certain utility systems, very efficient combined cycle units are also operated as baseload units.)[14]

The terms "baseload," "peaking," and "intermediate" are meaningful because their use helps to quickly give a sense of how many hours of the year these units will operate or what "roles" these units will play in meeting load during the year. The roles that baseload units and peaking units will play in meeting customers' demand for electricity are simple. They function at opposite ends of the operation spectrum. At one end of the range, baseload units will be operated during virtually every hour of the year in which these units are available to operate. At the opposite end of the range, peaking units will only operate a relatively few hours of the year in which electrical demand is very high and/or a large amount of other generating capacity is unavailable.

However, the role the intermediate units will play is actually much more interesting in regard to utility system operation. (And, as we shall see later in this book, these units play a very important role in analyses that are conducted in order for the utility to plan for its future resource additions.) The role of the intermediate units for our utility system (*i.e.*, combined cycle and steam-natural gas units) can be seen graphically by returning to Figure 2.7 and putting a finger on the load duration curve itself.

---

[13] The example also points out the large fuel-saving benefits that can be obtained from continuing to operate existing nuclear generating units.

[14] Note that solar and wind turbine facilities are generally not considered baseload units due to the intermittency of their energy sources, *i.e.*, sunshine and wind. They are better thought of as types of intermediate generating units.

Then, as one "follows" the curve up or down the load level (*i.e.*, to the left or to the right on the load curve line), one would readily see that during almost the entire length of the curve (except for the extreme bottom right-hand portion of the curve), the intermediate units are the means by which the utility meets load that is continually varying from one hour to the next. This is carried out by using more or less capacity from the intermediate units.

Because the intermediate units for our utility system are assumed to primarily burn natural gas, the volume of natural gas that is burned varies from hour to hour to meet electrical demand that is continually changing. From virtually any point on the load duration curve, if we go up the curve (toward the upper left-hand corner) to the next higher load point, more natural gas is being burned in either combined cycle or steam-natural gas units. Likewise, if we go down the curve (toward the lower right-hand corner) to the next lower load point, less natural gas is being burned in these same types of units.

In utility parlance, natural gas can be referred to as our hypothetical utility system's primary "marginal" fuel because the amount of natural gas used varies from hour-to-hour as the load goes up or down across almost all of the hours in the year. (The sole exception is during the relatively few hours of low load shown on the extreme right-hand side of the curve when coal serves as the marginal fuel.) This is not the case for nuclear fuel, or for coal during most of the hours in the year, because the nuclear- and coal-fueled units that use these fuels operate as many hours as they are available to operate because of their low operating costs.[15]

As we shall see later in this book, the addition of new resources to our hypothetical utility system will have the most impact, in terms of fuel use, on our utility's natural gas usage. Stated another way, the addition of these new resources will have the greatest impact on the number of hours that our utility's existing natural gas-fueled generating units will be operated. This will be true regardless of whether the new resources to be added are Supply options (new generating units) or DSM options. These effects will be seen in three different areas: (i) the utility's system fuel usage, (ii) the utility's system air emissions, and (iii) the economics of selecting one new resource option versus another resource option.

But we are again getting ahead of ourselves. As I said, we will return to these topics in later chapters. Now it is time to back up and see what we have learned about the basic operation of our utility system and then peek around the corner to see what is coming next.

SO WHAT HAVE WE LEARNED AND WHERE DO WE GO NEXT?

We have actually covered a lot of ground so far. We have discussed three facets of electric utility systems: (i) the electrical load, (ii) the type of conventional generating units that are likely to exist on a utility system, and (iii) how those generating

---

[15] Another utility with a larger amount of coal-fired generation would likely have coal as a marginal fuel for more hours in the year than is shown for our hypothetical utility system. For our discussion, the key point is that fossil fuels—natural gas or coal, primarily—will typically be a utility's marginal fuels. For our hypothetical utility, natural gas is the predominant marginal fuel.

units will be operated to meet the electrical load both annually and on our utility's summer peak day. We have illustrated these three facets through the use of our hypothetical utility system. These three facets are fundamental to understanding both how utility systems work and, as we shall see later, how a utility decides what new resource options should be added to the system.

From the previous discussions that made use of our hypothetical utility and its electrical load, we can summarize what we have learned in the following five points:

1. Electrical load varies significantly for an electric utility both from hour to hour on a given day (as shown in the load curve for the summer peak day) and over the course of a year (as shown in the annual load duration curve).
2. There are five basic types of conventional generating units (steam-natural gas, steam-coal, combined cycle, gas turbine, and nuclear) that have historically made up the bulk of the electrical generating capacity for electric utilities.
3. The operating costs for the five types of conventional generating units vary significantly and utilities operate these units to the extent possible (subject to equipment maintenance, contractual, environmental, etc. constraints) so that the generating units with the lowest operating costs are used the most, and the units with the highest operating costs are used the least.
4. As a consequence, nuclear units (with the lowest operating costs) will operate as much as possible followed closely by coal-fueled units. Conversely, gas turbines (with the highest operating costs) will operate as little as possible. This leaves our hypothetical utility with combined cycle natural gas units, and steam-natural gas units, as the primary units whose operation is dictated by the amount of load the utility must serve at any given hour.
5. Because the operation of both steam-natural gas units, and the combined cycle units that also burn natural gas, play the role of "ramping up" or "ramping down" to meet ever-changing load levels over almost all of the hours in a year, the amount of natural gas that is used is also constantly changing as the load changes. For this reason, natural gas is referred to as the primary marginal fuel for our hypothetical utility system. (As previously mentioned, other utilities may have coal as a marginal fuel for more hours in the year than is assumed for our hypothetical utility system.)

These summary points provide a good overview picture of what our hypothetical utility system looks like regarding its electrical load and generating units, plus how it will operate its generating units to meet the electrical load. Stated another way, we know how our utility system will meet its electrical load, at least for the year we have just examined.

In this chapter, we created a hypothetical utility system that we will use as a tool to assist us in subsequent chapters. In those chapters, we will use our utility system to help discuss and illustrate various electric utility principles, concepts, and analytical approaches. These discussions will be assisted by actually performing calculations using our hypothetical utility system.

We should also again note that what is really important in the discussions that follow are these principles, concepts, and analytical approaches that will be introduced and discussed, not the results of the actual calculations. This is because the calculation results will almost certainly differ if we had constructed our hypothetical utility's system of generating units and load patterns in a different way and/or used significantly different heat rate and fuel cost assumptions for our hypothetical utility system.

This leads me to introduce the first of my "Fundamental Principles of Electric Utility Resource Planning":

Fundamental Principle #1 of Electric Utility Resource Planning: "All Electric Utilities are Different"

Each electric utility is different in regard to (at least) its electrical load characteristics and its existing generating units. Therefore, when faced with a particular problem or issue such as *"Which resource option is the best selection?"*, the correct answer for one electric utility may not be the correct answer for another electric utility.

Of my Fundamental Principles that are introduced in this book, I believe this may be the most important to keep in mind, particularly for those who seek to influence the future direction of the electric utility industry through legislation and/or regulation.

In Chapter 3, we begin to place our hypothetical utility system in a situation in which it will need to make a decision regarding the need for additional resources in the future. We then begin an overview discussion of how resource planning is actually performed by introducing three basic questions that utility resource planning must always answer.

3 Overview of Utility Resource Planning

In this chapter, an overview of the various aspects of utility resource planning is presented. The main objective of this chapter is to simply introduce basic concepts that will be used in the remainder of the book. These concepts will then be discussed in more detail, and applied, in subsequent chapters.

ONE MORE ASSUMPTION REGARDING OUR HYPOTHETICAL UTILITY SYSTEM

When discussing resource planning for an electric utility, it is helpful to use a specific example of why the utility may need to make a resource decision. There are a variety of reasons why a particular utility may need to make a decision regarding one or more resource options. A partial list of these reasons would include the following possibilities:

- A growing number of customers, and the accompanying growth in electrical demand and usage, that the utility must serve;
- Recognition that one (or more) existing electric generating unit is nearing the end of its useful life, necessitating replacement of this soon-to-be-retired resource in order to maintain system reliability;
- Increasing costs of fuel for a utility system that has older, less fuel-efficient generating units that result in significantly increasing energy costs for the utility's customers; and
- The introduction of new environmental regulations that will result in unacceptably high environmental compliance costs for the utility's customers unless new or more fuel-efficient resources are added to the utility system.

For purposes of the discussions that follow, we make the assumption that the reason our hypothetical utility system needs to make a resource decision is that the number of customers, and their accompanying electrical demand and usage which the utility must serve, is forecasted to increase.

This assumption has been chosen for two reasons. First, it allows a somewhat simpler discussion of the concept of "reliability analyses" (that will be discussed shortly) than might be possible if another reason for having to make a resource decision was chosen for use in our discussion. Second, virtually all utilities have faced increasing growth in customers and electric load in the past. Therefore, the use of this assumption is helpful in understanding how an electric utility "grew" into its present (or recent) system of electric generating units and other resources.

DOI: 10.1201/9781003301509-4

However, it is important to note that the resource planning principles, concepts, and analytical approaches discussed in the remainder of this book for our hypothetical utility system that is assumed to face increasing growth are also applicable when a utility faces other reasons for having to make a resource decision. For example, regardless of the reason for why a utility must make a resource decision, the utility resource planner is faced with three basic questions.

THREE QUESTIONS UTILITY RESOURCE PLANNING MUST ALWAYS ANSWER

As just stated, our utility system now forecasts that it is facing an increasing number of customers in the coming years. Assuming all else equal, the electrical load that the utility must serve will also increase. Does the utility need to take any actions in order to meet this increased load? If so, when must these actions be taken? In utility parlance (you're starting to feel like a utility insider, aren't you?), the utility is trying to determine if it has a "resource need," *i.e.*, a need to add new resources in the future to meet this increased load.

In looking at this situation, a utility planner is faced with the following three questions which must always be answered:

1. When (in what year) does the utility need to add new resources?
2. What is the magnitude (MW) of the new resources that are needed?
3. What is the best resource option(s) with which to meet this need?

The first and second questions are not directly concerned with economics.[1] Instead, the focus is solely on the timing and magnitude of the new resource need. Only if the answers to the first and second questions—"When are new resources needed?" and "What is the magnitude of the need?"—indicate that a significant MW amount of new resources are needed in a near enough time frame so that the utility must make a decision soon, does the resource planner need to move on to the third question. The answer to the third question is focused on making the best choice for all of the utility's customers from both an economic perspective and a non-economic perspective.

We will provide an overview of how a utility approaches finding answers to these three questions in this chapter. Then, in subsequent chapters, we will provide examples of how these questions might actually be answered using our hypothetical utility system. We will start by examining the first and second questions. And, as you might guess, these two questions ("When does a utility need to add new resources?" and "What is the magnitude of the needed resources?") are related.

In the case of a utility facing increasing electrical demand, as is the case for our utility system, utilities usually address these two questions jointly and often refer

---

[1] As we shall soon discuss, the first two questions relate to whether the utility system meets one or more pre-established "reliability" criteria. Economic considerations may enter the picture in the course of establishing what the reliability criteria themselves should be. However, once these criteria are established, the first two questions are typically answered by performing calculations that do not involve economic considerations.

to the analyses designed to answer these two questions as "reliability analyses." In conducting reliability analyses, utilities use certain standards or criteria to gauge how reliable their utility system is projected to be. These reliability standards/criteria are frequently mandated by the utility's regulatory authority, but utilities can utilize other standards/criteria as well. As we shall see, because our utility system's electrical load is projected to increase, these criteria will be examined in light of the forecasted load growth to see if/when new resources need to be added to ensure that the utility continues to meet its reliability criteria.

RELIABILITY ANALYSIS: WHEN DOES A UTILITY NEED TO ADD NEW RESOURCES AND WHAT IS THE MAGNITUDE OF THOSE NEEDED RESOURCES?

Utilities typically make use of at least two perspectives in reliability analyses to ensure that a comprehensive picture of the utility's future resource needs is captured. Our discussion will focus on two perspectives commonly used at the time this book is written. Both of these perspectives are concerned with projections of the utility system in future years.

The first of these perspectives typically focuses on a specific hour for each of two days in the year: the hour on the one summer day in which the electrical load during the summer is projected to be the highest, and the hour on the one winter day in which the electrical load during the winter is projected to be the highest. These two hours are called the summer peak hour and winter peak hour, respectively. Analyses that use this perspective are generally referred to as "deterministic" analyses. The most commonly used form of a deterministic analysis is a "reserve margin" analysis.[2] This perspective has the advantage of being easy to explain, and the analyses can be performed on a simple spreadsheet or even with a calculator.

The second perspective examines all of the days in the year to determine the probability that generating units on the utility system might fail in a way that results in the utility not having enough generation available to meet all of its electrical load at some point during the year. These analyses are generally referred to as "probabilistic" analyses. These analyses are more complicated than deterministic analyses and require the use of fairly sophisticated spreadsheets or computer models.

RESERVE MARGIN PERSPECTIVE (SIMPLE TO CALCULATE)

As previously mentioned, a reserve margin perspective looks at the highest hourly load (the peak hourly load) that is projected to be experienced in the summer and in the winter. The intent is to determine if the utility can safely meet these two peak loads with the resources it is projected to have at that time.

---

[2] There are variations of reserve margin analysis that are also in use. (One such variation is termed capacity margin analysis.) These variations are similar to reserve margin analysis in their basic concept and there is little difference in the actual calculations. For these reasons, and for the sake of simplicity, our discussion will focus solely on reserve margin analysis.

The criterion used in this type of analysis is called the "reserve margin criterion," and it is expressed as a percentage. The percentage represents the amount of generation capacity (MW) that a utility has on its system that is in excess of the highest projected load that the utility is expected to serve in both the summer and winter seasons for a given year. Values used as the reserve margin criterion for utilities often fall in the range of 15% to 20%.[3] This criterion is set at a level designed to ensure that utilities will have more generation capacity than their highest projected load, so the utility can still serve the load even if some generating units break on these peak load days and/or the actual load is higher than projected.[4]

A reserve margin criterion is a <u>minimum</u> threshold criterion. In other words, a utility with a 20% reserve margin criterion is deemed to need additional resources if its projected reserve margin value for a given year drops below 20%. However, if their projected reserve margin value for a given year is 20% or greater, the utility is viewed to be reliable from a reserve margin perspective and there is no need for additional resources to be added by that given year.

Using our hypothetical utility system, a basic reserve margin calculation is presented in Table 3.1.

TABLE 3.1
Example of a Basic Reserve Margin Calculation

(1)	(2)	(3) = (1)−(2)	(4) = (3)/(2)
Total Generating Capacity (MW)	Peak Electrical Demand (MW)	Reserves (MW)	Reserve Margin (%)
12,000	10,000	2,000	20.0

(And you thought I was kidding when I said this calculation was simple.)

In this calculation, we see that our hypothetical utility system has a projected reserve margin of 20% for this particular year. If our hypothetical utility system has a reserve margin criterion of 20%, the utility's projected reserve margin value for this year of 20% meets that criterion. If a lower reserve margin criterion of 15% was used by the utility and/or its regulatory authority, the utility's projected 20% reserve margin value for this year would not only meet, but also exceed the 15% minimum criterion. In either case, our hypothetical utility system is deemed to be reliable from a reserve margin perspective for this year. Conversely, if the projected

---

[3] This range of values for the reserve margin criterion has been established over the years based both on actual utility operating experience and economic analyses. The economic analyses examined the inconvenience/cost to customers from interruptions in electric service if the criterion is set too low and the costs to customers from building more power plants if the criterion is set too high. Utility experience and the results of economic analyses have often indicated that a range of 15% to 20% is sufficient. A variety of utility-specific considerations are taken into account when determining the actual criterion value to use.

[4] In regard to this first consideration, you may recall that we earlier looked at something similar with our hypothetical utility system in Figure 2.6. Our utility system has 12,000 MW of generating capability and a system peak load of 10,000 MW. We saw that our utility could still meet the 10,000 MW peak load if 1,500 MW of its steam—oil/gas units were broken on that day.

reserve margin value for this year was lower than the reserve margin criterion—for example, if it was projected to have a 17% reserve margin, but the utility had a 20% reserve margin criterion—then the utility would not be projected to be reliable from a reserve margin perspective. In other words, the utility would fail to meet its 20% minimum reserve margin criterion.

We now turn our attention to the second perspective commonly taken in utility reliability analyses.

PROBABILISTIC PERSPECTIVE (NOT SO SIMPLE TO CALCULATE)

This perspective requires analyses that are not nearly as simple as reserve margin calculations. Sophisticated computer models, or complex spreadsheets, are typically used to perform these analyses. The basic approach is to first obtain a variety of pro-jections for each day of the year(s) that is being analyzed. This information includes, but is not limited to, the following: (i) the projected highest hourly load for that day, (ii) the amount of generation that will be out of service due to planned maintenance on that day, and (iii) the projected likelihood that each individual generating unit will break on that day in which the unit is available to operate (*i.e.*, any day that the unit is not already out of service due to planned maintenance).

The likelihood that a generating unit will break is typically discussed in terms of a "forced outage rate" (FOR) for the generating unit and the FOR value is expressed as a percentage. For example, a 2% FOR value for a given generating unit means (roughly speaking) that, after accounting for the annual hours needed for planned maintenance, the unit has a 2% chance of breaking on any day it is expected to be available to operate. All else equal, the lower the FOR, the more reliable the genera-tor is. Therefore, a unit with a 2% FOR is more reliable than a unit with a 3% FOR, one with a 3% FOR is more reliable than one with a 4% FOR, etc.[5]

Using this information, the probability that the utility will not have enough gen-eration capacity available (after accounting for planned maintenance and the likeli-hood of breakage as expressed by each unit's FOR) to fully serve the electrical load for the highest load hour for each day in the year is calculated. Then the probabilities of not being able to serve the load for each day in the year are summed to derive an annual value. This calculation is often referred to as a loss-of-load-probability (LOLP) calculation.[6]

The criterion by which the results of LOLP calculations are judged is usually an annual probability value of 0.10 day per year that a utility may not be able to fully provide all of the electricity that customers demand. Loosely speaking, the LOLP criterion of 0.10 day per year means there is a 10% probability that the utility will not be able to meet all of its electrical demand at some point during the year. (The LOLP

---

[5] Note that if a generating unit's FOR were to begin to approach double digits, a utility's thoughts may turn to converting that generating unit into an artificial reef.

[6] Just as there are variations of reserve margin analysis in common use, there are also variations of the probabilistic reliability analysis approach that are referred to by names other than LOLP. These variations are similar to LOLP analysis in their basic concept and there is little difference in the actual calculations. For these reasons, and for the sake of simplicity, our discussion focuses solely on LOLP analysis.

criterion is sometimes also discussed in terms of an equivalent 10-year perspective. In this case, the same LOLP criterion is referred to in terms of a probability of 1 day per 10 years in which the utility will not fully meet its customers' load. Both ways of expressing the LOLP criterion are typically used to refer to an equal level of reliability for the utility system.)

In contrast to the reserve margin criterion, the LOLP criterion is a <u>maximum</u> threshold criterion. In other words, as long as the LOLP value calculated for a given year is *less than* 0.10 day per year, the utility is deemed to be reliable from a probabilistic perspective. However, if the calculated LOLP value is higher than 0.10 (e.g., 0.15 or 0.25) day per year, the utility is no longer deemed to be reliable for that year from a probabilistic perspective and new resources need to be added.

As previously mentioned, a full LOLP calculation is a complicated one that is performed on sophisticated spreadsheets or computer models. Therefore, a full LOLP analysis is not a simple calculation as is the case with the reserve margin reliability criterion. However, it is possible to provide a relatively simple example of how the LOLP calculation process essentially works.

In our example, let's assume that we have a very small utility with three generating units (creatively labeled as Unit 1, Unit 2, and Unit 3). We also assume that each generating unit is 50 MW in size and each has a 2% FOR. We also assume that on the particular day we shall be examining, the utility has a peak load of 100 MW and that none of the three generating units is scheduled to be on planned maintenance that day. The relevant question is: "*How likely is it that the utility will be able to serve the peak load of 100 MW that day with its three generating units?*"

We begin by taking a look at the possible "states" the utility could find itself in that day in regard to the operational status of the three generating units. There are eight possible states:

1. Units 1, 2, and 3 are all operational (*i.e.*, a total of 150 MW are operational);
2. Only Units 1 and 2 are operational (100 MW are operational);
3. Only Units 1 and 3 are operational (100 MW are operational);
4. Only Units 2 and 3 are operational (100 MW are operational);
5. Only Unit 1 is operational (50 MW are operational);
6. Only Unit 2 is operational (50 MW are operational);
7. Only Unit 3 is operational (50 MW are operational); and
8. None of the three units are operational (0 MW are operational).

At first glance, it is clear that if the utility has a peak load that day of 100 MW, then it will be able to meet that load in four of the eight states listed above because states (1) through (4) would result in the utility having either 150 or 100 MW of generation operational. But this does not tell us how likely it is that the utility will be able to meet the 100 MW load that day. In order to determine this, we need to make use of the FOR values for each of the three generating units.

Our assumption of a 2% FOR value for each unit essentially means that, for any specific unit, there is a likelihood of 2% that the unit will be broken at any point in time (such as the one day in question). Conversely, it also means that there is a likelihood of 98% that the unit will be operational on that same day. We use this information to see what the likelihood is of state (1) listed above occurring in which all three generating units are operational. This is calculated by multiplying the likelihood of Unit 1 being operational, times the likelihood of Unit 2 being operational, times the likelihood of Unit 3 being operational. In other words, the calculation becomes the following: $0.98 \times 0.98 \times 0.98 = 0.9412$. This means that there is a likelihood of 94.12% that all three units, or all 150 MW, will be operational on that day.

We next determine how likely it is that two of the three units will be operational, and the third unit will be broken, as is the case in states (2), (3), and (4) mentioned earlier. For any one of these three situations, the calculation consists of the multiplication of the likelihood of one unit being operational times the likelihood that another unit is operational times the likelihood that the third unit is broken. This calculation is as follows: $0.98 \times 0.98 \times 0.02 = 0.01921$ or 1.92%. Because there are three such possible states, this value must be multiplied by a factor of 3 which results in a value of $0.0576 (= 0.01921 \times 3)$, or a likelihood of 5.76% that only two units, or exactly 100 MW, will be operational on that day.

By looking at the likelihood that *at least* 100 MW will be operational on the day, we see that there is a likelihood of 94.12% of 150 MW being operational, and a likelihood of 5.76% of exactly 100 MW being operational. Therefore, the likelihood of the utility being able to meet its 100 MW load on this particular day is 99.88% (= 94.12% + 5.76%). Conversely, there is a likelihood of approximately 0.12% (= 100% − 99.88%) of the utility not being able to meet its 100 MW load on this particular day; *i.e.*, that one of the states (5) through (8) will occur on that day. From an LOLP perspective, the important value is the projected likelihood of 0.12% of the utility *not* being able to meet its 100 MW load on this particular day, or an LOLP value for this one day of 0.0012.

A full LOLP calculation essentially repeats this calculation for all 365 days in the year. These 365 calculations account for the highest load expected on each day and for whether one (or more) generating unit is scheduled for planned maintenance on each day. The probability values of the utility not being able to meet its projected load for each day are then summed together to obtain an annual probability of not being able to meet its projected load at some point in the year. This sum is then compared to the LOLP criterion which is typically 0.1 (or 0.100) day per year, as previously mentioned.

In our simple one-day example above, the LOLP value for that one particular day was 0.0012. If one were to assume that all 365 days had the identical LOLP value (which is highly unlikely, but useful for illustration purposes), then the projected annual LOLP for this utility would be $0.438 (= 0.0012 \times 365)$. In this case, the projected annual LOLP value of 0.438 exceeds the annual LOLP reliability criterion of a maximum 0.100 value and the utility would be projected to not be reliable from an LOLP perspective.

As this simple example suggests, a full LOLP calculation for a utility system for every day of the year is a complicated calculation. This is especially true when one considers that utility systems typically: (i) have many more than three generating units (recall that our hypothetical utility system has 10,000 MW of generating compared to the 150 MW used in this simple example), (ii) will have a different peak load value virtually every day, (iii) will need to have generating units out for planned maintenance on certain days, and (iv) individual generating units will often have different FOR values. Therefore, actual calculations of LOLP for our hypothetical utility system would be difficult to present concisely for the purpose of this book.

For this reason, the rest of this book will utilize the simpler-to-use reserve margin calculation when discussing a reliability analysis for our hypothetical utility system. But we are not quite ready to leave our discussion of the probabilistic perspective to reliability analyses. This is because we now have two reliability analysis perspectives, deterministic and probabilistic. This leads to a logical question.

WHICH RELIABILITY PERSPECTIVE IS MORE IMPORTANT?

We have two reliability analysis perspectives that a utility can, and often does, use. A logical question is whether one perspective is more important than the other when a utility is performing its reliability analyses.

Let's assume that a utility uses both perspectives in its reliability analyses of future years. Therefore, the utility develops projections of both reserve margin and LOLP for each year in its reliability analysis. For our hypothetical utility with its growing customer base and increasing electrical load, one of these perspectives will eventually show that its criterion (perhaps a reserve margin criterion of 20% or an LOLP criterion of 0.100 days per year) is not met, or is "violated," beginning in some future year. Regardless of which perspective is violated for that year, the utility will need to add resources in that year of a magnitude sufficient to either increase its generating capacity and/or lower its projected peak demand, so that the criterion for this reliability perspective is no longer violated.

Let's also assume that the utility's projections show that its reserve margin criterion of 20% will be violated in a particular year in the future, but that the LOLP criterion will be met for that same year. In this case, the reserve margin perspective can be said to be "driving" the utility's need for resources. Conversely, if the LOLP criterion was projected to be violated in that year, but the reserve margin criterion was projected to be met, then the LOLP perspective would be said to be driving the utility's need for resources. It may be helpful to think of the two criteria as two reliability "trip wires" and the purpose of reliability analyses is to see which of these two trip wires is triggered first. The one that is triggered first is the criterion that is driving the reliability analysis for that utility.

All else equal, a utility whose generating units have relatively high FORs is likely to find that reliability analysis based on the LOLP perspective is driving its need for resources. On the other hand, a utility whose generating units have relatively low FORs is more likely to be driven by the reserve margin criterion. Therefore, it is not unusual for either perspective to drive the need for additional resources for a particular utility depending upon the characteristics of the utility system.

Consequently, there is no universally correct answer to the question of which of the two reliability analysis perspectives is the more important perspective. Either of the two perspectives could be the one that drives a given utility's need for new resources.

Although it is hard to definitively answer which perspective is more important, it is not hard to answer which perspective is more commonly quoted. The reserve margin perspective is most often quoted by utilities when their reliability analyses are discussed. There are a couple of reasons for this.

First, as mentioned before, reserve margin analyses are much simpler to perform than LOLP analyses and are simpler to explain to utility regulators and the public.[7] Second, even if LOLP analyses are driving a utility's need for additional resources, a utility can use the results of its LOLP analyses to reset its reserve margin criterion at an appropriate higher level.

For example, a utility might change its reserve margin criterion from 15% to 18% if it is consistently found that the LOLP criterion was being violated with a 15% reserve margin criterion, but the LOLP criterion would not be violated if it were to use a higher 18% reserve margin criterion. In this way, its LOLP-based need for new resources may be "resolved" by the new, higher reserve margin criterion. This allows the utility to focus its reliability analyses on the simpler-to-calculate, and, perhaps more importantly, simpler-to-explain, reserve margin perspective. (However, there is a potential downside to such an adjustment: the reserve margin criterion is now being driven by the LOLP criterion. As a result, the two reliability criteria are no longer independent of each other. In my opinion, I believe there is value in having two independent criteria and not base one criterion on the results of the other criterion when attempting to project future reliability of a utility system.)

Finally, many utilities have undertaken significant efforts over the years to improve the reliability of their generating units, thus driving the FORs of their electric generating units lower and lower. This effort has often resulted in lower LOLP projections, thus making it less likely that LOLP will be driving the utility's need for new resources.

For these reasons, the reserve margin perspective is the analytical approach most often discussed in regard to utility reliability analyses that are used to answer the first and second of the three basic questions a utility resource planner must answer. Therefore, this book will use only reserve margin calculations in subsequent chapters that look at the future resource needs of our hypothetical utility system.

With this overview of how utilities attempt to answer the first and second of the three basic questions, we now turn our attention to a general discussion of the third question.

RESOURCE OPTION EVALUATION AND SELECTION: WHAT IS THE BEST RESOURCE OPTION TO SELECT FOR A GIVEN UTILITY?

Assuming that a utility has completed its reliability analyses, and that the results of the reliability analyses show that the utility needs to add a certain amount of new resource MW by a certain year, the obvious next question is "what is the best resource option(s) to add?" This is where the "fun" really begins in utility planning.[8] What the utility decides is the best answer to this third question is often the topic

---

[7] I am tempted to add "certain utility executives" to this list (but that would be wrong).
[8] This assumes a loose (and possibly perverse) definition of "fun."

of regulatory approval hearings.[9] For those readers who have never had the pleasure of attending or participating in such events, these hearings typically take place over several days (at least). The hearings come after months of writing testimony and answering written and oral questions under oath. The hearing itself involves oral testimony in which one verbally answers even more questions before that state's regulatory commission (such as the Public Service Commission or similar organization). These hearings frequently involve input from various organizations who advocate for, and/or against, certain types of resource options. This is what we call "fun," especially if you are a witness in these hearings.

But we have a bit more to discuss before you're ready to dress up nicely and take the witness stand as a participant in these hearings. In this section, we will provide an overview of this aspect of utility planning and decision-making. We will do so by briefly discussing four topics that are crucial to an understanding of the third question itself and how it may be answered. These four topics are as follows:

1. The two basic types of resource options;
2. The concept of integrated resource planning (IRP);
3. Economic evaluations; and
4. Non-economic evaluations.

TWO BASIC TYPES OF RESOURCE OPTIONS: SUPPLY AND DEMAND SIDE MANAGEMENT OPTIONS

In Chapter 2, we briefly introduced the fact that there are two basic types of resource options that utilities can select to meet a future need for additional resources. The first type of resource option includes all resource options that generate, or supply, electricity. For this reason, these options will be referred to in this book as "Supply" options. The list of Supply options includes the previously discussed five types of conventional generating units that exist on utility systems across the country: steam units that burn natural gas (with oil as a backup fuel), steam units that burn coal, gas/combustion turbines, combined cycle units, and nuclear units. The technology for some of these types of units (such as combined cycle units) has steadily advanced to the point where the currently available generating units of that type are significantly more fuel-efficient than the existing units of the same type currently found on a utility system.

The list of potential Supply options also includes a number of other types of generating units which are often renewable energy-based generators. A partial list of these, in alphabetical order, includes the following:

1. Battery/Energy storage
2. Biomass

---

[9] The first and second questions "when does the utility need to add new resources?" and "what is the magnitude (MW) of the new resources that are needed?" are less frequently the topics of regulatory hearings. This is primarily because the reliability standards/criteria that the utility uses to answer those questions are often mandated by the utility's regulatory authority and/or have been well established in prior regulatory hearings.

3. Solar photovoltaic (PV)
4. Solar thermal
5. Waste-to-energy
6. Wind turbines

In total, there are usually more than a dozen types, or variations of types, of Supply options that may be potentially applicable for a given utility. Each of these types of Supply options shares one common trait: they generate/provide electricity that can, at least theoretically, be used to meet a utility's need for additional resources. From that common point, these options differ significantly in regard to size (MW), cost, fuel use, air emissions, etc. In subsequent chapters, we will select a few of these Supply options for further examination as we work through analyses of how our hypothetical utility system might meet its growing need for electricity with a new Supply option.[10]

The second basic type of resource option with which a utility may meet its need for additional resources includes all resource options with which the utility can change its customers' demand for electricity. Traditionally, these programs have lowered customers' demand for electricity, particularly at the system's summer peak hour or winter peak hour (because it is the peak hour demand for electricity that is used in reserve margin calculations as previously seen in Table 3.1).

There are at least three "paths" by which utility customers' demand for electricity can be directly lowered:

1. Federal and/or state government-mandated appliance and lighting efficiency standards and building codes;
2. Voluntary actions taken by customers that may involve forms of electricity use not directly addressed by government-mandated standards/codes or by utility programs; and
3. Programs offered by the utility to its customers that are designed to help participating customers reduce their electrical demand and usage.

The projected effects of government-mandated appliance and lighting efficiency standards and building codes, plus projections of voluntary customer efforts to conserve (often in reaction to higher electric rates and bills), are typically recognized up front by electric utilities in their resource planning efforts. This is frequently done by incorporating these projected effects into the utility's forecasts of future electric demand (typically referred to as the utility's "load forecast"[11]). These projected impacts, especially from efficiency standards and codes, have become an increasingly

---

[10] Supply options can take the form of either utility-owned generating units or purchased power contracts. In the latter, another party owns the generating unit and sells capacity (MW) and/or energy (MWh) to the utility under contractual terms. For simplicity's sake only, our discussions throughout the book will assume that the new Supply options being considered would be utility-owned.

[11] The subject of "load forecasting" for electric utilities is a complex one that would take many pages (if not a separate book) to adequately describe. Consequently, we will simply use a load forecast in our discussion of resource planning for our hypothetical utility system and not attempt to discuss how this load forecast was developed.

large factor in utilities' load forecasts, lowering the projected load below what it otherwise would have been.

Having already attempted to incorporate the projected energy reduction effects of the first and second paths listed above in their load forecasts, utilities then turn to the third path: utility programs. These utility programs will be referred to as demand side management (DSM) programs/options. There are two basic types of DSM programs: load management programs and energy conservation programs.

There are two basic types of DSM programs that have traditionally been offered by utilities. Both types of DSM programs are designed to reduce the participating customers' electric load (kW) at the utility's summer and/or winter peak hours and to lower the total amount of electricity (kWh) the participating customers will use over the course of a year. The two types of DSM programs can be described as follows:

1. *Load management programs* generally take one of two forms: (i) direct utility control ("load control") of customers' appliances/equipment (*i.e.*, the utility can remotely turn off, or otherwise regulate, the equipment's use of electricity during high load periods); or (ii) special electric rate structures (that feature higher prices during peak load hours, and lower prices during low load hours) that encourage a reduction of electricity usage during, and/ or a shift of electricity usage away from, the utility's peak hours. In either case, load management programs often result in relatively small impacts on the total amount of energy (kWh) the customer consumes over a year compared to energy conservation programs.[12]

2. *Energy conservation programs* typically do not feature direct control of the customer's load by the utility or offer special electric rate structures. Instead, financial incentives are typically offered by the utility to encourage customers to purchase a more efficient appliance, install higher levels of insulation, etc. Compared to load management programs, energy conservation programs generally result in relatively larger impacts on the total amount of energy a participating customer consumes over the course of a year. On the other hand, energy conservation programs may result in lower peak hour demand (kW) savings than is the case with load management programs.[13]

It is appropriate at this time to mention that there is a "flip side" to utility DSM programs designed to reduce electric usage. These "flip side" utility programs result in increasing, not decreasing, the demand for electric usage. An example of such a DSM program would be one that encourages the use of electric vehicles (EVs). Such a program would certainly result in increased annual energy (kWh) use and, in all likelihood, also increase electrical demand (kW) in the utility's system peak hour. However, in Part I of this book, the discussion of utility DSM programs will focus solely on DSM programs that reduce electrical demand. We will return to discuss utility DSM programs that increase electrical demand in Part II of this book.

---

[12] Load management type DSM programs are also referred to as "demand response" programs.
[13] Similarly, energy conservation-type DSM programs are also referred to as "energy efficiency" programs.

Just as there are numerous Supply options a utility will have to choose from, there is also a very long list of DSM options that may be applicable to a utility based on its location, climate, customers' electric usage patterns, etc. In fact, the list of potential DSM options may include hundreds, if not thousands, of possibilities.

Even from this brief overview of the two basic types of resource options (Supply options and DSM options) a utility has to choose from in order to meet its resource needs, it is readily apparent that these two basic types of resource options are fundamentally different. One type of option is designed to supply electricity and one type of option is designed to lower the demand for, and usage of, electricity. So how should a utility proceed to compare these two very dissimilar types of resource options in order to determine which option(s) is best for its customers? I'm glad you asked.

INTEGRATED RESOURCE PLANNING (IRP)

The concept of "integrated resource planning" (IRP) has been around since at least the late 1980s. The basic concept behind an "integrated" resource planning approach is to ensure that both types of resource options, Supply and DSM, are evaluated on what I will term a "level playing field." Let's explain what we mean by this term. This term means that analyses are performed in a manner that shows no preference or bias toward either type, Supply or DSM, of resource option. In more recent years, the term has broadened to encompass unbiased analyses of fossil-and-nuclear-fueled Supply options versus renewable energy-based Supply options.

Utilizing an unbiased IRP approach for analyses of resource options helps ensure that a wide variety of resource options is examined, thus increasing the likelihood that the best possible choice of resource options will be selected for a utility's customers. The logic inherent in using such an analytical approach is readily apparent, and by the early 1990s, utilities across the country regularly performed analytical approaches that were designed to implement the IRP concept.

However, also about that time, a number of states and utilities began a move toward an unregulated utility environment (and its underlying premise of lower electric rates that would, in theory, be brought about by greater competition). As a result, these states and utilities largely abandoned the IRP concept because these utilities had been "split" apart into separate companies. A number of companies would generate electricity and offer that electricity to customers; another company or organization would be in charge of transmitting electricity from the generators to specific geographic regions; then yet another company would distribute electricity to individual customers within the regions.

In such cases, there was less need to perform any type of planning except for the specific function (generation, transmission, or distribution) each company was responsible for. Furthermore, the two different types of resource options, Supply and DSM, would now typically be handled by two different companies. Typically, Supply options would be handled by a generation company, and DSM options would be handled by a distribution company or other entity. As a consequence, the IRP approach to evaluating the two types of resource options was usually abandoned by those states that had decided to pursue the path of an unregulated utility structure.

However, as previously mentioned, I believe that this "experiment" of an unregulated utility environment has had mixed results to-date. As a consequence, some of these states and utilities have moved back in a direction in which they attempt to incorporate at least some facets of an IRP approach.

At this point, it is useful to provide my definition of IRP:

Integrated resource planning (IRP) is an analytical approach in which both types of resource options, Supply and DSM, are analyzed on a level playing field. For each resource option, an IRP analysis accounts for all known cost impacts on the utility system that are passed on to its customers through the utility's electric rates. In addition, non-economic impacts to the utility system from the resource options are also evaluated. In this way, IRP analyses result in a comprehensive competition among resource options.

The concept that resource options must *compete* with each other is explicit in this definition of IRP. In fact, this concept is a key principle of IRP analysis. Only in this way can a utility truly identify the best option(s) for its customers. It is necessary to include all of the cost impacts to the utility system, that will be accounted for in electric rates, that will result from the selection of a resource option. This is because, if one does not account for all of these cost impacts for a particular resource option, the competition can no longer be unbiased and the results will not be accurate.

It is my belief that the IRP approach is the best way to analyze all resource options for electric utilities. That stated, it is instructive to point out that utilities in a number of states that profess to use an IRP approach actually use a modified version of my definition of IRP. This situation can occur for a variety of reasons, but the situation occurs most often because of state regulations that mandate a certain amount of a particular type(s) of resource option(s) to be included in a utility's resource plan. Examples of types of resource options that are currently most frequently mandated at certain levels are DSM and/or renewable energy options. In cases where specific levels of selected resource options such as these are mandated, the utility is no longer utilizing a true IRP approach according to the IRP definition just given.

However, a utility may then apply a modified IRP approach in which it "works around" these mandated levels of selected resource options to determine the optimum selection of Supply and DSM options to meet the *remaining* resource needs of the utility not addressed by the mandated resources. In such cases, the utility is seeking to utilize as much of an IRP approach as possible given the mandates it faces. Another way to state this is that a restricted IRP approach is used that accounts for certain constraints (*i.e.*, the mandates).

We will return to aspects of my IRP definition at various times in the book as we discuss various types of evaluations or analytical approaches. In Chapter 8, we will again discuss the issue of mandates/constraints that prevent a true IRP approach from being used and why a decision to step away from an IRP analytical approach can create problems for utilities and, most importantly, their customers.

With this basic definition of IRP analyses in hand, we now turn our attention to the first of two basic types of evaluations a utility undertakes in its IRP work: economic evaluations.

ECONOMIC EVALUATIONS

Yes, we have now gotten to the point at which we talk about what these resource options cost. And when discussing resource options for electric utilities, especially Supply options, we are talking about very large sums of money.

We start the discussion of economic evaluations by returning briefly to the IRP definition just provided. In the definition, an IRP analysis is described as an analysis that "... *accounts for all known cost impacts on the utility system that are passed on to its customers through the utility's electric rates.*"

Inherent in that description is the concept that any resource option will have a variety of cost impacts on the utility system as a whole. Obviously, the actual cost to build or add a particular resource option will need to be accounted for. However, there are a number of other cost impacts. For example, the addition of a new generating unit will result in changes in how the utility dispatches its current generating units. For example, if our hypothetical utility were to add a new nuclear unit and operate it as a baseload unit, our utility's marginal natural gas-fueled units would likely not operate as much. A change in the dispatch order of a utility's existing system of generating units will result in changes in system fuel costs and system environmental compliance costs.

Similarly, the addition of a DSM option will, by design, result in changes to the electrical load that the utility must serve. This changing electrical load shape, or pattern of electrical use, will also result in changes in how the utility dispatches its current generating units, as well as changes in system fuel costs, system environmental compliance costs, and the number of sales with which utilities recover their costs through their then-current electric rates. Therefore, economic analyses of resource options must account for all such changes (and related cost impacts) in the operation of the utility system as a whole.

Between the many Supply and DSM options that a utility can choose from, the utility may be faced with the daunting task of having to analyze hundreds of potential resource options. For that reason, a utility may utilize a two-step approach in its economic evaluation.

First, a *preliminary* economic evaluation may be used to eliminate some of the options. (Note that this preliminary economic evaluation is not a necessary step in an IRP approach, but it may be a useful step to take.) This step is often called a preliminary economic "screening" of options in which the less economically competitive options are "screened out." This preliminary economic screening work can sometimes be performed using a spreadsheet approach. However, because of the inherent differences between Supply and DSM options, and the very different impacts each type of option may have on a utility system, the preliminary economic screening approaches for each of the two types of resource options typically differ as we shall discuss shortly.

Second, those options that survive the preliminary economic screening then compete with each other in what we will term a *final* or *system* economic evaluation to identify the option or options that are truly the best economic choice for the specific utility system.[14] In order to perform this work correctly, sophisticated resource planning computer models are used to evaluate Supply and DSM options in a way that accounts for all system cost impacts. This ensures that a level playing field exists for the analysis of resource options.

Continuing with our "overview" theme of this chapter, we will briefly discuss these preliminary and final economic evaluation approaches. In subsequent chapters, we will flesh out these economic evaluation approaches with examples that utilize our hypothetical utility system in the analyses of specific resource options.

PRELIMINARY ECONOMIC SCREENING
EVALUATION OF SUPPLY OPTIONS

Supply options typically can vary in size from less than 1 MW to approximately 2,000 MW. Intuitively, it makes sense that a Supply option of, for example, 2,000 MW may have very different impacts on a utility system and its costs than a 1 MW option. This difference in size is not a problem when evaluating these options in the final economic evaluation, because the final evaluation will utilize a sophisticated computer model that can accurately accommodate these different sizes in the evaluation. But if the utility is considering a large number of Supply options, it is often very time consuming to run these computer models for so many options. It would be helpful if some of these options could be screened out up front using a simpler analytical approach.

The question is how to address such huge differences in size without incurring the cost and staff person-hours necessary to use the sophisticated computer models. The most commonly used approach is called a "screening curve"[15] approach that is performed on a spreadsheet. The screening curve approach can be described as follows for an analysis of a generating unit:

1. The perspective taken in this approach is that of placing each Supply option to be compared "out in a field" by itself, *i.e.*, the Supply options are <u>not</u> connected to the utility system. (As a consequence of this basic assumption that the Supply option is not connected to a utility system, a number of cost impacts which the addition of the Supply option will actually have on the utility system are omitted in this analytical approach. One example of such an omitted cost impact is the fact that the new Supply option will cause the dispatch of, and thus the costs of operating, other generating units already existing on the utility system to change.)

---

[14] "Final" economic evaluations are also referred to as "system" economic evaluations because such evaluations ensure that all of the system cost impacts are accounted for. These system cost impacts are often not fully accounted for in preliminary economic screening evaluations as will be discussed later.
[15] The most commonly used screening curve approach is also referred to as a "levelized cost of electricity (LCOE)" approach.

2. The Supply option is assumed to operate for a number of hours per year.[16] Typically, a range of potential operating hours per year is examined.
3. All of the annual costs of building, maintaining, and operating the Supply option over each year of the option's projected life (typically 25 or more years) are then calculated.
4. These annual costs are first converted into what is termed "present value" costs. Then the present value costs are used to develop a levelized annual cost value.[17]
5. These levelized annual costs are then divided by the number of MWh of electricity the generating unit is assumed to produce annually. (The annual MWh value is simply the product of multiplying the MW of the unit by the number of hours per year the generating unit is assumed to operate.)
6. The result is a projected levelized cost to build and operate the generating unit and the result is typically expressed in terms of $/MWh or (equivalently) cents/kWh.
7. As previously mentioned, the calculation is often repeated using a range of potential annual operating hours (*i.e.*, capacity factors). These results are often graphed. The shape of the graph is a curve, thus leading to the "screening curve" label commonly applied to this analytical approach.

Figure 3.1 shows the results of a typical screening curve analysis for two competing Supply options.

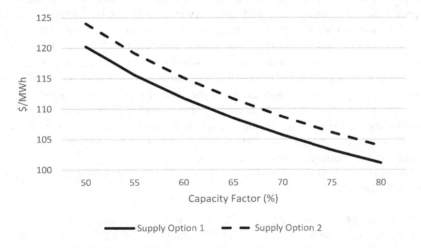

FIGURE 3.1 Preliminary economic analysis: Screening curve approach with levelized $/MWh costs for two CC supply options.

---

[16] The number of hours a Supply option (generating unit) operates per year is referred to as its "capacity factor" which is expressed in terms of the percentage of hours in the year the generating unit runs. The number of hours the generating unit runs is divided by the 8,760 annual hours to derive its capacity factor. As we discussed in Chapter 2, the actual number of hours a generating unit will operate is determined primarily by its operating cost; the lower the operating cost, the more hours the unit will run (thus the higher its capacity factor will be).

[17] Please see Appendix C for a discussion of present value costs and levelized costs.

This figure shows that the projected levelized costs for Supply Option 1 are less than the projected levelized costs for Supply Option 2 throughout the range of capacity factors examined (from 50% to 80% of the hours in a year). Therefore, the cost of producing an MWh *solely from each of these generating units*, without any consideration of the impacts that either of the new generating units will have on the existing generating units on the utility system, will be lower with Supply Option 1 than with Supply Option 2.

The theory is that the lower the $/MWh value for a given Supply option, the more economic the option is. *However, the theory only holds true if the Supply options being compared are identical, or at least very similar, in regard to each of four key characteristics.* (For illustrative purposes at this point, we will mention that one of these key characteristics is the size (MW) of the resource options being considered.[18]

In rare cases in which the Supply options being compared are identical, or very similar, in regard to all four key characteristics, a screening curve approach can be useful in identifying the more economical of the Supply options being compared. However, the value of a screening curve analytical approach *becomes virtually meaningless* when any of the four key characteristics are dissimilar from the other Supply options being compared.

This is because when any of the four key characteristics are dissimilar, the fact that a screening curve approach does *not* take into account a number of cost impacts of the Supply option being connected to the utility system becomes of overriding importance. (Recall that the screening curve perspective is of a generating unit "alone in a field," unconnected to the utility system.) Therefore, a screening curve approach not only can, but also will, give very inaccurate results in regard to which of these dissimilar Supply options is really the most economical choice for the utility.

Unfortunately, this fact is often overlooked and/or not understood. As a consequence, one may perform an analysis of vastly dissimilar resource options using a screening curve approach and believe that this analysis provides the answer of which resource option is more economical. Over the years, I have seen this mistake made numerous times by novice utility resource planners and by others who are trying to analyze resource options (or promote a selected resource option.) This mistake is usually a fatal one in terms of meeting the objective of accurately determining the most economic resource option for a utility system.

Therefore, we will now introduce my second Fundamental Principle of Electric Utility Resource Planning:

Fundamental Principle #2 of Electric Utility Resource Planning: "System Cost Impacts of Producing or Conserving Electricity are of Upmost Importance. Individual Resource Option Costs of Producing or Conserving Electricity are of Little or No Importance When Considered Separate from the Utility System as a Whole."

The projected costs of producing/conserving electricity for any *individual* resource option by itself, often expressed in terms of cents/kWh or $/MWh, whether

---

[18] We will discuss in detail what each of these four key characteristics is in Chapter 5. We will also discuss the fundamental problems inherent in a screening curve approach in Appendix D.

a Supply or DSM option, is of little/no consequence when performing economic evaluations, the objective of which is to select the most economic resource option for the utility system as a whole. When selecting a resource option for a particular utility, the objective is to identify the resource option that results in the lowest electric rates that are charged to customers. Only an analysis that accounts for all of the cost impacts that a resource option will have on the entire utility system can determine the most economic resource option.

As previously mentioned, we will return to a discussion of the screening curve approach, the four key characteristics of resource options, and the limitations to the use of screening curve analyses, in Chapter 5. In that chapter, we will flesh out this overview discussion by performing analyses that utilize our hypothetical utility system to examine Supply options. For now, simply keep in mind that a screening curve can be useful *only if each of four key characteristics of the Supply options being compared is identical or very similar.* If this is not the case, the results of a screening curve approach have little or no value. And, most importantly, a screening curve approach should *never* be used to make a final decision of which resource option to add to a utility system.

PRELIMINARY ECONOMIC SCREENING EVALUATION OF DSM OPTIONS

The size of DSM options also offers a challenge in the evaluation of these options, but the challenge is a bit different than the type of "size challenge" previously discussed when evaluating two (or more) competing Supply options. With Supply options, the challenge was how to compare one generating unit that could supply 1 MW with a second generating unit that could supply approximately 2,000 MW.

However, the challenge when evaluating competing DSM options is not due to a significant size difference from one DSM option to another. That is because the vast majority of DSM options will reduce load at a single participating customer's premises (home or business) by a relatively small amount. Consequently, there are seldom large differences in the amount of demand reduction per participating customer between two DSM options. For example, DSM options that are designed for residential and small commercial customers often are projected to have demand reduction values per participating customer of 1 kW or less. (Recall that 1,000 kW = 1 MW, so DSM's impact for one participating customer is typically much smaller than even the smallest Supply option.) Therefore, many DSM options will all have roughly the same demand reduction impact (approximately 1 kW) for a single participating customer.[19]

The challenge in evaluating DSM options is how to determine if signing up large numbers of customers to participate in any DSM option that reduces load by 1 kW (or less) per participating customer is a better economic decision than adding a new generating unit that, for example, may provide 1,000 MW of electrical generating capacity.

---

[19] DSM options for large commercial and industrial customers can have significantly larger demand reduction values per participating customer. However, as a general rule, individual DSM options typically vary much more in regard to the amount of energy (kWh) reduction they achieve for a single participating customer than they do in regard to the amount of demand (kW) reduction.

This problem is often addressed in final (or system) economic analyses by combining a number of DSM options that each is projected to have thousands of participating customers.[20] This approach results in a large amount of MW reduced by the DSM options competing against a comparable amount of MW of generating capacity from a Supply option. But how does one decide which DSM options are worth pursuing in an effort to sign up thousands of participating customers? In other words, how do we perform preliminary economic screening evaluations that can be used to screen all potential DSM options?

The answer is at once both similar to and different from the approach used in the screening curve analyses used to perform preliminary economic screening analyses of Supply options. The DSM preliminary evaluation approach is somewhat similar to a screening curve approach for Supply options in the sense that it also seeks to evaluate options on a comparable size basis. Recall that in screening curve analysis, each Supply option was compared on a $/MWh basis. In other words, what is the cost of producing a comparable amount of electricity (1 MWh)? This approach can also be viewed as essentially evaluating Supply options from a 1 MW size perspective.[21]

However, the screening curve analysis approach used for Supply options is based on comparing one Supply option versus another Supply option. The DSM preliminary economic screening approaches that are most in use are different in that respect. This DSM approach does not compare one DSM option versus another DSM option, but instead compares one DSM option versus one Supply option.[22]

The actual DSM preliminary economic analysis approach combines these two facets by comparing the DSM option to a comparable-sized generating unit. In practice, this means that one must "shrink" the generating unit down to a size comparable to the DSM option. This is typically done by assuming that the generating unit is 1 kW in size. One then calculates the costs for this 1 kW "mini" generating unit by dividing the total costs of the full-size generating unit by the total capacity (in terms of kW) to derive a $/kW cost. For example, assume that the cost to build the full-sized generating unit is $200 million and its capacity is 400 MW. The 400 MW equates to 400,000 kW.

Therefore, by dividing the $200,000,000 cost by 400,000 kW capacity, one derives a "pro-rata" $/kW cost to build the generating unit of $500/kW. One then assumes that a 1 kW-sized "mini" generating unit can, hypothetically, be built for the same cost.[23] A 1 kW mini-sized generating unit with a cost of $500/kW can then be readily compared to similar-sized DSM options.

---

[20] In Chapter 6, we will simplify our discussion by assuming just two DSM options, each with many thousands of participants.

[21] An equally valid way to view this approach is to say that the approach completely ignores the size of the Supply options being evaluated.

[22] The Supply option selected for this comparison should be the type of generating unit the utility would actually add if it were going to address the need for new resources with a new generating unit. To choose any other type of generating unit would render the subsequent analysis relatively meaningless.

[23] In reality, it is not possible to do this. However, this simplifying assumption is useful in performing preliminary economic screening of individual DSM options.

There are three preliminary economic evaluation "tests" that are commonly applied when evaluating potential DSM options. One of these DSM preliminary "cost-effectiveness tests" (as they are commonly called) is the Participant test. The Participant test is designed to provide the perspective of a customer who might participate in the DSM option if it were offered by the utility. The other two commonly used tests, the Rate Impact Measure (RIM) Test and the Total Resource Cost (TRC) Test, are, in theory, designed to provide the perspective of whether the utility should offer the DSM option.[24] (We will examine the three tests in detail in Chapter 6.)

The results of each of the three tests are presented in terms of a benefit-to-cost ratio. In regard to the Participant test, the sum of the various benefits the participant receives is divided by the sum of the various costs the participant must pay. (The participating customer's benefits include annual savings on the electric bill, incentive payments received from the utility, and government tax credits, if applicable. The participating customer's costs include the up-front costs of buying and installing the more efficient appliance or installing more insulation, etc., plus any ongoing maintenance costs, if applicable.) The result of this division, using the present value of benefits and the present value of costs, produces a benefit-to-cost ratio.

If this benefit-to-cost ratio is 1.0, then the projected benefits exactly equal the projected costs, and the customer would theoretically be indifferent to participating in this DSM program if it were offered by the utility. If the ratio is greater than 1.0, then benefits are greater than costs, and, all else equal, it is in the customer's best economic interest to participate in the DSM program. Conversely, if the ratio is lower than 1.0, then benefits are less than costs and, all else equal, it is not in the customer's best economic interest to participate in the DSM program.

The RIM and TRC tests also use a benefit-to-cost ratio to present the results of each test. However, with these tests, the benefits represent the benefits to all of the utility's customers, participants, and non-participants alike, from implementing the DSM option and not building and operating the generating unit that otherwise would have been built. Likewise, the cost values accounted for by these two tests represent DSM-related cost impacts to all of the utility's customers from implementing the DSM option. And, just as with the Participant Test, both the benefits and the costs are presented in present value dollars.

In general terms, the benefits of implementing DSM are the costs that are *avoided* by not building and operating the generating unit and by not serving as much energy to customers. (In other words, DSM benefits in these two tests are avoided costs.) The DSM cost impacts are the cost impacts associated with implementing and operating the DSM program.

Just as there were cautionary notes presented regarding using a screening curve analytical approach to perform preliminary economic screening evaluations of Supply

---

[24] These two tests, RIM and TRC, can be thought of as the "basic" DSM cost-effectiveness tests. There are other tests, including the Utility Cost test and the Societal test. These additional tests are essentially variations of the RIM and TRC tests. By understanding the RIM and TRC tests, one will also gain an understanding of the key components of these test variations as well. Consequently, our discussion will focus solely on the RIM and TRC tests.

options,[25] a cautionary note should also be made at this point regarding the use of DSM cost-effectiveness tests. However, the cautionary note does not apply to the Participant test. This test is widely recognized as a valid and necessary test to undertake when a utility is considering offering a DSM program. After all, if a customer won't benefit from the DSM program, why would/should a customer participate? Therefore, the Participant test is an essential test to undertake when evaluating DSM options.

The cautionary note is aimed at one of the other two tests that are designed to provide the perspective of whether the utility should offer the DSM option. A number of intelligent people reading this (and I am sure you are one of them) will be quick to ask the following: *"Why does one need two tests that are designed to provide the same perspective?"* The answer is that you don't need both, you only need one of these tests. But which of these two tests, RIM and TRC, should really be used?

This has been the subject of debate at one time or another in virtually every state, and for many electric utilities, during the last several decades. In short, no universal consensus has yet been reached. This is unfortunate because one of these tests is clearly the logical choice in regard to the IRP concept of evaluating all resource options on a level playing field in which all cost impacts to a utility's customers that are reflected in the utility's electric rates are accounted for.

However, this conclusion is more easily explained using examples. Therefore, we will return to discuss these two tests in Chapter 6 when we flesh out the cost-effectiveness tests in the course of examining DSM options for our hypothetical utility system. For now, keep in mind that although these tests are commonly used for performing preliminary economic screening of DSM options, these preliminary tests (just as is the case with screening curve analysis results for Supply options) should *never* be used to make a final decision regarding which resource options should be added to a utility system. A more comprehensive analysis is needed to make a final decision regarding the resource options, *i.e.*, a final (or system) economic evaluation.

FINAL (OR SYSTEM) ECONOMIC EVALUATIONS

A final (or system) economic evaluation is one in which those resource options, Supply and DSM, that have survived the preliminary economic screening analyses (if preliminary economic analyses have been performed) are evaluated using analyses that account for all cost impacts to the utility system. This ensures that the competing resource options are evaluated on a level playing field to determine which options are the best choices for the utility system to meet the utility's future resource needs.

A utility's final (or system) economic evaluation is an IRP approach that includes several attributes. First, the evaluation should address all of the potential resource options that remain after any preliminary economic screening evaluation has been

---

[25] Some unfortunate attempts have been made to use a screening curve (cents/kWh) analysis approach to compare DSM options to Supply options. Unfortunately, the fact that DSM and Supply options are very dissimilar options violates the rule that this analytical approach can provide meaningful results *only* if the resource options being compared are identical, or very similar, in regard to each of four key characteristics. In other words, a screening curve approach cannot be used to produce meaningful results in comparing DSM and Supply options.

completed, and which could realistically meet the utility's projected resource needs. As previously discussed, if preliminary economic screening evaluations are utilized, a number of Supply and/or DSM options may have been screened out in these preliminary evaluations.

Second, the evaluation should include all of the cost impacts that each resource option will have on the utility system as a whole. Third, the evaluation should ensure that the reliability criterion (*i.e.*, the minimum reserve margin percentage and/or the maximum LOLP value) the utility utilizes is not violated by the selection of a particular resource option.

For these reasons, IRP final economic evaluations typically take the form of a comparison of different "resource plans" with each resource plan, including one (or more) of the competing resource options. A resource plan is simply a projection of possible combinations of Supply and/or DSM resources, including those that the utility already has, that the utility could add to meet its projected resource needs over a number of years.

For example, suppose a utility is considering whether to add a new combined cycle generating unit to address a projected need for new resources in 5 years, or to add a comparable amount of DSM by the end of that same 5-year period. The utility would construct two resource plans that met their reliability criterion for each year in the analysis period. In one resource plan, the new combined cycle unit would be added, but not the DSM. In the second resource plan, the needed amount of DSM would be added, but not the combined cycle unit.

If the projected economic life of a combined cycle unit is assumed to be 25 years, the analysis would address the 5 years from the present to when the new combined cycle unit would be added, plus the 25 years of economic life for the unit from that point. In other words, the analysis would address, at a minimum, a 30-year time period.

The first resource plan, Resource Plan A, would add the combined cycle unit in year 5. The second plan, Resource Plan B, would add the needed amount of DSM[26] so that by year 5, both resource plans would meet the utility's reliability criterion. Then additional resource options would be added to both resource plans in years 6 through 30 so that the reliability criterion would continue to be met for both plans for all years in the evaluation.[27] This ensures that both of the competing resource plans will meet the utility's reliability criteria in all years, thereby making the economic evaluation results more meaningful than would be the case if the two plans differed significantly in one or more years in regard to how reliable the utility system would be.

The idea is to eliminate any significant variations in the reliability of the two resource plans to ensure that the economic differences between the two resource plans are meaningful. In practice, utilities usually attempt to further eliminate significant variations between the two plans after its "decision year." (In this example, year 5 is the decision year because it is the year in which the utility first needs new resources.) This can be done by assuming that all of the new resources that are added

---

[26] The term "needed amount of DSM" simply refers to the amount of DSM that allows the utility to meet its reliability criterion. This concept is discussed in more detail in Chapters 4 and 6.

[27] As will be discussed in Chapter 6, DSM options typically have shorter life "terms" than do new generating units. We will discuss how this difference between the two types of resource options should be accounted for in developing resource plans involving DSM options in Chapter 6.

to the resource plan in years <u>after</u> the decision year to maintain reliability are the same type of resource. For example, the utility might assume only combined cycle generating units would be added in these latter years. This approach, often termed a "filler unit" approach, minimizes/eliminates other variables that could complicate the economic evaluation results and make it harder to be sure of which of the two resource options is the best choice for the decision year.

Once two resource plans are constructed that are of comparable reliability, the two plans are evaluated using sophisticated computer models that determine all utility system costs for each resource plan over the 30-year analysis period.[28] The economic impacts that will result from each resource plan are then compared in order to determine which resource plan is the best economic choice for the utility's customers. The competing resource option that is included in the most economic resource plan in the decision year is thus identified as the most economic resource option for the utility's customers for the decision year.

The economic impacts of the resource plans can be compared on two economic bases: a total cost basis or an electric rate basis (in which the costs are divided by the total kWh of electricity used by customers to derive the cents/kWh electric rate that customers are charged). The most meaningful way to compare two resource plans is on an electric rate basis.

To see why this is the case, we first assume that we are comparing two resource plans that contain only Supply options (e.g., a new combined cycle unit and a new combustion turbine unit). The two resource plans are typically compared in regard to total costs that are referred to in terms of the "cumulative present value of revenue requirements (CPVRR)."[29] Suppose that Resource Plan X (the plan with the new combined cycle unit) was $100 million CPVRR less expensive than Resource Plan Y (the plan with the new combustion turbine unit). Resource Plan X is clearly the more economical choice from a cost basis. But is it also the economical choice on an electric rate basis?

The answer is "yes." To better understand this, think of an electric rate as a simple fraction. The numerator (*i.e.*, the top value in the fraction) is the cost of the resource plan. The denominator (*i.e.*, the bottom value) is the number of kWh the utility serves. If the denominator is identical in two fractions, then the fraction with the higher numerator will result in the larger value. For example, 3/4 is a larger value than 1/4.

When only Supply options are being compared in resource plans, the resource plans will serve the same number of kWh (*i.e.*, the denominators are identical) because Supply options do not change the amount of electricity that customers use. Consequently, the resource plan with the lowest cost (*i.e.*, the smallest numerator) will also be the resource plan with the lowest electric rate.

To flesh out this example using a simple example, suppose a (really small) utility would serve 1,000 kWh with either of two resource options. If the utility's total cost is projected to be $90 with one resource option, and $100 with another option.

---

[28] At the time this book is written, there are a number of computer models that are commonly used in IRP analyses. The number and names of such models are subject to change from time-to-time. For that reason, the names of applicable computer models will not be used in the book.

[29] "Revenue Requirements" and CPVRR are simply "utility-speak" terms for the utility's total costs for each resource plan. See Appendix B for a definition of each term and see Appendix C for numerical examples of these terms.

Assuming all else equal, from a total cost perspective, the option resulting in the $90 cost clearly is the economic choice. The same result occurs if we switch to an electric rate perspective because this option will result in an electric rate of $90/1,000 kWh = $0.09/kWh (or 9 cents/kWh), while the other option will result in an electric rate of $100/1,000 kWh = $0.10/kWh (or 10 cents/kWh).

Therefore, once it has been determined which resource plan (among the resource plans in which only Supply options are being evaluated) is more economical from the perspective of resource plan total costs, that same resource plan will also be the more economical plan from an electric rate perspective as well.

Consequently, when evaluating resource plans that examine only Supply options, it is *not necessary* to take the additional step of examining the resource plans from an electric rate basis because we already know the outcome. (However, utilities may choose to perform a rate analysis in order to more clearly indicate the level by which a utility's electric rates will change.)

This is obviously not the case when comparing resource plans in which at least one of the resource plans contains a different amount of DSM. This is due to the fact that different amounts of DSM will result in differing amounts of energy (kWh) reductions occurring, thus changing the amount of kWh that the utility will serve. Or, in other words, the denominators for the two resource plans will no longer be identical as is the case when considering only Supply options.

To continue our earlier example, suppose that Resource Plan X (which features the combined cycle unit in the decision year) is now compared to Resource Plan Z, which features a comparable amount of DSM being added by the decision year. In this case, if one were to stop work after determining the total costs of each resource plan, might one be selecting a resource plan that resulted in higher electric rates (which would clearly not be the customers' choice for the best resource plan)? The answer is "yes," one might be doing just that unless one takes the next step of actually evaluating the projected electric rates for each resource plan.

Continuing our simple example, suppose that one were to stop work after determining that a DSM-based Resource Plan Z would result in a lower total cost than Resource Plan X (e.g., $89 for Resource Plan Z versus $90 for Resource Plan X). However, after taking into the account the change in the denominator caused by the DSM-induced reduction in kWh served by the utility (let's assume a reduction of 50 kWh), the electric rates for Resource Plan Z would be $89/(1,000 − 50 = 950 kWh) or $0.94/kWh (or 9.4 cents/kWh). The utility's customers would clearly prefer Resource Plan X because the electric rates they will be charged would be lower, 9 cents/kWh versus 9.4 cents/kWh.[30]

Therefore, the correct way to evaluate competing resource plans in which even one resource plan contains a new DSM option is *to compare them from an electric rate perspective* in the final (or system) economic evaluation. As we have seen, it is generally acceptable to take a shortcut and look only at the resource plan total costs when only Supply options are being compared (because for Supply options,

---

[30] Just as we utilized a cumulative present value perspective to discuss utility costs over a number of years, we will also use a cumulative present value perspective of electric rates that will be introduced and discussed in Chapters 5, 6, and 7.

the lowest cost resource plan is automatically the plan with the lowest electric rates), but never when DSM options are being compared to Supply options.

We will return again to the topic of IRP final (or system) economic evaluations in Chapters 5 through 7 where we will flesh out our discussion of conducting a final or system economic evaluation of both Supply and DSM options utilizing our hypothetical utility system.

Before leaving this overview of the subject of final economic evaluations, another point needs to be made. The discerning reader (and you should definitely know if you are one by now) will recall that we included cautionary notes when discussing the preliminary economic screening evaluations for both Supply and DSM options. The cautionary notes basically said that final resource decisions should *never* be based on the results of preliminary economic evaluations; one needs to perform a complete final (or system) economic evaluation which accounts for all impacts on the utility system. There is still one more cautionary note to make.

The resource plan and/or resource option that emerges from the final economic evaluation as the best economic choice may not always be the best *overall* choice. This can occur if there are important non-economic considerations by which the resource plans and/or resource options will also be judged. We will discuss a few of these non-economic considerations next.

NON-ECONOMIC EVALUATIONS

We start by first discussing what we will refer to as "non-economic" considerations. The items that we will discuss as examples of non-economic considerations may also have been considered in part in the final economic evaluations, but not always. We will briefly discuss three examples of such considerations.

I refer to these as non-economic considerations because they are usually referred to in non-economic terms such as "years," "tons," "mmBTU," or "percentages." In addition, the non-economic evaluation results for a given resource plan in regard to at least a few of these considerations are often compared to limits or standards that are themselves not discussed (at least directly) in terms of economics.

The three examples of considerations that we will discuss in this chapter are as follows:

1. The length of time it takes before a resource plan and/or resource option becomes the economic choice;
2. The utility system's fuel usage for a resource plan; and
3. The utility system's air emissions for a resource plan.

These three considerations are often used, to varying degrees, by electric utilities in their resource planning work.[31] We will discuss each of these considerations separately.

---

[31] There are other non-economic considerations that are used in utility resource planning. Two of these will be discussed in Chapter 8 in our discussion of "constraints" on resource planning. These two additional considerations are relatively new at the time this book is written and, therefore, are not yet in as widespread use as the three considerations discussed in this chapter. In addition, other non-economic considerations beyond those discussed in this book exist currently and new considerations are continually emerging.

NON-ECONOMIC CONSIDERATION EXAMPLE (I): THE LENGTH OF TIME IT TAKES BEFORE A RESOURCE PLAN BECOMES THE ECONOMIC CHOICE

Suppose one has decided that a particular resource plan is the economic choice, *i.e.*, it results in the lowest electric rate for the utility's customers. But is it the resource plan that most of the utility's current customers would actually choose today? If the resource plan provides the lowest electric rate overall during the time period addressed by the analyses, why wouldn't it be the choice of the utility's current customers?

To answer that, let's take a look at "*when*" the resource plan that is deemed to be the most economic actually becomes the most economic plan. Note that there are several ways in which one could look at a "time" question in regard to the economics of two (or more) resource plans. One could look either at costs (when examining only Supply options) or at electric rates (when comparing DSM and Supply options). And one could look at either nominal or present value costs or electric rates. Each different way of looking at this economic "timing" issue can provide a somewhat different perspective from which to view the resource plans. The value of each perspective may vary on a case-by-case basis.

For the sake of simplicity in our discussion, we will select and examine only one such perspective: cumulative present value costs. We now take a look at the results of two resource plans, Resource Plan A and Resource Plan B, which are used to evaluate two competing Supply options. These two resource plans are compared in regard to their CPVRR costs.[32] We start with Figure 3.2.

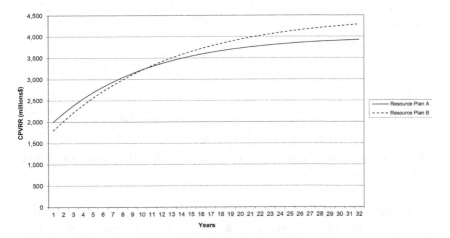

FIGURE 3.2   "Cross over" graph of two hypothetical resource plans: with cross over point in 10 years.

---

[32] The graph used in the example would be essentially the same if we had chosen to compare the two resource plans in regard to the present value of electric rates. However, I have found it is easier for most people to readily understand a graph of cumulative present value costs from year-to-year than of cumulative present value electric rates from year-to-year. And, as previously discussed, when resource plans featuring only new Supply options are evaluated, the resource plan with the lowest costs will also have the lowest electric rates. (Note, however, that if the perspective had been nominal costs, instead of present value costs, the "cross over" point will usually be different.)

In Figure 3.2, the resource plan that is judged to be the more economic plan of the two from a CPVRR cost perspective over the entire time period addressed in the evaluation is Resource Plan A. This is seen by looking at the ending (*i.e.*, the last cost value on the right-hand side of the graph) CPVRR cost value. Resource Plan A has a CPVRR value of less than $4,000 million, while Resource Plan B has a CPVRR value of greater than $4,000 million.

Now that the outcome is known, a relevant question is: *"How long does it take before Resource Plan A becomes the less expensive plan?"* From looking at the graph, we see that Resource Plan A becomes the less expensive (or the more economic) plan starting in year 10 when the cost curves for the two plans "cross over" each other (*i.e.*, Resource Plan A becomes the lower cost curve in the figure in year 10). Resource Plan A then remains the more economic plan for the remaining 20 years of the approximately 30-year analysis period.

However, the "cross over" picture changes considerably if we assume different cost values. We do so and present the results in Figure 3.3 below.

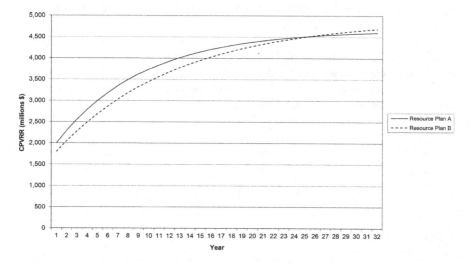

FIGURE 3.3 "Cross over" graph of two hypothetical resource plans: with cross over point in 25 years.

In Figure 3.3, we have changed the costs for the two resource plans (as shown by the fact that the final CPVRR values for both plans now exceed $4,500 million). Resource Plan A remains the more economical resource plan, but it doesn't earn this designation until year 25 when the cost curves of the two plans cross over. Thus, in this particular case, customers must wait 25 years to see net cumulative present value benefits from Resource Plan A.

Is Resource Plan A still the correct selection in this case? There are no hard-and-fast rules regarding this, but it is unlikely that the utility's current customers would be thrilled to pay higher costs (on a present value basis) for 25 years and then begin to see present value net benefits that would largely be realized by the utility's customers 25 years in the future, *i.e.*, by the next generation of the population served by the utility.

For this reason, this particular non-economic consideration has been called a "generational equity" (or "intergenerational equity") issue in utility parlance. In "customer parlance," it might be called an issue of: *"I have to wait how many years before I see the net present value benefits of lower costs?"*[33]

It is relatively rare for a resource plan (or resource option, regardless of whether one is considering a Supply or DSM option) to immediately become the economic choice, and remain the economic choice for the full period of time addressed in the evaluation, when multiple resource options are being evaluated. In other words, there is usually some number of years before the "cross over" point in the projected economics of competing resource plans is reached. How long it takes for a resource plan to cross over is frequently of interest to the utility (and its regulators) in regard to the economics of the resource options in question.

However, the importance of the generational equity issue may be significantly reduced, or even eliminated, if a particular resource option in question has clear non-economic advantages in comparison to the competing resource options. Such non-economic advantages might include, but are not necessarily limited to, the ability of the particular resource option to significantly reduce the utility's dependence upon one or more types of fossil fuel and to significantly reduce the risk of future environmental regulations that would require comprehensive reductions in air emissions from the utility system.

Before we leave this discussion of how long it takes before a resource plan becomes the more economic resource plan, let's introduce a different way to present the information shown by the two figures shown earlier. In certain cases, it can become difficult to easily distinguish the cross over point if presented in this type of graph. This can be especially true in cases in which curves for more than two resource plans are presented.

For that reason, we will use a tabular format to present cross over information in the remainder of this book. Table 3.2, that is presented on the next page is an example of this tabular format. It presents the same basic information in regard to when Resource Plan A crosses over in comparison to Resource Plan B (*i.e.*, in 10 years) as shown in Figure 3.2. The table uses a designation of "1st" to denote the resource plan with the lower CPVRR cost through that year, and a designation of "2nd" to denote the resource plan with the higher CPVRR cost through that year.

---

[33] Generational equity issues also arise with other utility projects that do not involve Supply and DSM options. For example, the high cost incurred today in burying overhead distribution lines when projected benefits may only be seen a number of years later could be seen as a generational equity issue. Non-utility, multi-year projects such as the building of new canals and tunnels can also be viewed as involving generational equity issues.

TABLE 3.2

"Cross Over" Table for Two Hypothetical Resource Plans: CPVRR Ranking of Resource Plans by Year

Year	Resource Plan A	Resource Plan B
1	2nd	1st
2	2nd	1st
3	2nd	1st
4	2nd	1st
5	2nd	1st
6	2nd	1st
7	2nd	1st
8	2nd	1st
9	2nd	1st
10	1st	2nd
11	1st	2nd
12	1st	2nd
13	1st	2nd
14	1st	2nd
15	1st	2nd
16	1st	2nd
17	1st	2nd
18	1st	2nd
19	1st	2nd
20	1st	2nd
21	1st	2nd
22	1st	2nd
23	1st	2nd
24	1st	2nd
25	1st	2nd
26	1st	2nd
27	1st	2nd
28	1st	2nd
29	1st	2nd
30	1st	2nd

Note: 1st = lowest CPVRR cost.

As shown in Table 3.2, Resource Plan A is shown to cross over in year 10 to become the lower CPVRR cost resource plan and it maintains that position for the remaining years of the analysis.

Because this tabular approach is easier to interpret when multiple resource plans are discussed, we will utilize this tabular format when discussing the economic cross over points for CPVRR costs and/or cumulative present value of electric rates in later chapters of this book.

NON-ECONOMIC CONSIDERATION EXAMPLE (II): THE UTILITY SYSTEM'S FUEL USAGE DUE TO A RESOURCE PLAN

The cost of fuel will already have been accounted for in all aspects of economic evaluations. However, for fuel usage, the non-economic consideration is often the question of how much a utility relies upon a particular type of fuel. In other words, it is an issue of fuel supply reliability or security, plus an issue of cost volatility (*i.e.*, how rapidly a fuel's cost may increase in a relatively short time). Concerns by both utilities and regulators tend to increase when a utility becomes increasingly reliant on any one particular type of fuel.

Therefore, utilities frequently strive to maintain and/or enhance their fuel "diversity" to avoid overdependence on any one fuel type. Consequently, a resource plan that emerges from the final economic evaluation as the most economic choice, but which results in significantly more reliance upon a fuel that the utility is already heavily dependent on, may not be the best overall choice. In such a case, another resource plan that was not the best in terms of system economics, but which is the resource plan that is the next-most economical and provides a desired level of fuel diversity, may emerge from the overall evaluation of both economics and non-economics as the resource plan of choice.

NON-ECONOMIC CONSIDERATION EXAMPLE (III): THE UTILITY SYSTEM'S AIR EMISSIONS DUE TO A RESOURCE PLAN

This consideration is another example in which the associated currently known costs will have already been accounted for in the final economic evaluation. At the time of this writing, the types of air emissions that are most often included in IRP evaluations are: sulfur dioxide (SO_2), nitrogen oxides (NO_x), and carbon dioxide (CO_2).

Resource plans might be evaluated non-economically by simply projecting the amounts of system air emissions of interest for each resource plan. This is one way in which one might gauge the viability of resource plans over many years in the future if environmental regulations were to be tightened over the analysis period. Another way would be to account for the projected costs of complying with either current or potential environmental regulations in economic evaluations. This can be done by accounting for the projected costs for pollution control equipment, the projected costs of emission allowances, and/or the use of projected environmental compliance costs.[34]

We now have an idea of the types of non-economic considerations that may be used to judge resource plans. But how do utilities address these non-economic considerations in IRP evaluations? One way in which utilities could attempt to address concerns about not only system air emissions, but also concerns about cross over time and/or fuel usage, would be to utilize self-imposed constraints to all resource

---

[34] In subsequent chapters, the analyses that will be presented in Chapters 5, 6, and 7 for our hypothetical utility system will use a projection of environmental compliance costs for air emissions. We then return to the subject of potential tightening of environmental regulations in Chapter 8 in our discussion of potential constraints on the utility resource planning process.

plans that are being evaluated.[35] In other words, no resource plan will be considered to be acceptable unless it meets these pre-set constraints. For example, the following self-imposed constraints could be used to address the three non-economic considerations we have just discussed:

1. A maximum length-of-time-to-cross over (e.g., x years or less) to a resource plan becoming the most economical resource plan;
2. The annual percentage of the amount of total energy supplied by a particular type of fuel cannot exceed pre-set annual limits;
3. The annual tonnage of a particular air emission cannot exceed pre-set annual limits.[36]

In summary, the non-economic evaluation portion of IRP analyses could result in a resource plan that is the most economic choice *not* being selected as the best overall plan. In order to maximize efficiency and minimize time in their IRP evaluations, utilities may use pre-set constraints to address these non-economic considerations in their IRP final economic evaluation. However, the use of such constraints will have economic consequences.

As we conclude Chapter 3, we note that the chapter has presented an overview discussion of key concepts important in electric utility resource planning. In the next chapter, we will now return to our hypothetical utility system and begin to apply these concepts in evaluations of resource options and resource plans.

---

[35] In Chapter 8, we will also discuss a number of other constraints on utility resource planning. These will include both constraints that are self-imposed by the utility and constraints that are not self-imposed.
[36] Note that utilities which set a low- or zero-carbon goal are self-imposing exactly this type of constraint.

4 Reliability Analyses for Our Utility System

At this point, let's recap where we are in our examination of electric utility systems. In Chapter 2, we discussed some fundamental concepts of how the electric load for utilities varies on both a daily and an annual basis. We also briefly examined five types of conventional generating units that are commonly found in utility systems. We also discussed how a utility uses the operating cost of each type of generating unit in deciding which type of generating unit to operate at any given time to meet the continually varying electric demands of its customers.

In Chapter 3, we discussed the fact that utilities plan for the future by finding answers to three basic questions. We then discussed how reliability analyses, using both reserve margin and probabilistic perspectives, are utilized to answer two of these three basic questions: *"When does the utility need to add new resources?"* and *"What is the magnitude (MW) of the new resources that are needed?"* We then examined how an integrated resource planning (IRP) approach is the best way to determine the answer to the third question: *"What is the best resource option with which to meet this future need for resources?"*

We also briefly discussed how both economic and non-economic evaluations are often required before the final answer to this third question can be determined.

In summary, the previous two chapters have provided an overview of how utilities operate their existing generating units (and why they do it this way), plus provided an overview of the IRP process that electric utility systems can (and should) use to plan for the future.

It is now time to move beyond the overview presentations of these concepts and examine these concepts in more detail. In so doing, we will see how these concepts actually work in practice. Starting with this chapter, and continuing for several more chapters, we shall begin to "flesh out" these concepts using our hypothetical utility system. We will begin by returning to our hypothetical utility system and conducting a reliability analysis for our utility system to see when it needs new resources and what the magnitude of the future resource need is.

WHEN DOES OUR UTILITY SYSTEM NEED NEW RESOURCES?

When we last left our hypothetical utility system, we had conducted a reliability analysis that showed, for a certain year, a reserve margin projection that was presented in Table 3.1. We now provide this same projection again in Table 4.1 that is presented on the next page.

As shown in the table, the projected reserve margin value for this particular year is 20%. Let's now assume that the reserve margin perspective is the one driving the need for new resources for our utility system. In other words, the reserve margin

DOI: 10.1201/9781003301509-5

TABLE 4.1
Example of a Basic Reserve Margin Calculation

(1)	(2)	(3) = (1) − (2)	(4) = (3)/(2)
Total Generating Capacity (MW)	Peak Electrical Demand (MW)	Reserves (MW)	Reserve Margin (%)
12,000	10,000	2,000	20.0

criterion will be triggered or violated in future years for our utility before the LOLP criterion is violated.

Let's also assume that our utility's reserve margin criterion is 20%. Clearly, for the particular year addressed in Table 4.1, our utility system is deemed to be reliable because its projected reserve margin value is equal to the minimum reserve margin criterion of 20%.

Now let's expand our view to look at a few more years. We do so in Table 4.2.

TABLE 4.2
Reserve Margin Analysis Projection for the Hypothetical Utility

Year	(1) Total Generating Capacity (MW)	(2) Peak Electrical Demand (MW)	(3) = (1) − (2) Reserves (MW)	(4) = (3)/(2) Reserve Margin (%)
Current Year	11,600	9,600	2,000	20.8
Current Year + 1	12,000	9,700	2,300	23.7
Current Year + 2	12,000	9,800	2,200	22.4
Current Year + 3	12,000	9,900	2,100	21.2
Current Year + 4	12,000	10,000	2,000	20.0
Current Year + 5	12,000	10,100	1,900	18.8

Instead of a reserve margin projection for just 1 year, we now have a projection for 6 years. The first thing to notice is that the projection for the one particular year that is presented in Table 4.1 is for the year now labeled as "Current Year + 4." In other words, this projection is for 4 years from now (*i.e.*, 4 years from the "Current Year").[1]

In looking back to the Current Year, we see in Column (1) that our utility system actually has 400 MW less generating capacity than what it is projected to have in Current Year + 4: 11,600 MW in the Current Year instead of the 12,000 MW of generating capacity in Current Year + 4 that we just discussed. Our utility's peak electrical demand, shown in Column (2), is also lower in the Current Year by 400 MW:

---

[1] The term "Current Year" is used throughout this book to denote the year our hypothetical utility is conducting its analysis. This approach avoids the use of a specific year (such as the year this book was written), which will, of course, no longer be the actual "current year" when someone (hopefully) reads this book years after it is written.

9,600 MW instead of 10,000 MW. This combination of lower generating capacity and lower peak demand results in a reserve margin value for the Current Year of 20.8% as shown in Column (4). Because this reserve margin value is higher than our utility's 20% reserve margin criterion, our utility is deemed to be reliable in the Current Year.[2]

However, by examining the projected annual load values in Column (2) in this table, we see that our utility system is projecting growth in its peak demand of 100 MW/year for each of the next 5 years. Because of this projected growth, our utility's projected reserve margin 1 year in the future (Current Year + 1) would drop below the 20% criterion to 19.6% unless new resources are added. (Although this value is not shown in the table, it is easily calculated: 11,600 MW of generation capacity minus 9,700 MW of peak load would result in 1,900 MW of reserves. Then 1,900 MW of reserves divided by 9,700 MW of peak load yields a reserve margin of 19.6%.)

Although a utility might decide not to add new resources in that year because its reserve margin criterion of 20% is projected to be violated by a relatively small amount, the projection of continued growth in peak load each year going forward would cause our utility's projected reserve margin value of 19.6% 1 year in the future to drop to 18.4% two years in the future. The reserve margin would then continue to drop each year thereafter. Consequently, our utility had previously decided to add 400 MW of generating capacity in Current Year + 1 to bring its total generating capacity to 12,000 MW. This increased amount of generation capacity is projected to ensure that the 20% reserve margin criterion would be met for 4 years (Current Year + 1 through Current Year + 4).

Yet the 100 MW per year growth in peak load is projected to continue past this 4th year into the future, causing the projected reserve margin to drop significantly below the 20% criterion, to 18.8%, in the 5th year (Current Year + 5).[3] Our utility now knows the answer to the first question: *"When does it need to add new resources?"* The answer is 5 years from the Current Year (or, in Current Year + 5).

We now turn our attention to how reserve margin analyses can be used to answer the second question: *"What is the magnitude (MW) of the resources needed?"* Perhaps unexpectedly, we shall see that the answer depends upon what type of resource the utility chooses to add.

WHAT IS THE MAGNITUDE OF THE NEW RESOURCES NEEDED BY OUR UTILITY SYSTEM?

From Table 4.2, it is clear that, because the peak load is projected to grow at 100 MW/year, the utility needs to add new resources by Current Year + 5 because the projected reserve margin has dropped to 18.8% for that year. But what is the magnitude (MW) of the new resources that are needed? Table 4.2 does not provide an answer to this question.

---

[2] I'm relieved to hear this. Clearly, this indicates superior resource planning was carried out for our utility system in prior years.

[3] Just a reminder that our utility system could have also found itself in a similar situation of needing new resources in Current Year + 5, even without increased load growth, if, for example, an existing generating unit was scheduled to be retired in that year, thus decreasing the Total Generating Unit Capacity (MW) value in Column (1).

Also, as we discussed in Chapter 3, there are two basic types of resource options that the utility could add to meet a need for future resources: Supply options and DSM options. Our answer(s) to the question of *"What is the magnitude (MW) of new resources needed?"* must be responsive to both of these types of resource options that the utility could choose. This will ensure that the correct amount of either type of resource option has been chosen so that the utility's reserve margin criterion of 20% continues to be met.

We shall start by looking at one of the two types of resource options the utility might choose: Supply options. Then we shall see how we would answer the question regarding the magnitude of resources needed in Current Year + 5 if the utility chooses to meet the need by adding new generating capacity. In order to do this, we turn our attention to Table 4.3.

TABLE 4.3

Reserve Margin Analysis Projection for the Hypothetical Utility: MW Needed If Only New Generation Is Added

Year	(1) Total Generating Capacity (MW)	(2) Peak Electrical Demand (MW)	(3) = (1) − (2) Reserves (MW)	(4) = (3)/(2) Reserve Margin (%)	(5) = ((2)*1.20) − (1) Generation Only MW Needed to Meet Reserve Margin (MW)
Current Year	11,600	9,600	2,000	20.8	(80)
Current Year + 1	12,000	9,700	2,300	23.7	(360)
Current Year + 2	12,000	9,800	2,200	22.4	(240)
Current Year + 3	12,000	9,900	2,100	21.2	(120)
Current Year + 4	12,000	10,000	2,000	20.0	(0)
Current Year + 5	12,000	10,100	1,900	18.8	120

Table 4.3 is an expanded version of Table 4.2 in which one new column, Column (5), has been added. In Column (5), we calculate how the generating capacity of the utility would have to change in order for a 20% reserve margin criterion to be met each year. The calculation formula to derive the values shown in Column (5) is provided in the table under (5) in the column heading.

This formula is straightforward. The projected peak load for a given year (shown in Column (2)) is multiplied by 1.20 to determine how many MW of generating capacity would be needed to have total generating capacity that would be exactly 20% higher than the projected peak load. This amount of generating capacity would allow the utility to meet its 20% reserve margin criterion. One can refer to this value as the "needed" capacity value to ensure the 20% reserve margin criterion is met.

Then the expected annual total generating capacity of the utility for that year (shown in Column (1)) is subtracted from the "needed" capacity value to determine how far off the utility is in regard to the generating capacity needed to meet the 20% criterion.

As shown in Column (5), this calculated value is negative (*i.e.*, values presented in parentheses indicate negative values) for the Current Year and through 3 more years in the future (Current Year through Current Year + 3). These negative values denote that the utility has more generating capacity than it needs to meet its 20% reserve margin criterion. This is not exactly news because, as shown in Column (4), the utility's projected reserve margin exceeds the 20% criterion in these same years, but it does tell one how much "excess" generating capacity the utility has in each year. (This information can be useful if the utility is interested in making a short-term sale of generating capacity to another utility or is considering taking a generating unit out of service for an extended time to perform modifications or enhancements.)

Consequently, the utility does not need to add generating capacity in these years. Looking ahead one more year to Current Year + 4, we see that the value calculated for this year in Column (5) is 0 MW of needed new generating capacity. This is again expected because the utility's projected reserve margin for that year is exactly 20%.

However, for Current Year + 5, Column (5) presents a calculated value of 120 MW of needed new generation capacity to return the utility to a projected reserve margin level of 20%. From the formula presented for Column (5), we see that this value is derived by multiplying the projected load for that year, 10,100 MW, by 1.20. The result is a total of 12,120 MW of resources needed to meet the 20% reserve margin criterion. Then, after subtracting the current generation resources of 12,000 MW, the difference is 120 MW. The formula tells us that this is the magnitude of new resources that need to be added if the new resource is new generating capacity. But is 120 MW of new generation really the correct answer?

We can check that calculation by altering the table to allow the 120 MW of new generating capacity to be entered back into the reserve margin calculation. We do so in Table 4.4, which is a modified version of Table 4.3 that we just discussed. In order to focus better on this calculation "check," only information for the last 2 years that are shown in Table 4.3 is presented in Table 4.4.

TABLE 4.4

Reserve Margin Analysis Projection for the Hypothetical Utility: MW Needed If Only New Generation is Added (Checking)

	(1a)	(1b)	(1c) = (1a) + (1b)	(2)	(3) = (1c) − (2)	(4) = (3)/(2)
Year	Previously Projected Generating Capacity (MW)	Additional Projected Generating Capacity (MW)	Total Projected Generating Capacity (MW)	Peak Electrical Demand (MW)	Reserves (MW)	Reserve Margin (%)
Current Year + 4	12,000	0	12,000	10,000	2,000	20.0
Current Year + 5	12,000	120	12,120	10,100	2,020	20.0

In Table 4.4, Column (1) from the previous table has now been expanded into three columns. Column (1a) presents the same information as Column (1) from the previous table, *i.e.*, our utility's previously projected generating capacity. Two new columns have also been added. Column (1b) shows the newly projected generating capacity that will be added to our utility system in each of those 2 years and Column (1c) shows the sum of the previously projected and newly projected generating capacity. Columns (2) through (4) are essentially unchanged from the previous table.

As expected, there is no change for Current Year + 4 because no new generating capacity is added. However, the calculation for Current Year + 5 shows that adding 120 MW of new generation in that year will indeed return our utility's projected reserve margin to 20%. This check confirms the original projection presented in Table 4.3: 120 MW of new generation is needed if the utility decides to meet its projected resource needs 5 years in the future with a Supply option.[4]

But is this projection of 120 MW of new resources also valid if the utility decides to meet its need for new resources with a DSM option? The answer is "no" as shown in Column (5) of Table 4.5.

TABLE 4.5

Reserve Margin Analysis Projection for the Hypothetical Utility: MW Needed If Only New DSM Is Added

Year	(1) Total Generating Capacity (MW)	(2) Peak Electrical Demand (MW)	(3) = (1) − (2) Reserves (MW)	(4) = (3)/(2) Reserve Margin (%)	(5a) = ((2)*1.20) − (1)/1.20 DSM Only MW Needed to Meet Reserve Margin (MW)
Current Year	11,600	9,600	2,000	20.8	(67)
Current Year + 1	12,000	9,700	2,300	23.7	(300)
Current Year + 2	12,000	9,800	2,200	22.4	(200)
Current Year + 3	12,000	9,900	2,100	21.2	(100)
Current Year + 4	12,000	10,000	2,000	20.0	0
Current Year + 5	12,000	10,100	1,900	18.8	100

This table is a DSM-based version of the previously presented Table 4.3, which addressed the magnitude of resources needed if the new resource is new generating capacity. Table 4.5 is modified in Column (5) to perform a new calculation that is necessary if the new resource is DSM. This table shows that the projected value for

---

[4] The alert reader might also have noticed that, starting with the 20% reserve margin value in Current Year + 4, the utility is adding 100 MW of load in the next year. Because the utility's reserve margin is 20%, the utility would need to add 1.20 × 100 MW of new load = 120 MW to "keep pace" with the new load in Current Year + 5. (However, the question of how much generating capacity is needed is rarely answered so easily in real life because a utility's total MW of existing generating capacity often changes from 1 year to the next due to recent changes in one or more of these generating units.)

the magnitude of new resources needed to meet a 20% reserve margin criterion is only 100 MW if the need is met by new DSM resources.[5] As before, let's check this by inserting the 100 MW of DSM back into the calculation. We do so in Table 4.6, which again focuses only on the last 2 years.

TABLE 4.6
Reserve Margin Analysis Projection for the Hypothetical Utility: MW
Needed If Only New DSM Is Added (A Check)

	(1)	(2a)	(2b)	(2c) = (2a) − (2b)	(3) = (1) − (2c)	(4) = (3)/(2c)
Year	Total Generating Capacity (MW)	Previously Forecasted Electrical Demand (MW)	Newly Projected DSM (MW)	Firm Electrical Demand (MW)	Reserves (MW)	Reserve Margin (%)
Current Year + 4	12,000	10,000	0	10,000	2,000	20.0
Current Year + 5	12,000	10,100	100	10,000	2,000	20.0

Similar to Table 4.4, one of the columns from the original table (Table 4.3) has again been expanded. However, in this case, Column (2) has been expanded. Column (2a) presents the previously forecasted peak demand and two new columns have been added to this table. Column (2b) shows the amount of projected new DSM that will be added to the utility system. Column (2c) then shows the result of a calculation of how the new DSM will lower the forecasted peak demand to provide a lower projection of electrical demand. This projection of load after DSM is accounted for is commonly called the "firm" demand for the utility. Columns (1), (3), and (4) are essentially unchanged from those presented earlier in Table 4.5.

As expected, there is no change for Current Year + 4 because no new DSM needs to be added in that year. However, the calculation for Current Year + 5 shows that adding 100 MW of new DSM in/by that year will indeed return the utility's projected reserve margin to 20%.[6] This check confirms the original projection presented in Table 4.5 that 100 MW of new DSM is needed if the utility decides to meet its projected resource needs 5 years in the future with new DSM only.

---

[5] The alert reader may have already determined that the answer to this question is 100 MW. Table 4.1 shows that, in Current Year + 4, our utility had exactly a 20% reserve margin. Then the load increased to 100 MW in the next year. Therefore, the addition of new DSM that lowers the peak load by 100 MW would return the utility to a 20% reserve margin. (However, the question of how much DSM is needed is rarely answered so easily because generation values typically change each year which we have not allowed to happen in this simple example.)

[6] In reality, a utility would almost certainly begin adding DSM at least several years earlier instead of waiting until the year the full DSM resource is actually needed. The earlier implementation of DSM is driven both by economics and by the practical problems of signing up thousands of participating customers for the DSM program(s). Therefore, the picture presented here is a simplified one in which we were solely concerned with establishing the magnitude of the needed DSM resources by Current Year + 5.

WHAT HAVE WE LEARNED AND WHAT IS NEXT?

We have now taken the first step in fleshing out our discussion of resource planning concepts by determining that our hypothetical utility system has a resource need 5 years in the future, *i.e.*, in Current Year + 5. We have also determined that the resource need is either 120 MW if the need is met solely by a new Supply option or 100 MW if the need is met solely by a new DSM option.

The fact that there is a difference in the amount of new resources needed depending on the type of resource option that is selected, new Supply or new DSM, is a basic concept of utility resource planning. A reserve margin-driven utility (such as our hypothetical utility system) will always have a greater resource need if that need is met by a new Supply option than if the resource need is met by a new DSM option.[7] The difference between the two amounts is determined by a utility's reserve margin criterion, which is 20% for our hypothetical utility system, and which is representative of the reserve margin criterion used by certain utilities (and which should be easier for a reader to follow when reading this book than if a reserve margin criterion of, for example, 18% was used).

In other words, if a utility's new resource needs can be met by 100 MW of new DSM, it will take 120 MW of a new Supply option to meet the utility's projected resource needs, assuming that the utility's reserve margin criterion is 20%. Looked at another way, if the utility's projected peak demand will increase in the future by 100 MW, and it currently exactly meets its 20% reserve margin criterion, then it will need 120 MW of new generation to maintain the 20% reserve margin level.

At this point, you may be asking yourself the following question: *"Because the utility only needs 100 MW of a DSM option to meet its reliability criterion, versus 120 MW of a Supply option, doesn't a DSM option have a built-in advantage over a Supply option when it comes time to choose a resource option?"*

Perhaps surprisingly, the answer is "yes" and "no." (I hate these types of answers, so let's explain.) From an economic perspective of just the initial cost of installing or acquiring each resource option, DSM does have an advantage. Assume for a moment that the costs of acquiring 1 MW of both the DSM and Supply options are equal. In such a case, and assuming all else equal, DSM has an economic advantage in regard to the cost of acquiring the resource because the utility only has to obtain 100 MW of the DSM resource versus 120 MW of the Supply resource.

In practice, the acquisition costs of DSM and Supply options will likely vary considerably. If the acquisition cost of DSM is lower than that of the competing Supply option, then the DSM advantage is increased due to the fact that fewer MW of DSM compared to Supply are needed. Conversely, if the acquisition cost of DSM is higher than that of the competing Supply option, this DSM disadvantage is lessened for the same reason. In either case, the initial or installed cost of acquiring a resource option provides an advantage to DSM because the utility doesn't need as much of that type of resource. This particular advantage is accounted for in economic evaluations of resource options.

---

[7] The situation may be less clear cut if LOLP is driving a utility's resource needs.

However, because the utility doesn't need to acquire as many DSM MW as it would if it chose a Supply option, DSM may also have less of an impact on the operation of the utility system than would be the case if the same MW amount of DSM resource had been acquired as would be acquired if a Supply option had been chosen (120 MW in our example). Assuming all else equal, the impacts to the utility system operation from acquiring only 100 MW of DSM will be less than the impacts would be if 120 MW of the same DSM had been acquired. Two of the types of impacts to system operation will be: (i) changes to the utility system's fuel usage, and (ii) changes to the utility system's air emissions. Assuming all else equal, the implementation of 100 MW of DSM will have less of an impact on system fuel use and system emissions than if 120 MW of the same DSM had been implemented.

Consequently, DSM's potential impact on either system fuel usage or system air emissions will be less than it would have been if the same amount of DSM were added to the utility system as would be added with Supply options. This can actually work to DSM's disadvantage when evaluating the utility's system fuel usage and/or system air emissions. These particular operational disadvantages are also accounted for in both economic and non-economic evaluations of resource options.

In summary, the fact that a smaller amount of a DSM option is needed, compared to the amount of a Supply option, to ensure the same reserve margin level for a utility system results in both advantages and disadvantages for DSM versus Supply options.

A utility could attempt to remove the system operation-based disadvantages inherent in implementing a smaller amount of DSM by implementing the exact same amount (MW) of DSM as would be needed if a Supply option had been chosen. However, by doing so, the initial or installed cost advantage of having to acquire a smaller amount of DSM would be lost. Consequently, utilities typically do not attempt such an approach, and we will not take such an approach, as we continue to work with our utility system in evaluating Supply and DSM options in the next several chapters.

In Chapter 5, we turn our attention to some Supply options that our utility system will consider for meeting its resource needs 5 years in the future (in Current Year + 5). We will then follow these Supply options through an economic evaluation to determine which one of these Supply options is the best economic choice for our utility. In Chapter 6, it will be DSM's turn. We will follow some DSM options through an economic evaluation to determine which DSM option is the best economic choice for the utility. Then, in Chapter 7, we will pull together the Supply and DSM options and evaluate them from both an economic and a non-economic perspective.

5 Resource Option Analyses for Our Utility System: Supply Options

In this chapter, we follow our utility system through economic analyses of four different Supply options it will consider as possible choices with which to meet its projected resource need 5 years in the future (in Current Year + 5). In practice, a utility would likely consider more than four Supply options. However, in order to simplify the discussion, while still illustrating the important points regarding economic analyses of Supply options, our utility system has (graciously) condensed its list to only four Supply options.

TYPES OF SUPPLY OPTIONS UNDER CONSIDERATION

The four Supply options our utility system is considering include the following: two natural gas-fueled combined cycle (CC) units (imaginatively labeled as CC Unit A and CC Unit B), one natural gas-fueled combustion turbine (CT) unit, and a photovoltaic (PV) facility that uses sunlight to generate electricity. We will make the simplifying assumption that the operating life for each of these four Supply options is 25 years. The rest of the assumptions for the key inputs needed for an economic evaluation of these four Supply options are presented in Table 5.1 that is presented on the next page.

For discussion purposes, the inputs presented in the table will be grouped into general categories that are directly or indirectly related to three types of costs: (i) capital costs, (ii) other fixed costs, and (iii) operating costs.

CAPITAL COSTS: ROWS (1) THROUGH (3)

We start the discussion of inputs related to capital costs with the capacity (MW) shown in Row (1) of the table. Most fossil-fueled generating units, such as the CC and CT options, come in pre-determined sizes or capacities (MW).[1] The capacities of new generating units being offered usually vary from year-to-year as changes in the technologies occur. The CT size of 160 MW is representative of certain CT units available at the time this book is written. Because CC units use CT units as their basic "building block," CC units can be built in various sizes depending on how

---

[1] In this book, the term "capacity" is being used to denote the projected maximum MW output of a generating unit based on a given set of conditions (such as air temperature, humidity, and type of fuel being used). This projected maximum output value is also referred to as "nameplate capacity."

DOI: 10.1201/9781003301509-6

TABLE 5.1

Key Inputs for Economic Evaluation: Supply Options

Input	Measurement	CC Unit A	CC Unit B	CT	PV
(1) Capacity (Nameplate)	MW	500	500	160	120
(2) Capital Cost	$ (millions)	$475	$520	$104	$180
(3) Capital Cost	$/kw	$950	$1,040	$650	$1,500
(4) Fixed O&M	$/kw	$6.00	$6.00	$4.00	$5.00
(5) Capital Replacement	$/kw	$10.00	$10.00	$8.00	$1.00
(6) Firm Gas Transportation	$/mmBTU	$2.10	$2.10	$2.10	N.A.
(7) Firm Gas Needed	mmBTU/day	50,000	50,000	10,000	N.A.
(8) Variable O&M	$/MWh	$0.10	$0.10	$0.20	N.A.
(9) Type of Fuel	---	Natural Gas	Natural Gas	Natural Gas	Sunlight
(10) Heat Rate	BTU/kwh	6,600	6,700	10,400	N.A.
(11) Availability	% of hours per year	92%	92%	92%	50%
(12) Capacity Factor	% of hours per year	80%	80%	5%	20%
(13) Firm Capacity Value	% of nameplate rating considered as firm capacity	100%	100%	100%	50%
(14) SO_2 Emission Rate	lbs/mmBTU	0.006	0.006	0.006	0.000
(15) NO_x Emission Rate	lbs/mmBTU	0.010	0.010	0.033	0.000
(16) CO_2 Emission Rate	lbs/mmBTU	117	117	117	0

many CTs are used. CC units being built at the time this book is written typically use from one to four CT units, plus an equal number of heat recovery steam generators. The 500 MW capacity assumed for both of our CC options is representative of some of these CC units. Although the size of the two CC options, CC Unit A and CC Unit B, are identical (500 MW), the capital costs for the two CC units are slightly different as shown in Row (2) of the table.

Note that our utility system will be considering CC and CT options of a size considerably larger than its next projected resource need of 120 MW for Current Year + 5 that was determined in the prior chapter. This is not unusual. Conversely, because PV options consist of very small (approximately one to several kW) modules that are grouped together, a PV option can be built that more exactly matches a utility's projected resource need. For our purposes, we will assume that our utility will add 120 MW of PV. The 120 MW size of the PV option was specifically chosen to illustrate an aspect of solar (and other renewable) resource options: the assumed contribution of such intermittent resources at the time of a utility's peak hour that is used in reliability analyses.

At first glance, one might assume that the selection of this 120 MW PV option would exactly meet the projected 120 MW resource need in Current Year + 5 if this Supply option is chosen. However, that is not the case. With fossil-(and nuclear-) fueled generators, the "nameplate" capacity rating is considered to be 100% available at the time of the utility's system peak hour (assuming the generator is not out for planned maintenance or unexpectedly broken). This is commonly referred to as having a "firm capacity value" of 100%. Using either of the CC options as an example, all of their 500 MW of nameplate capacity rating is assumed to be available at the system peak hour when performing reliability analyses such as reserve margin calculations. As a result, the nameplate capacity rating and the firm capacity value of the CC units are identical (500 MW).

However, the same is not true for solar options such as the 120 MW PV option our utility is considering. The PV option's energy source (sunlight) is intermittent and, therefore, not as reliable as an essentially constant fuel supply (such as coal or natural gas) of a conventional generating unit. For example, the sun does not shine at night, the sun is at different positions in the sky at different hours of the day (thus changing the intensity of the sunlight that hits the solar panels), and the amount of sunshine at a given moment can differ due to clouds, rain, snow, etc.

For these reasons, utilities develop a projection of firm capacity value for solar options that is almost always less than 100% of the PV nameplate rating. These percentages are typically determined by computer modeling of expected capacity and energy outputs and/or by actual operating experience at the specific site(s) where the PV modules will be installed. As shown in Table 5.1 in Row (13), our utility has assumed a firm capacity value of 50% for the PV option. What that value means is that, at the utility's system's peak hour, the 120 MW PV option is projected to only supply 60 MW (= 120 MW × 0.50).

In the analyses to follow, this means that only 60 MW of the PV option's 120 MW nameplate rating is considered firm capacity in the utility's reserve margin calculations. Therefore, the utility will have to add two 120 MW PV facilities, or 240 MW of PV in total, to meet the projected 120 MW resource need in Current Year + 5. This will result in both disadvantages and advantages for the PV option. An obvious disadvantage is that the installed cost of the PV option will double because 240 MW, not 120 MW, will need to be added. On the other hand, the advantage is that doubling the amount of PV installed to meet a reserve margin target will also result in doubling the amount of fuel cost savings and emission savings from the PV option because PV's annual output of fuel-and emission-free MWh will be doubled. (We will return to the concept of firm capacity value for solar, and discuss it more fully, in Part II of this book.)

Before leaving Rows (1) and (13), we note that the differences in the capacity (MW) and firm capacity offered by these four Supply options will play an important role in the economic evaluation. We shall see why this is so later in this chapter.

Rows (2) and (3) of Table 5.1 provide the capital cost for the options. By capital cost, we mean the cost of building the Supply option. The assumed capital cost for each option is presented in the table in two ways. First, in Row (2), the capital cost is presented in terms of the actual cost in millions of dollars.[2] For example, CC Unit A is assumed to have a cost of $475 million.

---

[2] These assumed capital costs account for the costs of the generating facility itself, including assumptions made for interest charges during construction, the cost of the land upon which the facility will be built, and the cost of transmission additions needed both to connect the generating facility to the rest of the utility system, and to upgrade other areas in the overall transmission system to accommodate handling the output of the new generating unit. Obviously, variations in these cost aspects of any of the resource options could result in changes in the assumed overall cost for the options.

Second, in Row (3), the capital cost is presented in terms of cost per a common unit of capacity (*i.e.*, cost/kW) for the option. This value is derived by dividing the capital cost (in millions of dollars) by the capacity (MW) value in Row (1). (The capacity MW value is first converted to a kW by multiplying the MW value by 1,000 because 1,000 kW = 1 MW.) The result is expressed in terms of $/kW as shown in Row (3). The reason for expressing the capital cost in terms of $/kW is that Supply options can vary so widely in terms of their capacity that it is difficult to gauge the relative cost of building units if one looks only at the cost in terms of millions of dollars. The use of $/kW values allows one to better gauge the relative costs of generating units on a common size (1 kW) basis.

As shown in Table 5.1 in Row (2), the installed costs for the two CC units are $475 million and $520 million, respectively. The cost of the CT unit is $104 million and the cost of the PV option is $180 million (for one 120 MW PV facility). Just as we noted in Chapter 1, and in our introductory discussion of assumed fuel costs in Chapter 2, we again point out that the specific assumptions used in this book for capital (and other) costs of these types of generating units assuming are not important to this discussion of resource planning concepts and analytical approaches. Although the assumed generating unit cost values we are using are representative of costs at the time this book was written, at any point in the future, these cost (and other) values will certainly be different because they are continually changing. However, the principles, concepts, and analytical approaches we are discussing will remain valid regardless of the assumption values used to illustrate these concepts and analytical approaches.

From the information presented in Row (2), the CT appears to be an inexpensive choice relative to the CC and PV units. However, because the capacity (MW) of each of these four options varies so much, *i.e.*, from 500 MW to 120 MW, it is difficult at first glance to gauge the relative costs of building a set amount of capacity with each option. As previously mentioned, converting the capital costs into $/kW values in Row (3) allows one to easily see the cost on a "cost per unit" (/kW) basis.

After performing this conversion, we see that the cost/kW value for the CC and CT options has narrowed significantly because the two CC units have costs of $950/kW and $1,040/kW, respectively, while the CT option's cost is $650/kW. Conversely, this "cost per unit" perspective shows that PV is a more expensive option to build because it has an installed cost of $1,500/kW compared to a cost range of $650/kW to $1,040/kW for the CC and CT options.

Therefore, a first impression of the economics of the four Supply options is that the renewable energy option, PV, is more expensive to construct compared to the fossil-fueled units, CC and CT. Although this impression is correct based on the assumed costs, we must remember that our overall cost picture is far from complete.

OTHER FIXED COSTS: ROWS (4) THROUGH (7)

We turn next to what we will refer to as other "fixed costs" which include costs not related to building or directly operating the unit, but which are incurred essentially to keep the unit available for operation. Table 5.1 lists three such costs: Fixed operation and maintenance (O&M) in Row (4), capital replacement in Row (5), and firm gas transportation-related information in Rows (6) and (7).

Fixed O&M costs typically refer to costs incurred for salaries of the generating unit's operating personnel and for some regular maintenance activities. These costs are typically expressed in terms of $/kW. As shown in Row (4), the assumed Fixed O&M costs for these four options vary little with costs ranging from $4/kW to $6/kW. Consequently, this cost category is likely to play a relatively minor role in the economic evaluation of these four Supply options.

The second fixed cost category is termed "capital replacement." We are using this term to refer to expenditures that are incurred to periodically replace major equipment that wears out over time as the generating unit is operated. (A non-utility analogy might be an automobile tires that wear out and must be replaced after some threshold number of miles driven has been reached.) These values are also typically expressed in terms of $/kW. The assumed capital replacement cost values shown in Row (5) for the CC and CT units again vary little with values ranging from $8/kW to $10/kW. However, PV has no moving parts and, therefore, is typically perceived to have only minor capital replacement costs; and a value of $1/kW is assumed for the PV option. In this case, the PV option has a clear advantage in regard to this cost category.

This is also the case when discussing the third of the fixed cost categories, "firm gas transportation." This term refers to the cost that is incurred to ensure that a natural gas-fired generating unit (or an entire utility system that has one or more natural gas-fueled generators) will have sufficient natural gas to fuel a generating unit when it is operating. Because of its gaseous nature, large quantities of natural gas are not typically stored next to a generating unit as readily as one can store large quantities of coal or oil. Natural gas must be delivered by pipeline, often by a company separate from the electric utility, and the electric utility will want to ensure that a sufficient amount of gas is readily available at all times the generating unit is likely to be operating.

Therefore, the electric utility frequently enters into a contract with the natural gas supplier to "reserve" a set amount of natural gas per day for use by the utility. In exchange, the utility pays a type of reservation charge, often referred to as a "firm gas transportation" charge, as shown in Row (6). This charge is typically expressed in terms of dollars per million BTUs ($/mmBTU). These costs vary but, at the time this book is written, are in the general vicinity of approximately $1.75/mmBTU to $2.50/mmBTU. In Row (6), we have assumed a firm gas transportation cost of $2.10/mmBTU for the CC and CT options. Because PV does not need natural gas (or any fuel/energy source other than sunlight), this cost is not applicable ("N. A.") for the PV option.

In regard to the set amount of gas, the utility is reserving for the two different types of gas-fueled generating units, and this can vary significantly from one utility system to another. In addition, a utility system may already have a sufficient volume of firm natural gas already reserved so that no additional firm natural gas would be needed when considering a new natural gas-fueled resource option such as a new CC or CT. This becomes more likely if the new CC option is much more fuel-efficient than the existing natural gas-fueled generating units on the utility's system. In this case, the new CC would likely be run for many hours of the year, and existing, less fuel-efficient natural gas-fueled units are run for considerably fewer annual hours than would otherwise be the case, resulting in a decrease in the utility's annual natural gas usage.

However, for our utility system, we have assumed that additional firm natural gas supplies will be needed if a new natural gas-fueled Supply option is chosen. These additional volumes are assumed to be 50,000 mmBTU/day for the CC options and 10,000 mmBTU/day for the CT option, as shown in Row (7).[3] For the PV option, consideration of a needed amount of natural gas per day assumption is again not applicable.

When the assumed firm gas transportation charge of $2.10/mmBTU is combined with the assumed amount of gas that is reserved per day, and this is then extended to all 365 days/year, the overall annual firm gas transportation cost is significant. Using the CC option as an example, our calculation for the annual firm gas transportation cost becomes:

Annual firm gas transportation cost

$= \$2.10/\text{mmBTU} \times 50,000 \text{ mmBTU/day} \times 365 \text{ days/year}$

$= \$38.3$ million per year for the CC options

The calculation for the CT option is identical except for the lower 10,000 mmBTU/day value and the result of the calculation for the CT option is approximately $7.7 million/year. Consequently, the PV option again has an economic advantage over both the CC and CT options in regard to this fixed cost category, and the CT option has an advantage over the CC options.

OPERATING COSTS: ROWS (8) THROUGH (16)

The remaining rows of Table 5.1 provide information related to the costs and other impacts of actually operating these Supply options on our utility system.[4] The first of these operating costs is shown in Row (8): the Variable O&M cost that is expressed in terms of $/MWh. These costs are directly tied to the number of hours the generating unit operates and may include expenses for water and lubricants but typically do not include the cost of the fuel.

Variable O&M costs typically are in the range of approximately $2.00/MWh or less. Accordingly, we have assumed that the two CC options have variable O&M costs of $0.10/MWh and the CT option's variable O&M cost is $0.20/MWh. Because the PV option has no moving parts and uses virtually no water except for cleaning the modules, we assume that a variable O&M cost is again not applicable for this option (*i.e.*, it has a variable O&M cost of $0.00/MWh).

---

[3] The amount of incremental firm natural gas that a utility will actually require if it chooses to acquire a new natural gas-fueled generating unit is completely utility-specific. The amount will be dependent upon a number of factors, including the amount of firm natural gas the utility already has under contract, the amount of gas the new generating unit is projected to require, etc. In our discussion, we are assuming (pessimistically) that our utility system has to acquire a significant amount of additional natural gas to operate the new gas-fueled units, particularly for a new CC unit.

[4] Recall that the fuel-based operating costs were briefly discussed in Chapter 2.

The table next lists the type of fuel used by each option in Row (9). Natural gas is used by the CC and CT options, while sunlight is used by the PV option. (The costs of natural gas were briefly discussed in Chapter 2. We will further discuss this assumption later in this chapter.)

The heat rate (or efficiency of converting fuel into electricity) of the generating unit is then listed in Row (10) and is expressed in terms of BTU/kWh. The two CC options are assumed to have relatively low heat rates of 6,600 BTU/kWh and 6,700 BTU/kWh, respectively, which indicate a high level of efficiency in converting fuel into electricity. By contrast, the CT option's heat rate of 10,400 BTU/kWh shows that the CT option is a much less efficient type of generating unit.

For purposes of this part of our discussion, the PV option's "heat rate" is not applicable. This is because the PV option's energy source (sunlight) is free and, therefore, the efficiency with which the PV option converts its fuel into electricity is simply not a factor in economic analyses in regard to heat rate. (However, PV's conversion efficiency is indirectly accounted for in other inputs for PV, including the projected $/kW cost and capacity factor.)

The next two input categories shown in Rows (11) and (12), respectively, are the availability and capacity factor. Availability, shown in Row (11), refers to how many hours of the year a generating unit is capable of operating after accounting for planned maintenance and projected unplanned maintenance (*i.e.*, the unit's forced outage rate, or FOR).[5] The other input, capacity factor, shown in Row (12), refers to how many hours of the year a generating unit is expected to operate on a particular utility system. As we have seen in Chapter 2, this value is typically driven by economics pertaining to the utility system in question. Therefore, the less expensive the operating cost of a generating unit is, the more hours the unit will typically operate as we previously discussed.

In Row (11), we have assumed annual availability factors of 92% for the CC and CT options based on an assumption that planned and unplanned maintenance will require approximately 8% of the hours in the year. Consequently, the CC and CT options are available to run the remaining 92% of the hours in a year.

Because PV options require relatively little planned maintenance (because there are no moving parts, at least with PV installations that do not include machinery that moves the PV modules throughout the day to track the sun's movements), the availability of PV is primarily limited by the number of expected daylight hours.[6] Consequently, our utility assumes an availability of 50% for the PV option.

In Row (12), we have assumed annual capacity factors of 80% for the CC options, 5% for the CT option, and 20% for the PV option. The low heat rate values (indicative of high efficiency) for the CC options will result in low operating costs, thus leading to the CC options' high capacity factor value of 80%. Conversely, the high heat rate value (indicating low efficiency) for the CT option

---

[5] Generating unit availability is a facet of the electric utility industry that is actually quite complicated. Therefore, the approach we are using to discuss generating unit availability is a simplified one.

[6] Certain PV-related equipment, such as inverters that convert a PV's direct current (DC) output to alternating current (AC), can fail. Such equipment failure will also limit the output of PV facilities. However, the availability of sunlight is typically the overriding factor in estimating PV availability.

results in high operating costs, thus resulting in a low capacity factor value for the CT option of 5%.

In regard to PV, because there is no fuel cost and we are assuming zero variable O&M costs, PV has no operating cost. Therefore, it will generally operate as much as possible. Therefore, its assumed capacity factor is determined by projections of the number of annual hours of rain/precipitation and cloud cover, whether the PV modules are designed to remain fixed in place or track the sun's path across the sky each day, and by the assumed annual degradation of the solar panel's output (which is typically around 0.5% or less per year). Taking all of these factors into account, our utility assumes that this PV option has an annual average capacity factor of 20%.

Row (13) shows the assumed firm capacity value of each option which has already been discussed (and which will be further discussed in Part II of this book). The last three rows of the table, Rows (14), (15), and (16), provide inputs for the generating options' emission rates for three types of air emissions: sulfur dioxide (SO_2), nitrogen oxides (NO_x), and carbon dioxide (CO_2). These emission rates are typically expressed in terms of pounds of emission per million BTUs of fuel burned (lbs/mmBTU).

Row (14) shows the SO_2 emission rates for the four Supply options. For the CC and CT options, we have assumed an emission rate is 0.006 lb/mmBTU, and for the PV option, this emission rate is zero because no fossil fuel is burned. Row (15) shows the assumed NO_x emission rate is 0.010 lb/mmBTU for the CC options, 0.033 lb/mmBTU for the CT option, and zero for the PV option. Row (16) shows the assumed CO_2 emission rate is 117 lb/mmBTU for the CC and CT options, and zero for the PV option. The emission rate inputs shown in these rows are representative values for these types of new generating units at the time this book is written. (Later in this chapter, we will discuss "environmental compliance cost" values, *i.e.*, projected costs associated with compliance with existing and/or potential legislation or regulation regarding emissions, for each of these three types of air emissions.[7])

After reviewing the list of key inputs, including cost assumptions, for our utility's four Supply options in Table 5.1, one might inquire about what is assumed for federal tax credits for our resource options. The second edition of this book is being written just after the federal Inflation Reduction Act (IRA) was signed into law in 2022. Among the many facets of this legislation were significant new tax credits that apply to selected resource options that are essentially zero carbon emitting resources. In regard to the four resource options our utility has selected for its analyses, the PV option would be eligible for tax credit treatments offered by the 2022 IRA.

In Part I of this book, the decision was made to ignore these tax credits in the discussion of the resource options, including the PV option. There are several reasons for this. First, the inclusion of the tax credits adds a layer of complexity to the discussion of our hypothetical utility's analyses of resource options that I believe is unnecessary for a basic understanding of how the economic analysis aspect of

---

[7] At the time this book is written, there are both state and federal regulations for SO_2 and NO_x in the United States, and there has been an established market for allowances for these two emissions. The same is not true at this time for CO_2. However, due to the expectation that U.S. federal regulations for CO_2 will eventually be established, we will also be projecting compliance costs for CO_2 emissions in our discussion.

integrated resource planning (IRP) works. Second, these tax credits did not exist as utilities made their pre-2022 IRA decisions that led them to add the resources they added to that point in time. Third, I view the tax credits as something that is not fundamental to the IRP process, but as a financial assumption that did not exist before the 2022 IRA (and which could conceivably disappear in the future under different federal administrations).

For these reasons, Part I of this book will not account for these tax credits in the analyses that are discussed. However, these tax credits will have a significant impact on utilities' resource planning work for as long as the tax credits exist. Therefore, Part II of this book will address these tax credits, including a discussion of their impacts, particularly as to how they impact the economic analyses of the PV option.

This completes the introduction of the inputs for the four Supply options that will be used in our utility's economic evaluations. However, before we leave this section, let's look back and see what appears to be obvious and, perhaps more importantly, what is not obvious, from looking at these inputs.

First, from looking at the inputs, the following statements about these four Supply options appear to be obvious:

- In regard to the capital costs on a $/kW basis, the CC and CT options are noticeably less expensive than the PV option. Similarly, the CT option has a capital cost advantage versus the CC options.
- In regard to the other fixed costs, the converse is true; the PV option has lower costs than either the CC or CT options. The CT option again has an advantage in regard to these other fixed costs compared to the CC options.
- In regard to the operating costs, the PV option again has lower costs than either the CC or CT options, simply because it has no fuel costs and a lower Variable O&M cost. As for the CC and CT options, it is somewhat difficult to judge at a glance which of these options has an operating cost advantage. The CT option has lower Variable O&M costs, a lower CO_2 emission rate, and will run much less of the time (a 5% capacity factor) than will the CC options (with an 80% capacity factor). Therefore, the fuel cost to operate the CT option during a year will be lower than with either of the CC options. Conversely, the CC options are much more efficient, and have lower NO_x emission rates, than the CT options. (There is no difference in the SO_2 or CO_2 emission rates between the CC and CT options.)

At this point, one might be tempted to make several other statements. The first of these statements is that it appears obvious that the CT option has a capital and other fixed cost advantage over the CC options because (i) the CT option is lower in both capital and other fixed costs, and (ii) a comparison of the various operating costs versus the CC options are a mixed bag with the CT option being lower in regard to some of these costs and higher in regard to other costs.

The second other statement one might be tempted to make is that it is not obvious whether the significant capital cost disadvantage of the PV option is overcome by PV's clear advantages in regard to other fixed costs and operating costs. This becomes more complicated when accounting for the fact that PV's firm capacity factor is 50% which results in not only a doubling of installed and O&M costs,[8] but also a doubling in the volume of MWh that will be delivered to the utility system at zero fuel cost and zero environmental compliance cost.

Finally, one might also be tempted to make a third statement which is that, at this point, it is not obvious how much effect the truly significant differences in both the capacity (MW) and the capacity factor (% of hours/year the options will operate), of the four Supply options, will have in regard to determining which option is the best economic choice for our utility system.[9]

We will see how all of this actually plays out in the next three sections of this chapter. We will start by performing a preliminary economic screening evaluation of the four Supply options.

PRELIMINARY ECONOMIC SCREENING EVALUATION OF THE SUPPLY OPTIONS

We shall next perform a preliminary economic screening evaluation of these four Supply options and we will use a screening curve analytical approach that was briefly discussed in Chapter 3. However, before we can conduct a screening curve analysis, there are a couple of inputs that are needed for this analysis that are not directly tied to these four specific Supply options.

These inputs are the projected costs of natural gas (the fuel that the CC and CT options utilize) and the projected environmental compliance costs for the three types of air emissions that our utility system is including in its analyses: SO_2, NO_x, and CO_2. Recalling that the screening curve analytical approach is a multi-year analysis, we will need projected cost values for fuel and environmental compliance for a number of years. Assuming, for purposes of our discussion, that the four Supply options have an operating life (or, perhaps more accurately, an accounting "book life") of 25 years, and knowing that our utility's resource needs begin 5 years in the future (in Current Year + 5), our projections for fuel cost and environmental compliance costs will need to address 30 years (from the Current Year through Current Year + 29).[10]

At this point, a brief discussion of environmental compliance costs may be helpful. We earlier described these costs as projected costs associated with compliance with existing and/or potential legislation or regulation regarding emissions. These costs are typically developed by third parties (*i.e.*, a consultant) who starts with a projection of future projected legislation and/or regulation that sets a limit on one

---

[8] Because our utility will need to install 240 MW of PV to meet its 120 MW need due to the 50% firm capacity value for the PV option.

[9] If one could accurately pick out the most economic resource option just by glancing at the inputs that will be used in economic (and non-economic) analyses, then there would be no need for this book. (And I could have spent a lot more time playing more golf.) Sadly, it is not that easy.

[10] We again offer a reminder that the assumed projected values for fuel costs and environmental compliance costs would, in reality, certainly change many times in the years after this book is written. However, it is necessary to assume values for these costs in order to illustrate resource planning analytical approaches. The analytical approaches will remain valid regardless of the assumed values used to illustrate them.

or more specific type of emissions in a future year. The third party then develops, through computer modeling, a projection of the costs it will take for utilities to comply with this legislation and/or regulation. Those costs are typically presented in annual \$ /ton values.

Although these compliance costs may not be an actual cost at in the Current Year, if a utility believes there is a strong likelihood that the projected legislation and/or regulation will be enacted, then it makes sense for a utility to include environmental compliance costs in its resource planning analyses for two reasons. First, the utility can perform scenario analyses both with and without these compliance costs (or use a \$0/ton cost) to see how the results of its resource planning might change between the two scenarios. The utility can then make a judgment call regarding what its resource plan should be. Second, if the projected legislation and/or regulation is enacted, then there will be actual costs for complying with the legislation/regulation that will be recovered from the utility's customers through the utility's electric rates.[11]

––––––––

(Author's Note: At this point, an opinion of the author is about to enter the discussion.)

(It is my opinion that the use of projected environmental compliance costs is a better way to account for potential cost impacts of emissions than the use of environmental "externalities" which is an approach taken by some states in their utilities' resource planning. I see four disadvantages with using externalities, instead of compliance costs, in resource planning work. First, an 'externality' typically refers to an action or impact that is not addressed by either existing or projected specific legislation or regulation. Conversely, the development of environmental compliance costs is based on specific proposed legislation or regulation.

Second, regarding electric utilities, values assigned to externalities are typically projected values for costs that are not typically recovered from utility customers through electric rates. Therefore, these values are not in the normal purview of utilities and their regulators whereas compliance costs will be recovered through electric rates if the legislation/regulation goes into effect. Third, externality values can be difficult to estimate/project with accuracy. For that reason, it is not uncommon to find that different estimates for the same externality that vary significantly (and can differ by an order of magnitude). On the other hand, projections of compliance costs are simpler to estimate using a normal resource planning approach. Fourth, the use of externality values has often focused primarily (if not solely) on projected negative impacts with little (or no) attention on projected positive impacts that may also occur that would provide a more complete picture.

For these reasons, I believe that the use of projected environmental compliance costs is a better way to attempt to account for the potential cost impacts of emissions. Therefore, our hypothetical utility will be using environmental compliance costs in the analyses that will be utilized in analyses presented in the rest of the book.)

––––––––

These assumed cost projections for fuel and environmental compliance costs do not attempt to account for the projected timing of various market corrections, changes in economic conditions, and/or potential advancement in generating or

––––––––

[11] The subject of the use of environmental compliance costs is addressed again in Chapter 14.

emission control equipment. One reason for this is that our economic evaluation is not designed to look at an actual specific starting year (such as the year you are reading this book). Instead, the evaluation is presented in terms of a "Current Year" plus 29 years into the future from that year. For this reason, it made no sense to attempt to factor in the timing of such potential occurrences of the types listed above. (In real life, because the actual starting year for the forecast is known, utility forecasters of load, fuel costs, environmental compliance costs, etc. often do attempt to incorporate the impacts of projected specific events.)

And, as previously mentioned, utilities attempt to account for uncertainty in various forecasts by utilizing multiple forecasts in their resource planning work. For simplicity's sake in discussing the concepts and analytical approaches of utility resource planning, we are using single forecasts for load, fuel costs, environmental compliance costs, etc.

The annual costs for natural gas and the annual compliance costs for SO_2, NO_x, and CO_2 emissions that we are assuming for the economic evaluations of the four Supply options are presented in Table 5.2 on the next page. As previously mentioned, 30 years of projected costs, from "Current Year" through "Current + 29" year, are presented.

Armed with the Supply option-specific cost assumptions previously provided in Table 5.1, and the assumptions for natural gas and environmental compliance costs presented in Table 5.2, we can now conduct a preliminary economic screening evaluation of the four Supply options using a screening curve approach.[12] Figure 5.1 presents the results of the screening curve analysis for these four Supply options.

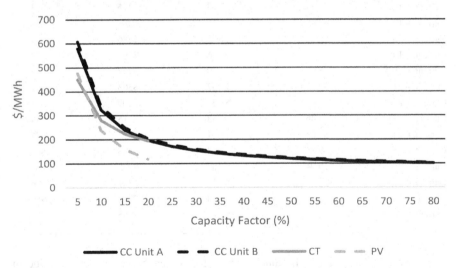

FIGURE 5.1 Preliminary economic screening analyses: Screening approach levelized $/MWh costs for all four supply options.

---

[12] Yes, I know that I said earlier that a screening curve approach cannot provide meaningful information when the resource options being compared are dissimilar in regard to four key characteristics. We are ignoring the differences in the Supply options for the moment solely for the purpose of allowing you to get more comfortable with how screening curve results "look." Fear not, we will return to discuss the significant limitations of screening curve analyses in short order.

TABLE 5.2
Assumed Costs for Natural Gas and Environmental Compliance

Year	Natural Gas Price ($/mmBTU)	Environmental Compliance Costs		
		SO$_2$($/ton)	NO$_x$($/ton)	CO$_2$($/ton)
Current	$6.00	$1.00	$200	$5
Current + 1	$6.12	$1.02	$204	$8
Current + 2	$6.24	$1.04	$208	$11
Current + 3	$6.37	$1.06	$212	$14
Current + 4	$6.49	$1.08	$216	$17
Current + 5	$6.62	$1.10	$221	$20
Current + 6	$6.76	$1.13	$225	$23
Current + 7	$6.89	$1.15	$230	$26
Current + 8	$7.03	$1.17	$234	$29
Current + 9	$7.17	$1.20	$239	$32
Current + 10	$7.31	$1.22	$244	$35
Current + 11	$7.46	$1.24	$249	$38
Current + 12	$7.61	$1.27	$254	$41
Current + 13	$7.76	$1.29	$259	$44
Current + 14	$7.92	$1.32	$264	$47
Current + 15	$8.08	$1.35	$269	$50
Current + 16	$8.24	$1.37	$275	$53
Current + 17	$8.40	$1.40	$280	$56
Current + 18	$8.57	$1.43	$286	$59
Current + 19	$8.74	$1.46	$291	$62
Current + 20	$8.92	$1.49	$297	$65
Current + 21	$9.09	$1.52	$303	$68
Current + 22	$9.28	$1.55	$309	$71
Current + 23	$9.46	$1.58	$315	$74
Current + 24	$9.65	$1.61	$322	$77
Current + 25	$9.84	$1.64	$328	$80
Current + 26	$10.04	$1.67	$335	$83
Current + 27	$10.24	$1.71	$341	$86
Current + 28	$10.45	$1.74	$348	$89
Current + 29	$10.66	$1.78	$355	$92

As we can see, a $/MWh value is presented for each capacity factor value in increasing increments of 5%, *i.e.*, at 5%, 10%, etc. (Note that no value for a capacity factor of 0% is shown because that would reflect only the capital and other fixed costs for the option; no operating hours—and therefore no operating costs—would be assumed.) Then, for each subsequent 5% increase in capacity factor, the option's operating costs are added to the capital and other fixed costs. Then that total cost is divided by the number of MWh that represents that option's capacity (500 MW for CC, 160 MW for CT, or 120 MW for PV) operating at the given capacity factor. For example, for the CT option, the $/MWh value shown at the 5% capacity factor mark represents the total costs divided by 70,080 MWh (= 160 MW × 8,760 hours/year × 5%).

As we can see from Figure 5.1, there is a wide range of values driven partly by the capital costs of the Supply options but driven mostly by the wide disparity in the assumed maximum capacity factors for each Supply option: 20% for PV and CT,[13] and 80% for CC. As a glance at this particular figure shows, the wide range of values typically presented in a screening graph can make it quite difficult to visually distinguish between the values for the different resource options. In order to clear up this picture a bit, we will take a closer look at the options, but this look will examine only pairs of options with similar maximum capacity factors. This will allow us to narrow the focus in regard to the capacity factors addressed in the figure.

We first look at the two options, CT and PV, which have the lower maximum capacity factors. This view is provided in Figure 5.2.

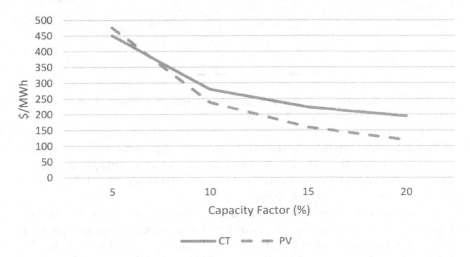

FIGURE 5.2 Preliminary economic screening analyses: Screening approach levelized $/MWh costs for CT and PV options.

In this figure, we narrow our focus to the range of capacity factors that the two resource options will operate in: 5% to 20%. This narrower focus allows us to examine these two lower capacity factor options more closely. From this figure, it is clear that the $/MWh cost for the PV option is definitely higher than the $/MWh cost for the CT option at a 5% capacity factor level. This is primarily due to the higher installed cost of PV ($1,500/kW) versus the CT ($650/kW). However, at a 10% capacity factor, the PV option's $/MWh cost would be lower than the CT option's $/MWh cost (if the CT had a 10% capacity factor). The same result holds true for capacity factors of 15% and 20%.

---

[13] Note that, for illustrative purposes, the "maximum" capacity factor for the CT option has been extended from 5% to 20% to avoid having a single data point for this option. (CTs can operate at capacity factors of 20% or more, but such occurrences are relatively rare.)

In Figure 5.3, we now examine the projected $/MWh costs for the two CC options. We also have narrowed the focus for the two CC options as well.[14]

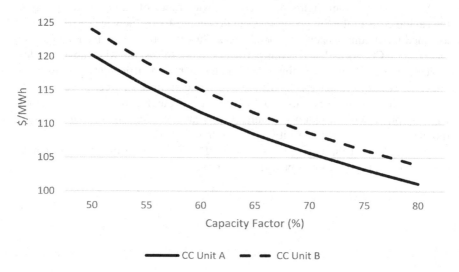

FIGURE 5.3 Preliminary economic screening analyses: Screening approach levelized $/MWh costs for CC Unit A and CC Unit B.

In this figure, our focus is narrowed to a range of capacity factors (50% to 80%) that is closer to how these CC units will actually operate on our utility system (80%). This view shows that CC Option A is clearly the less expensive of the two CC options throughout the range of capacity factors.

So what do we do with this screening curve information? Recall the discussion in Chapter 3 in which we said that a basic rule of using a screening curve approach is that the resource options being analyzed must be identical, or at least very similar, in each of four key characteristics. This seems like a good time to state what those key characteristics are.

The four key characteristics of resource options that must be identical, or very similar, for a screening curve approach to yield meaningful results for a comparison of those resource options are:

1. Capacity (MW);
2. Annual capacity factor;
3. The percentage of the option's capacity (MW) that can be considered as firm capacity at the utility's system peak hours; and
4. The projected life of the option.

We will soon see why these characteristics are so important in regard to screening curve analyses as we work through the remainder of this chapter.

---

[14] You may note that you have seen this figure before, only with slightly different labeling of the two resource options, in Figure 3.1.

Armed now with this knowledge of what the four key characteristics are, we return first to Figure 5.3 for the screening curve analysis of CC Option A and CC Option B. Both of these CC units have identical capacity (500 MW) and annual capacity factor (80%) assumptions, both CC units are assumed to have 100% of their capacity as firm capacity at the utility system's peak hour, and both CC units have identical projected life values (25 years).

In other words, our rule that all four key characteristics must be identical, or very similar, for a screening curve to provide meaningful results is met in regard to the two CC options. Therefore, we can safely eliminate CC Option B using the screening curve results because its $/MWh cost is definitely higher than CC Option A's $/MWh cost throughout the capacity factor range these units will be operating in. So we can safely say "goodbye" to CC Option B.

However, in regard to the CT and PV comparison shown in Figure 5.2, we note that two of the four key characteristics for CT and PV are neither identical, nor even very similar, to the CC options (nor to each other). Their capacities, 160 MW for the CT option and 120 MW for the PV option, are very dissimilar to the 500 MW capacity of the CC units. There are also very large differences in the capacity factors of the three types of Supply options: 5% for CT, 20% for PV, and 80% for CC. In addition, the firm capacity value for the PV option is 50% versus 100% for the CT and CC options, thus presenting a dissimilarity in a third of the four characteristics that must be identical or very similar for a comparison of resource options using a screening curve methodology to yield meaningful results.

Therefore, the basic rule for utilizing a screening curve approach is violated in regard to these three Supply options. As a result, a screening curve approach cannot be used to eliminate any of the three remaining Supply options: CC Unit A, the CT option, or the PV option. A screening curve approach is simply not adequate to provide meaningful information regarding such dissimilar resource options.

Consequently, the preliminary economic screening evaluation using a screening curve approach has eliminated only one Supply option, CC Option B. The remaining Supply options, CC Unit A, the CT option, and the PV option, are carried forward for more detailed economic analyses. We prepare for that evaluation in the next section. (And, near the end of this preparation, we shall see why dissimilarities in even one of the four key characteristics of resource options can severely limit the usefulness of a screening curve analytical approach.)

CREATING THE COMPETING "SUPPLY ONLY" RESOURCE PLANS

As discussed in Chapter 3, an IRP approach looks at long-term resource plans in order to ensure that all of the impacts that a proposed resource addition will have on the utility system over an extended period of time are captured in the analyses. We are now at a point where we will create three separate resource plans. Each resource plan will utilize one of the three remaining Supply options to address the utility's projected resource need in the first year the utility needs to add resources (Current Year + 5 as we saw in Chapter 4). And, for purposes of distinguishing these three resource plans from other resource plans that we will develop later when examining DSM options, we will refer to these three resource plans collectively as "Supply Only" resource plans.

In creating these Supply Only resource plans, we must first extend the projection of the utility's reserve margin beyond the initial 6-year period (Current Year through Current Year + 5) we examined in Table 4.3. This is necessary because the addition of any one of the three Supply options in Current Year + 5 will only meet the utility's projected needs for a few years at best if we assume (and we do) that the utility's load forecast continues to grow at 100 MW/year.

Our utility's reserve margin projection, assuming no resource additions after Current Year + 1, can be extended to encompass the assumed 25-year life of our three remaining Supply options. That projection is shown in Table 5.3 that is presented on the next page.

Note that the information contained in Table 5.3 for the Current Year through Current Year + 5 is identical to that shown previously in Table 4.3. Then information for 24 more years (through Current Year + 29) is presented. We see that the projection of continued load growth of 100 MW/year (shown in Column (2)) continues to decrease the utility's projected reserve margin each year (shown in Column (4)). In fact, the projected reserve margin values even turn negative starting in Current Year + 25.

Consequently, it is clear that the addition of any of the three new Supply options in Current Year + 5 to address the utility's first year of resource need will only carry our utility so far. Other resource options will need to be added at various times in later years to ensure that our utility's reserve margin criterion continues to be met. But when must these other resource additions be added and what do we assume these new resource additions will be?

For purposes of this discussion, we will refer to these later resource additions as "filler units" because their purpose is to "fill in" as needed in later years when our utility's reserve margin criterion would not otherwise be met. However, our utility is not making any definitive decision at this time regarding what type of resource addition would be made in these future years because it does not have to make that decision now. The only definitive resource option decision our utility has to make now is for Current Year + 5.

For that reason, we will refer to Current Year + 5 as the "decision year." And, as discussed in Chapter 3, the addition of the filler units will ensure that each resource plan meets the utility's reliability criterion in all years, thus ensuring that the comparison of competing resource plans is meaningful in regard to system reliability.[15]

We have three Supply options that we are discussing; CC Unit A, CT, and PV. (Recall that we will examine DSM options in the next chapter.) Rather than introduce a fourth Supply option to play the role of the filler units, we will simply select one of these three Supply options to also play the role of the filler unit. For the purpose of the calculations that will follow, we will select the 500 MW CC Unit A to play the role of filler unit.

We are now ready to take a look at how the introduction of any of the three remaining Supply options in Current Year + 5 will lead to the creation of a resource plan. This is done in two steps. First, we look at how the introduction of one of these three Supply options in Current Year + 5 impacts the utility's projected reserve

---

[15] The questions of how many years in the future one continues to add filler units, and what type(s) of filler unit is used, are often discussed by utilities in terms of "end effects" in resource planning. The real issue is to attempt to ensure that these choices are not unduly influencing the choice of the resource option being selected for the near-term decision year. Therefore, the number of years for which filler units are added, and/or the types of filler units, may be varied by the utility in additional analyses of resource plans in order to determine how much influence the filler unit assumptions may have.

TABLE 5.3

**Long-Term Projection of Reserve Margin for the Hypothetical Utility:
Assuming No Resource Additions after "Current Year + 1"**

	(1)	(2)	(3) = (1) − (2)	(4) = (3)/(2)	(5) = ((2)* 1.20) − (1)
Year	Total Generating Capacity (MW)	Peak Electrical Demand (MW)	Reserves (MW)	Reserve Margin (%)	Generation Only MW Needed to Meet Reserve Margin (MW)
Current Year	11,600	9,600	2,000	20.8	(80)
Current Year + 1	12,000	9,700	2,300	23.7	(360)
Current Year + 2	12,000	9,800	2,200	22.4	(240)
Current Year + 3	12,000	9,900	2,100	21.2	(120)
Current Year + 4	12,000	10,000	2,000	20.0	0
Current Year + 5	12,000	10,100	1,900	18.8	120
Current Year + 6	12,000	10,200	1,800	17.6	240
Current Year + 7	12,000	10,300	1,700	16.5	360
Current Year + 8	12,000	10,400	1,600	15.4	480
Current Year + 9	12,000	10,500	1,500	14.3	600
Current Year + 10	12,000	10,600	1,400	13.2	720
Current Year + 11	12,000	10,700	1,300	12.1	840
Current Year + 12	12,000	10,800	1,200	11.1	960
Current Year + 13	12,000	10,900	1,100	10.1	1,080
Current Year + 14	12,000	11,000	1,000	9.1	1,200
Current Year + 15	12,000	11,100	900	8.1	1,320
Current Year + 16	12,000	11,200	800	7.1	1,440
Current Year + 17	12,000	11,300	700	6.2	1,560
Current Year + 18	12,000	11,400	600	5.3	1,680
Current Year + 19	12,000	11,500	500	4.3	1,800
Current Year + 20	12,000	11,600	400	3.4	1,920
Current Year + 21	12,000	11,700	300	2.6	2,040
Current Year + 22	12,000	11,800	200	1.7	2,160
Current Year + 23	12,000	11,900	100	0.8	2,280
Current Year + 24	12,000	12,000	0	0.0	2,400
Current Year + 25	12,000	12,100	(100)	(0.8)	2,520
Current Year + 26	12,000	12,200	(200)	(1.6)	2,640
Current Year + 27	12,000	12,300	(300)	(2.4)	2,760
Current Year + 28	12,000	12,400	(400)	(3.2)	2,880
Current Year + 29	12,000	12,500	(500)	(4.0)	3,000

margin. Second, using that information, we add a 500 MW filler CC unit as needed in the appropriate years so that the 20% minimum reserve margin criterion is met in all years.

We will start with the introduction of CC Unit A in Current Year + 5 and see how the utility's projected reserve margin is impacted in Table 5.4 (which is based on Table 5.3).

TABLE 5.4
Long-Term Projection of Reserve Margin for the Hypothetical Utility: With New CC Unit Added in "Current Year + 5"

	(1a)	(1b)	(1c)	(1d)	(1e)	(1f)	(2)	(3)	(4)	(5)
	Previously Projected Generating Capacity (MW)	Cumulative No. of New Unit Additions				Total Projected Generating Capacity (MW) = (See Formula Below)	Peak Electrical Demand (MW)	Reserves (MW) = (1f) − (2)	Reserve Margin (%) = (3)/(2)	Generation Only MW Needed to Meet Reserve Margin (MW) = ((2)* 1.20) − (1f)
Year		CC Unit A (No. Units)	CT (No. Units)	PV (No. Units)	Filler Units (No. Units)					
Current Year	11,600	0	0	0	0	11,600	9,600	2,000	20.8	(80)
Current Year + 1	12,000	0	0	0	0	12,000	9,700	2,300	23.7	(360)
Current Year + 2	12,000	0	0	0	0	12,000	9,800	2,200	22.4	(240)
Current Year + 3	12,000	0	0	0	0	12,000	9,900	2,100	21.2	(120)
Current Year + 4	12,000	0	0	0	0	12,000	10,000	2,000	20.0	0
Current Year + 5	12,000	1	0	0	0	12,500	10,100	2,400	23.8	(380)
Current Year + 6	12,000	1	0	0	0	12,500	10,200	2,300	22.5	(260)
Current Year + 7	12,000	1	0	0	0	12,500	10,300	2,200	21.4	(140)
Current Year + 8	12,000	1	0	0	0	12,500	10,400	2,100	20.2	(20)
Current Year + 9	12,000	1	0	0	0	12,500	10,500	2,000	19.0	100
Current Year + 10	12,000	1	0	0	0	12,500	10,600	1,900	17.9	220
Current Year + 11	12,000	1	0	0	0	12,500	10,700	1,800	16.8	340
Current Year + 12	12,000	1	0	0	0	12,500	10,800	1,700	15.7	460

Current Year + 13	12,000	1	0	0	12,500	10,900	1,600	14.7	580
Current Year + 14	12,000	1	0	0	12,500	11,000	1,500	13.6	700
Current Year + 15	12,000	1	0	0	12,500	11,100	1,400	12.6	820
Current Year + 16	12,000	1	0	0	12,500	11,200	1,300	11.6	940
Current Year + 17	12,000	1	0	0	12,500	11,300	1,200	10.6	1,060
Current Year + 18	12,000	1	0	0	12,500	11,400	1,100	9.6	1,180
Current Year + 19	12,000	1	0	0	12,500	11,500	1,000	8.7	1,300
Current Year + 20	12,000	1	0	0	12,500	11,600	900	7.8	1,420
Current Year + 21	12,000	1	0	0	12,500	11,700	800	6.8	1,540
Current Year + 22	12,000	1	0	0	12,500	11,800	700	5.9	1,660
Current Year + 23	12,000	1	0	0	12,500	11,900	600	5.0	1,780
Current Year + 24	12,000	1	0	0	12,500	12,000	500	4.2	1,900
Current Year + 25	12,000	1	0	0	12,500	12,100	400	3.3	2,020
Current Year + 26	12,000	1	0	0	12,500	12,200	300	2.5	2,140
Current Year + 27	12,000	1	0	0	12,500	12,300	200	1.6	2,260
Current Year + 28	12,000	1	0	0	12,500	12,400	100	0.8	2,380
Current Year + 29	12,000	1	0	0	12,500	12,500	0	0.0	2,500

CC Unit A = 500 MW
CT = 160 MW
PV = 60 MW (firm)
Filler Units = 500 MW

Formula: $(1f) = (1a) + ((1b) \times 500) + ((1c) \times 160) + ((1d) \times 60) + ((1e) \times 500)$

Let's first discuss the two changes we have made to the table. First, Column (1) in Table 5.3 has been expanded to encompass Columns (1a) through (1f). The values in Column (1a) are identical to what appeared in Column (1) in the previous table. The values in Columns (1b) through (1e) represent the cumulative number of new Supply options of the various types that are assumed to be added: the number of CC options in Column (1b), the number of CT options in Column (1c), the number of PV options in Column (1d), and the number of filler units in Column (1e). Finally, Column (1f) then shows the total generating capacity for the utility for each year.

Second, the capacity (MW) for each of these types of Supply options is shown in a small separate section that appears at the bottom of the table below Columns (1b) through (1d). The number of Supply options of each type is then multiplied by the MW for each Supply option type shown in this section. For example, by looking at the row for Current Year + 5 and Column (1b), we see that 1 CC unit, CC Unit A, is added in that year. This number of CC units, 1 unit, is multiplied by the 500 MW for each CC unit from the small table to yield a gain of 500 MW in that year. Likewise, no (or zero) CT, PV, or filler units are added in that year so 0 MW are added in that year from these types of Supply options.

Column (1f) then adds the previously projected generating capacity in Column (1a) to the additional MW from each of the types of Supply options to derive the total projected generating firm capacity for that year. This total is presented in Column (1f). For Current Year + 5, the calculation is: 12,000 existing MW + (1 × 500 MW) + (0 × 160 MW) + (0 × 60 MW) + (0 × 500 MW) = 12,500 MW. Note that the firm capacity value assumed for PV in this formula is only 60 MW due to the assumption that the PV option has a 50% firm capacity value rather than the 100% firm capacity value assumed for the other options. (This formula is shown on the table starting at the bottom of Column (1f).)

By examining the reserve margin values in Column (4), we see that the addition of the 500 MW CC Unit A in Current Year + 5 allows the utility to not only meet its reserve margin criterion for that year but also for 3 more years. The first year in which the reserve margin criterion is not met is now Current Year + 9. That year is highlighted by shading the Current Year + 9 row in the table. Consequently, no filler unit addition is needed until Current Year + 9. In subsequent years, additional filler units will also be needed. We see how many filler units are needed, and when they are needed, in Table 5.5.

(This page/space is intentionally left blank.)

TABLE 5.5

Supply Only Resource Plan 1 with New CC Unit Added in "Current Year + 5" Plus Filler Units

Year	(1a) Previously Projected Generating Capacity (MW)	(1b) CC Unit A (No. Units)	(1c) CT (No. Units)	(1d) PV (No. Units)	(1e) Filler Units (No. Units)	(1f) = (See Formula Below) Total Projected Generating Capacity (MW)	(2) Peak Electrical Demand (MW)	(3) = (1f) − (2) Reserves (MW)	(4) = (3)/(2) Reserve Margin (%)	(5) = ((2) * 1.20) − (1f) Generation Only MW Needed to Meet Reserve Margin (MW)
			Cumulative No. of New Unit Additions							
Current Year	11,600	0	0	0	0	11,600	9,600	2,000	20.8	(80)
Current Year + 1	12,000	0	0	0	0	12,000	9,700	2,300	23.7	(360)
Current Year + 2	12,000	0	0	0	0	12,000	9,800	2,200	22.4	(240)
Current Year + 3	12,000	0	0	0	0	12,000	9,900	2,100	21.2	(120)
Current Year + 4	12,000	0	0	0	0	12,000	10,000	2,000	20.0	0
Current Year + 5	12,000	1	0	0	0	12,500	10,100	2,400	23.8	(380)
Current Year + 6	12,000	1	0	0	0	12,500	10,200	2,300	22.5	(260)
Current Year + 7	12,000	1	0	0	0	12,500	10,300	2,200	21.4	(140)
Current Year + 8	12,000	1	0	0	0	12,500	10,400	2,100	20.2	(20)
Current Year + 9	12,000	1	0	0	1	13,000	10,500	2,500	23.8	(400)
Current Year + 10	12,000	1	0	0	1	13,000	10,600	2,400	22.6	(280)
Current Year + 11	12,000	1	0	0	1	13,000	10,700	2,300	21.5	(160)
Current Year + 12	12,000	1	0	0	1	13,000	10,800	2,200	20.4	(40)
Current Year + 13	12,000	1	0	0	2	13,500	10,900	2,600	23.9	(420)

Current Year + 14	12,000	1	0	2	13,500	11,000	2,500	22.7	(300)
Current Year + 15	12,000	1	0	2	13,500	11,100	2,400	21.6	(180)
Current Year + 16	12,000	1	0	2	13,500	11,200	2,300	20.5	(60)
Current Year + 17	12,000	1	0	2	13,500	11,300	2,200	19.5	60
Current Year + 18	12,000	1	0	3	14,000	11,400	2,600	22.8	(320)
Current Year + 19	12,000	1	0	3	14,000	11,500	2,500	21.7	(200)
Current Year + 20	12,000	1	0	3	14,000	11,600	2,400	20.7	(80)
Current Year + 21	12,000	1	0	3	14,000	11,700	2,300	19.7	40
Current Year + 22	12,000	1	0	4	14,500	11,800	2,700	22.9	(340)
Current Year + 23	12,000	1	0	4	14,500	11,900	2,600	21.8	(220)
Current Year + 24	12,000	1	0	4	14,500	12,000	2,500	20.8	(100)
Current Year + 25	12,000	1	0	4	14,500	12,100	2,400	19.8	20
Current Year + 26	12,000	1	0	5	15,000	12,200	2,800	23.0	(360)
Current Year + 27	12,000	1	0	5	15,000	12,300	2,700	22.0	(240)
Current Year + 28	12,000	1	0	5	15,000	12,400	2,600	21.0	(120)
Current Year + 29	12,000	1	0	5	15,000	12,500	2,500	20.0	0

CC Unit A =   500   MW
CT =   160   MW
PV =   60   MW (firm)
Filler Units = 500   MW

Formula: (1f) = (1a) + ((1b) × 500) + ((1c) × 160) + ((1d) × 60) + ((1e) × 500)

By examining Column (1e) of this table, we see that a cumulative total of five filler units have been added through the 30-year analysis period. These filler units are added, one each, in the following years: Current Year + 9, Current Year + 13, Current Year + 18, Current Year + 22, and Current Year + 26.[16]

We now have a resource plan that adds CC Unit A in Current Year + 5 and then adds 5 filler units over the course of the next 25 years. We will (very creatively) name this resource plan Supply Only Resource Plan 1 (CC).

We will now construct similar resource plans based on adding either the CT option or the PV option in Current Year + 5, then adding the appropriate number of filler units as needed in subsequent years.

Table 5.6, presented on pages 104 and 105, shows the projected reserve margin if the CT option is added in Current Year + 5 instead of the CC option.

---

[16] The alert reader may ask why no filler unit is added in certain years such as Current Year + 17 in which the projected reserve margin is 19.5%. In practice, utilities, and their regulators, may allow the reserve margin to fall slightly below their reliability criterion for a particular year. Our hypothetical utility has taken that approach and will not add a filler unit unless the projected reserve margin drops below 19.50%, *i.e.*, below a value that will round up to the reserve margin criterion of 20%.

(This page/space is intentionally left blank.)

TABLE 5.6
Long-Term Projection of Reserve Margin for the Hypothetical Utility: With New CT Unit Added in "Current Year + 5"

	(1a)	(1b)	(1c)	(1d)	(1e)	(1f)	(2)	(3) = (1f) − (2)	(4) = (3)/(2)	(5) = ((2) × 1.20) − (1f)
	Previously Projected Generating Capacity (MW)	Cumulative No. of New Unit Additions				= (See Formula Below) Total Projected Generating Capacity (MW)	Peak Electrical Demand (MW)	Reserves (MW)	Reserve Margin (%)	Generation Only MW Needed to Meet Reserve Margin (MW)
Year		CC Unit A (No. Units)	CT (No. Units)	PV (No. Units)	Filler Units (No. Units)					
Current Year	11,600	0	0	0	0	11,600	9,600	2,000	20.8	(80)
Current Year + 1	12,000	0	0	0	0	12,000	9,700	2,300	23.7	(360)
Current Year + 2	12,000	0	0	0	0	12,000	9,800	2,200	22.4	(240)
Current Year + 3	12,000	0	0	0	0	12,000	9,900	2,100	21.2	(120)
Current Year + 4	12,000	0	0	0	0	12,000	10,000	2,000	20.0	0
Current Year + 5	12,000	0	1	0	0	12,160	10,100	2,060	20.4	(40)
Current Year + 6	12,000	0	1	0	0	12,160	10,200	1,960	19.2	80
Current Year + 7	12,000	0	1	0	0	12,160	10,300	1,860	18.1	200
Current Year + 8	12,000	0	1	0	0	12,160	10,400	1,760	16.9	320
Current Year + 9	12,000	0	1	0	0	12,160	10,500	1,660	15.8	440
Current Year + 10	12,000	0	1	0	0	12,160	10,600	1,560	14.7	560
Current Year + 11	12,000	0	1	0	0	12,160	10,700	1,460	13.6	680
Current Year + 12	12,000	0	1	0	0	12,160	10,800	1,360	12.6	800
Current Year + 13	12,000	0	1	0	0	12,160	10,900	1,260	11.6	920

Current Year + 14	12,000	0	0	1	0	12,160	11,000	1,160	1,040	10.5
Current Year + 15	12,000	0	0	1	0	12,160	11,100	1,060	1,160	9.5
Current Year + 16	12,000	0	0	1	0	12,160	11,200	960	1,280	8.6
Current Year + 17	12,000	0	0	1	0	12,160	11,300	860	1,400	7.6
Current Year + 18	12,000	0	0	1	0	12,160	11,400	760	1,520	6.7
Current Year + 19	12,000	0	0	1	0	12,160	11,500	660	1,640	5.7
Current Year + 20	12,000	0	0	1	0	12,160	11,600	560	1,760	4.8
Current Year + 21	12,000	0	0	1	0	12,160	11,700	460	1,880	3.9
Current Year + 22	12,000	0	0	1	0	12,160	11,800	360	2,000	3.1
Current Year + 23	12,000	0	0	1	0	12,160	11,900	260	2,120	2.2
Current Year + 24	12,000	0	0	1	0	12,160	12,000	160	2,240	1.3
Current Year + 25	12,000	0	0	1	0	12,160	12,100	60	2,360	0.5
Current Year + 26	12,000	0	0	1	0	12,160	12,200	(40)	2,480	(0.3)
Current Year + 27	12,000	0	0	1	0	12,160	12,300	(140)	2,600	(1.1)
Current Year + 28	12,000	0	0	1	0	12,160	12,400	(240)	2,720	(1.9)
Current Year + 29	12,000	0	0	1	0	12,160	12,500	(340)	2,840	(2.7)

CC Unit A = 500 MW
CT = 160 MW
PV = 60 MW (firm)
Filler Units = 500 MW

Formula: $(1f) = (1a) + ((1b) \times 500) + ((1c) \times 160) + ((1d) \times 60) + ((1e) \times 500)$

As shown in Table 5.6, the addition of the 160 MW CT option in Current Year + 5 results in the utility still having a resource need in the next year (Current Year + 6) because the projected reserve margin is 19.2% for that year. That row is again shaded in this table.

Table 5.7, presented on pages 108 and 109, then shows the subsequent addition of filler units that are needed starting in Current Year + 6.

By examining Column (1e) of this table, we see that a cumulative total of six filler units have been added through the 30-year analysis period. These filler units are added, one each, in the following years: Current Year + 6, Current Year + 10, Current Year + 15, Current Year + 19, Current Year + 23, and Current Year + 27.

Therefore, the addition of the CT option results both in (i) a greater number of filler units being added overall and (ii) a change in the timing of when filler units are added (*i.e.*, in what years they are added). This second Supply option resource plan, based on the addition of the CT option, is definitely different than the first Supply option resource plan which was based on the CC option. Continuing to be led by our highly creative instincts, we will refer to this second resource plan as Supply Only Resource Plan 2 (CT).

We will then turn to the PV option and Table 5.8 (presented on pages 110 and 111) shows the projected reserve margin if the PV option is added in Current Year + 5.

(This page/space is intentionally left blank.)

TABLE 5.7
Supply Only Resource Plan 2 with New CT Unit Added in "Current Year + 5" Plus Filler Units

Year	(1a) Previously Projected Generating Capacity (MW)	(1b) CC Unit A (No. Units)	(1c) CT (No. Units)	(1d) PV (No. Units)	(1e) Filler Units (No. Units)	(1f) Total Projected Generating Capacity (MW) = (See Formula Below)	(2) Peak Electrical Demand (MW)	(3) = (1f) − (2) Reserves (MW)	(4) = (3)/(2) Reserve Margin (%)	(5) = ((2) × 1.20) − (1f) Generation Only MW Needed to Meet Reserve Margin (MW)
Current Year	11,600	0	0	0	0	11,600	9,600	2,000	20.8	(80)
Current Year + 1	12,000	0	0	0	0	12,000	9,700	2,300	23.7	(360)
Current Year + 2	12,000	0	0	0	0	12,000	9,800	2,200	22.4	(240)
Current Year + 3	12,000	0	0	0	0	12,000	9,900	2,100	21.2	(120)
Current Year + 4	12,000	0	0	0	0	12,000	10,000	2,000	20.0	0
Current Year + 5	12,000	0	1	0	0	12,160	10,100	2,060	20.4	(40)
Current Year + 6	12,000	0	1	0	1	12,660	10,200	2,460	24.1	(420)
Current Year + 7	12,000	0	1	0	1	12,660	10,300	2,360	22.9	(300)
Current Year + 8	12,000	0	1	0	1	12,660	10,400	2,260	21.7	(180)
Current Year + 9	12,000	0	1	0	1	12,660	10,500	2,160	20.6	(60)
Current Year + 10	12,000	0	1	0	2	13,160	10,600	2,560	24.2	(440)
Current Year + 11	12,000	0	1	0	2	13,160	10,700	2,460	23.0	(320)
Current Year + 12	12,000	0	1	0	2	13,160	10,800	2,360	21.9	(200)
Current Year + 13	12,000	0	1	0	2	13,160	10,900	2,260	20.7	(80)
Current Year + 14	12,000	0	1	0	2	13,160	11,000	2,160	19.6	40

Current Year + 15	12,000	0	1	0	3	13,660	11,100	2,560	23.1	(340)
Current Year + 16	12,000	0	1	0	3	13,660	11,200	2,460	22.0	(220)
Current Year + 17	12,000	0	1	0	3	13,660	11,300	2,360	20.9	(100)
Current Year + 18	12,000	0	1	0	3	13,660	11,400	2,260	19.8	20
Current Year + 19	12,000	0	1	0	4	14,160	11,500	2,660	23.1	(360)
Current Year + 20	12,000	0	1	0	4	14,160	11,600	2,560	22.1	(240)
Current Year + 21	12,000	0	1	0	4	14,160	11,700	2,460	21.0	(120)
Current Year + 22	12,000	0	1	0	4	14,160	11,800	2,360	20.0	0
Current Year + 23	12,000	0	1	0	5	14,660	11,900	2,760	23.2	(380)
Current Year + 24	12,000	0	1	0	5	14,660	12,000	2,660	22.2	(260)
Current Year + 25	12,000	0	1	0	5	14,660	12,100	2,560	21.2	(140)
Current Year + 26	12,000	0	1	0	5	14,660	12,200	2,460	20.2	(20)
Current Year + 27	12,000	0	1	0	6	15,160	12,300	2,860	23.3	(400)
Current Year + 28	12,000	0	1	0	6	15,160	12,400	2,760	22.3	(280)
Current Year + 29	12,000	0	1	0	6	15,160	12,500	2,660	21.3	(160)

CC Unit A =	500	MW
CT =	160	MW
PV =	60	MW (firm)
Filler Units =	500	MW

Formula: (1f) = (1a) + ((1b) × 500) + ((1c) × 160) + ((1d) × 60) + ((1e) × 500)

TABLE 5.8
Supply Only Resource Plan 3 with 2 New PV Units Added in "Current Year + 5"

	(1a)	(1b)	(1c)	(1d)	(1e)	(1f)	(2)	(3)	(4)	(5)
	Previously Projected		Cumulative No. of New Unit Additions			= (See Formula Below)		= (1f) − (2)	= (3)/(2)	= ((2) × 1.20) − (1f)
										Generation Only
	Projected	CC Unit	CT	PV	Filler	Total Projected	Peak			MW Needed to
	Generating	A (No.	(No.	(No.	Units (No.	Generating	Electrical	Reserves	Reserve	Meet Reserve
	Capacity	Units)	Units)	Units)	Units)	Capacity	Demand	(MW)	Margin	Margin (MW)
Year	(MW)					(MW)	(MW)		(%)	
Current Year	11,600	0	0	0	0	11,600	9,600	2,000	20.8	(80)
Current Year + 1	12,000	0	0	0	0	12,000	9,700	2,300	23.7	(360)
Current Year + 2	12,000	0	0	0	0	12,000	9,800	2,200	22.4	(240)
Current Year + 3	12,000	0	0	0	0	12,000	9,900	2,100	21.2	(120)
Current Year + 4	12,000	0	0	0	0	12,000	10,000	2,000	20.0	0
Current Year + 5	12,000	0	0	2	0	12,120	10,100	2,020	20.0	0
Current Year + 6	12,000	0	0	2	0	12,120	10,200	1,920	18.8	120
Current Year + 7	12,000	0	0	2	0	12,120	10,300	1,820	17.7	240
Current Year + 8	12,000	0	0	2	0	12,120	10,400	1,720	16.5	360
Current Year + 9	12,000	0	0	2	0	12,120	10,500	1,620	15.4	480
Current Year + 10	12,000	0	0	2	0	12,120	10,600	1,520	14.3	600
Current Year + 11	12,000	0	0	2	0	12,120	10,700	1,420	13.3	720
Current Year + 12	12,000	0	0	2	0	12,120	10,800	1,320	12.2	840
Current Year + 13	12,000	0	0	2	0	12,120	10,900	1,220	11.2	960
Current Year + 14	12,000	0	0	2	0	12,120	11,000	1,120	10.2	1,080
Current Year + 15	12,000	0	0	2	0	12,120	11,100	1,020	9.2	1,200

Current Year + 16	12,000	0	0	2	0	12,120	11,200	920	8.2	1,320
Current Year + 17	12,000	0	0	2	0	12,120	11,300	820	7.3	1,440
Current Year + 18	12,000	0	0	2	0	12,120	11,400	720	6.3	1,560
Current Year + 19	12,000	0	0	2	0	12,120	11,500	620	5.4	1,680
Current Year + 20	12,000	0	0	2	0	12,120	11,600	520	4.5	1,800
Current Year + 21	12,000	0	0	2	0	12,120	11,700	420	3.6	1,920
Current Year + 22	12,000	0	0	2	0	12,120	11,800	320	2.7	2,040
Current Year + 23	12,000	0	0	2	0	12,120	11,900	220	1.8	2,160
Current Year + 24	12,000	0	0	2	0	12,120	12,000	120	1.0	2,280
Current Year + 25	12,000	0	0	2	0	12,120	12,100	20	0.2	2,400
Current Year + 26	12,000	0	0	2	0	12,120	12,200	(80)	(0.7)	2,520
Current Year + 27	12,000	0	0	2	0	12,120	12,300	(180)	(1.5)	2,640
Current Year + 28	12,000	0	0	2	0	12,120	12,400	(280)	(2.3)	2,760
Current Year + 29	12,000	0	0	2	0	12,120	12,500	(380)	(3.0)	2,880

CC Unit A =	500	MW
CT =	160	MW
PV =	60	MW (firm)
Filler Units =	500	MW

Formula: $(1f) = (1a) + ((1b) \times 500) + ((1c) \times 160) + ((1d) \times 60) + ((1e) \times 500)$

As previously discussed, the fact that the PV option has only a 50% firm capacity value, or 60 MW of firm capacity from the 120 MW nameplate PV option, results in our utility having to add <u>two</u> 120 MW PV options in Current Year + 5 in order to meet the 120-MW resource need in that year.

Similar to the projected reserve margin picture when the CT option is added, we see that our utility will also not meet its reserve margin criterion in Current Year + 6. That row is again shaded in this table. Table 5.9 (see pages 114 and 115) shows how the filler unit additions must occur if the PV option is selected for Current Year + 5.

We once again examine Column (1e) for this table and we see that six filler units have again been added through the 30-year analysis period. These filler units are added, one each, in the same years as filler units were added for the previous resource plan featuring the CT option with one exception; we now add a filler unit 1 year earlier in Current Year + 14 instead of in Current Year + 15 as was the case with the resource plan featuring the 160 MW CT. Therefore, although the addition of the PV option results in the same number of filler units being added as with the CT option, there is still a change in the timing of one of those filler units.

(This page/space is intentionally left blank.)

TABLE 5.9
Supply Only Resource Plan 3 with 2 New PV Units Added in "Current Year + 5" Plus Filler Units

Year	(1a) Previously Projected Generating Capacity (MW)	(1b) CC Unit A (No. Units)	(1c) CT (No. Units)	(1d) PV (No. Units)	(1e) Filler Units (No. Units)	(1f) = (See Formula Below) Total Projected Generating Capacity (MW)	(2) Peak Electrical Demand (MW)	(3) = (1f) − (2) Reserves (MW)	(4) = (3)/(2) Reserve Margin (%)	(5) = ((2) × 1.20) − (1f) Generation Only MW Needed to Meet Reserve Margin (MW)
		Cumulative No. of New Unit Additions								
Current Year	11,600	0	0	0	0	11,600	9,600	2,000	20.8	(80)
Current Year + 1	12,000	0	0	0	0	12,000	9,700	2,300	23.7	(360)
Current Year + 2	12,000	0	0	0	0	12,000	9,800	2,200	22.4	(240)
Current Year + 3	12,000	0	0	0	0	12,000	9,900	2,100	21.2	(120)
Current Year + 4	12,000	0	0	0	0	12,000	10,000	2,000	20.0	0
Current Year + 5	12,000	0	0	2	0	12,120	10,100	2,020	20.0	0
Current Year + 6	12,000	0	0	2	1	12,620	10,200	2,420	23.7	(380)
Current Year + 7	12,000	0	0	2	1	12,620	10,300	2,320	22.5	(260)
Current Year + 8	12,000	0	0	2	1	12,620	10,400	2,220	21.3	(140)
Current Year + 9	12,000	0	0	2	1	12,620	10,500	2,120	20.2	(20)
Current Year + 10	12,000	0	0	2	2	13,120	10,600	2,520	23.8	(400)
Current Year + 11	12,000	0	0	2	2	13,120	10,700	2,420	22.6	(280)
Current Year + 12	12,000	0	0	2	2	13,120	10,800	2,320	21.5	(160)
Current Year + 13	12,000	0	0	2	2	13,120	10,900	2,220	20.4	(40)
Current Year + 14	12,000	0	0	2	3	13,620	11,000	2,620	23.8	(420)

Current Year + 15	12,000	0	2	3	13,620	11,100	2,520	22.7	(300)
Current Year + 16	12,000	0	2	3	13,620	11,200	2,420	21.6	(180)
Current Year + 17	12,000	0	2	3	13,620	11,300	2,320	20.5	(60)
Current Year + 18	12,000	0	2	3	13,620	11,400	2,220	19.5	60
Current Year + 19	12,000	0	2	4	14,120	11,500	2,620	22.8	(320)
Current Year + 20	12,000	0	2	4	14,120	11,600	2,520	21.7	(200)
Current Year + 21	12,000	0	2	4	14,120	11,700	2,420	20.7	(80)
Current Year + 22	12,000	0	2	4	14,120	11,800	2,320	19.7	40
Current Year + 23	12,000	0	2	5	14,620	11,900	2,720	22.9	(340)
Current Year + 24	12,000	0	2	5	14,620	12,000	2,620	21.8	(220)
Current Year + 25	12,000	0	2	5	14,620	12,100	2,520	20.8	(100)
Current Year + 26	12,000	0	2	5	14,620	12,200	2,420	19.8	20
Current Year + 27	12,000	0	2	6	15,120	12,300	2,820	22.9	(360)
Current Year + 28	12,000	0	2	6	15,120	12,400	2,720	21.9	(240)
Current Year + 29	12,000	0	2	6	15,120	12,500	2,620	21.0	(120)

CC Unit A =	500	MW
CT =	160	MW
PV =	60	MW (firm)
Filler Units =	500	MW

Formula: $(1f) = (1a) + ((1b) \times 500) + ((1c) \times 160) + ((1d) \times 60) + ((1e) \times 500)$

Therefore, this third resource plan differs from the previous two resource plans in terms of the timing of the filler units added as well as in terms of the type of Supply option added in Current Year + 5. And, as you surely have guessed by now, we will refer to this third resource plan as Supply Only Resource Plan 3 (PV).

If we were to put these three resource plans side-by-side, we should more clearly see the differences between them. Therefore, we do so with Table 5.10.

TABLE 5.10
Overview of Three Supply Only Resource Plans

Year	Supply Only Resource Plan 1 (CC)	Supply Only Resource Plan 2 (CT)	Supply Only Resource Plan 3 (PV)
Current Year	------	------	------
Current Year + 1	------	------	------
Current Year + 2	------	------	------
Current Year + 3	------	------	------
Current Year + 4	------	------	------
Current Year + 5	CC Unit A (500 MW)	CT (160 MW)	2 PV (120 MW)
	Cumulative No. of Filler Units		
Current Year + 6	0	1	1
Current Year + 7	0	1	1
Current Year + 8	0	1	1
Current Year + 9	1	1	1
Current Year + 10	1	2	2
Current Year + 11	1	2	2
Current Year + 12	1	2	2
Current Year + 13	2	2	2
Current Year + 14	2	2	3
Current Year + 15	2	3	3
Current Year + 16	2	3	3
Current Year + 17	2	3	3
Current Year + 18	3	3	3
Current Year + 19	3	4	4
Current Year + 20	3	4	4
Current Year + 21	3	4	4
Current Year + 22	4	4	4
Current Year + 23	4	5	5
Current Year + 24	4	5	5
Current Year + 25	4	5	5
Current Year + 26	5	5	5
Current Year + 27	5	6	6
Current Year + 28	5	6	6
Current Year + 29	5	6	6

Table 5.10 shows the significant differences between the three Supply Only resource plans, particularly in regard to the timing and number of filler units being added in each resource plan. By looking at this table, one can see that in the years after Current Year + 5, there are only 5 years (such as Current Year + 9) in which the cumulative number of filler units is identical for the three resource plans.

These differences in the timing of filler units, plus the difference in the total number of filler units added in each resource plan, *is solely due to the difference in the amount of firm capacity that is added in the decision year of Current Year + 5*. As these three resource plans demonstrate, even a relatively small difference in firm capacity between two resource options (such as a difference of 40 MW [=160 MW firm capacity of the CT option versus 120 MW total firm capacity of the PV option] can result in a shift in the timing of future resource additions). And, as you might expect, this change in the timing of future resource additions in a given year will change the utility's total cost projection for that year (and each year thereafter). We shall see this in the next section.

This effect is magnified greatly when the difference in the capacity being added is more significant. This can be seen by comparing Supply Only Resource Plan 1 (CC) with either Supply Option Resource Plan 2 (CT) or Supply Only Resource Plan 3 (PV).

A discerning reader may see a light bulb go off right about now. Recall that the first and third of the four key characteristics that two competing resource options must be identical, or very similar, in regard to in order to allow a screening curve approach to give meaningful results are, respectively, (1) the capacity (MW) of the two resource options and (3) the percentage of the option's capacity that can be considered firm capacity at the utility's system peak hour.

We have just seen one reason why that is true. Even a relatively small difference of 40 MW in firm capacity between two Supply options, CT and PV, is enough to change the timing of future resource additions. This change in the timing of when future resource options are added will directly impact the costs for the utility from that year forward.[17] This change in the timing of future resource option additions, and the associated costs with this change, are simply not captured in a screening curve analysis approach.

If our utility had (incorrectly) eliminated one of these two resource options after completing only the screening curve-based preliminary economic evaluation, it would have missed accounting for this one impact to costs of the resource plans in future years. The impact of this seemingly small difference in firm capacity between the PV and CT options clearly shows up when comparing long-term resource plans based on these options. In turn, the impact of this difference in the resource plans can only be fully addressed in an economic evaluation of the resource plans themselves. We do so next.

---

[17] As we shall later see, there are other reasons why dissimilarities between resource options for even the first key characteristic (capacity (MW)) will result in significant impacts to the costs of resource plans that are not captured in a screening curve approach.

FINAL (OR SYSTEM) ECONOMIC EVALUATION
OF SUPPLY OPTIONS

Overview

Now that we have created a separate resource plan for each of the three remaining Supply options, the final (or system) economic evaluation can be carried out. This evaluation of each resource plan consists of two steps. The first step provides a perspective of total utility system costs with each resource plan. The second step provides a second perspective: the electric rates that our utility's customers will be charged to recover all of our utility's costs for each resource plan.

In the first step, the cumulative present value of annual revenue requirements (CPVRR) for our utility system with each resource plan is calculated.

The CPVRR value represents the total costs for each resource plan.[18] This calculation accounts for all of our utility system's costs that are driven by the Supply option selected for the decision year, plus all of the subsequent filler units for that resource plan. At a minimum, the following types of costs are accounted for in final (or system) economic evaluations:

- The capital, fixed O&M, variable O&M, and capital replacement costs for the selected Supply option and the subsequent filler units;
- The fixed costs associated with fuel delivery (such as firm gas transportation costs for all new Supply options that are fueled by natural gas);
- The administrative and incentive costs for any DSM options that are selected, plus the savings from any transmission and distribution costs that are avoided or deferred by the addition of a DSM option that reduces future load growth[19]; and
- The fuel costs and environmental compliance costs for the entire utility system, including costs for both the new Supply options and the utility's existing generating units.

All of these individual annual costs (or annual "revenue requirements") are then summed to derive the total annual revenue requirements for a specific resource plan. These total annual revenue requirements for each year in the analysis period are then "present valued" and each year's present value revenue requirement value is summed. The result is a single CPVRR number that represents the present value of the utility's revenue requirements for the entire analysis period that are driven by a specific resource plan.

The second step consists of first combining the resource plan-specific annual costs with other annual costs which our utility will have that are not affected by the addition

---

[18] If you are unfamiliar with the term CPVRR, please see Appendix C for a mini-lesson regarding various economic terms, such as CPVRR and revenue requirements, that we will be discussing in this chapter, and in subsequent chapters.

[19] Note that the DSM-related cost impacts are listed here in this chapter only for purposes of describing the types of costs that are included in an IRP final (or system) economic evaluation. In this chapter, we are not yet evaluating DSM resource options. Consequently, in this chapter, the DSM-related costs are zero. DSM options are addressed in the next chapter.

of a resource plan. For example, a list of these other costs will include, but not be limited to, the following: the remaining (after depreciation) capital costs of existing utility generating units, the remaining capital costs of transmission and distribution lines, buildings, and salaries for existing staff. Note that, although these other costs of the utility system are not affected by the selection of any of the resource options being evaluated, they are important in conducting final economic evaluations of resource options. This is because these other costs are factors in projecting what the utility's electric rates will be.

The sum of those system costs that are directly tied to resource options, and the other costs described earlier, represents the total annual cost of our utility that is passed on to its customers through electric rates. The annual total costs are then divided by the total number of annual sales (kWh) of the utility to derive an annual system average electric rate that is usually expressed in terms of cents per kilowatt-hour (cents/kWh).

Just as the total present value of the resource-plan-specific costs is represented by a single CPVRR value, a single value can also be calculated to represent the electric rate perspective. This is most often done by converting the annual system average electric rate values to a present valued electric rate value for each year and then summing these present valued annual electric rates. (We will refer to this sum as the "original" present valued sum of electric rates.)

Then a single, constant electric rate value is assumed for each year that, when present valued for each year and summed, results in the identical sum as the original present valued sum of electric rates. This single electric rate that is held constant in this calculation is referred to as the "levelized" system average electric rate.[20]

In summary, each resource plan can be described by two economic values: the CPVRR value that represents all of our utility system's costs that are driven of the resource plan itself, and the levelized system average electric rate value that represents the electric rates that will be charged to our utility's customers.

We now turn our attention to calculating these CPVRR and levelized system average electric rate values for each of the three Supply Only resource plans for our utility system.

TOTAL COST PERSPECTIVE (CPVRR) FOR THE SUPPLY ONLY RESOURCE PLANS

We begin with the total cost perspective for the three Supply Only resource plans. We will first calculate the CPVRR cost for the Supply Only Resource Plan 1 (CC). The total CPVRR cost for this resource plan is determined by a number of cost components. Rather than attempt to show all of these individual cost components on a single spreadsheet, the individual cost components will first be grouped into three groups of costs: "Fixed" Costs, "DSM" Costs, and "Variable" Costs. Each of these three groups of costs will be presented and discussed separately. The Fixed Costs for Supply Only Resource Plan 1 (CC) are presented in Table 5.11.

---

[20] I realize that both the total cost perspective (CPVRR) and the electric rate perspective (levelized system average electric rate) are likely a bit (or more) confusing at this point. Hang in there! Both perspectives should be much clearer as we work through an example of how these values are developed.

TABLE 5.11
Fixed Cost Calculations for Supply Only Resource Plan 1 (CC)

	(1)	(2)	(3)	(4)	(5)	(6) = Sum of Cols. (2) thru (5)
Year	Annual Discount Factor 8.00%	Generation Capital (Millions)	Generation Fixed O & M (Millions)	Generation Capital Replacement (Millions)	Firm Gas Transportation (Millions)	Total Generation Fixed Costs (Millions)
Current Year	1.000	0	0	0	0	0
Current Year + 1	0.926	0	0	0	0	0
Current Year + 2	0.857	0	0	0	0	0
Current Year + 3	0.794	0	0	0	0	0
Current Year + 4	0.735	0	0	0	0	0
Current Year + 5	0.681	95	3	5	38	141
Current Year + 6	0.630	91	3	5	38	138
Current Year + 7	0.583	87	3	5	38	134
Current Year + 8	0.540	84	3	5	38	130
Current Year + 9	0.500	188	6	11	77	282
Current Year + 10	0.463	180	7	11	77	274
Current Year + 11	0.429	172	7	11	77	266
Current Year + 12	0.397	164	7	11	77	259
Current Year + 13	0.368	273	11	18	115	416
Current Year + 14	0.340	260	11	18	115	404
Current Year + 15	0.315	247	11	18	115	391
Current Year + 16	0.292	234	11	19	115	379
Current Year + 17	0.270	221	11	19	115	367
Current Year + 18	0.250	338	16	26	153	533
Current Year + 19	0.232	320	16	26	153	516
Current Year + 20	0.215	302	16	27	153	498
Current Year + 21	0.199	284	16	27	153	481
Current Year + 22	0.184	406	21	35	192	654
Current Year + 23	0.170	382	21	36	192	631
Current Year + 24	0.158	359	22	36	192	609
Current Year + 25	0.146	335	22	37	192	586
Current Year + 26	0.135	463	27	45	230	766
Current Year + 27	0.125	434	28	46	230	738
Current Year + 28	0.116	404	28	47	230	710
Current Year + 29	0.107	374	29	48	230	681
Total CPVRR =		$1,658	$79	$131	$787	$2,656

The Fixed Costs component represents the following costs for the CC unit and the subsequent filler units: generation capital costs in Column (2), fixed O&M costs in Column (3), capital replacement costs in Column (4), and firm gas transportation costs in Column (5). The sum of these Fixed Costs is shown in Column (6). The cost values presented in the rows for all years are nominal costs, *i.e.*, actual costs incurred. Within each column, these costs are then present valued and summed with the result shown on the "Total CPVRR" line at the bottom of the table. The CPVRR Fixed Cost total value for this resource plan is $2,656 million CPVRR as shown at the bottom of Column (6).[21]

The second group of costs that will be discussed is that of DSM Costs associated with new DSM resource options that our utility has chosen. However, as we have discussed, no incremental DSM resources have been added for this Supply Only resource plan. Consequently, there are no incremental DSM costs for this resource plan. However, because the tabular cost format that is being introduced in this chapter will also be used later in Chapter 6 when we discuss resource plans with DSM resource options, we shall present the zero incremental DSM costs in order to introduce the complete tabular format.

Thus, the values for the DSM Costs grouping are presented on the next page in Table 5.12 which also "carries over" the sum of the Fixed Costs (Column (6)).

---

[21] In Table 5.11, note that after Current Year (CY) + 5, in which the CC option is installed, the fixed costs increase significantly in CY + 9, CY + 13, CY + 18, CY + 22, and CY + 26. These fixed cost increases are due to the addition of a filler unit in each of those years as previously depicted in Table 5.5.

TABLE 5.12
DSM Cost Calculations for Supply Only Resource Plan 1 (CC)

	(1)	(6) = Sum of Cols. (2) thru (5)	(7)	(8)	(9)	(10) = Col (7) + Col (8) − Col (9)
Year	Annual Discount Factor 8.00%	Total Generation Fixed Costs (Millions)	DSM Administrative Costs (Millions)	DSM Incentive Payments (Millions)	T & D Costs Avoided by DSM (Millions)	DSM Net Costs (Millions)
Current Year	1.000	0	0	0	0	0
Current Year + 1	0.926	0	0	0	0	0
Current Year + 2	0.857	0	0	0	0	0
Current Year + 3	0.794	0	0	0	0	0
Current Year + 4	0.735	0	0	0	0	0
Current Year + 5	0.681	141	0	0	0	0
Current Year + 6	0.630	138	0	0	0	0
Current Year + 7	0.583	134	0	0	0	0
Current Year + 8	0.540	130	0	0	0	0
Current Year + 9	0.500	282	0	0	0	0
Current Year + 10	0.463	274	0	0	0	0
Current Year + 11	0.429	266	0	0	0	0
Current Year + 12	0.397	259	0	0	0	0
Current Year + 13	0.368	416	0	0	0	0
Current Year + 14	0.340	404	0	0	0	0
Current Year + 15	0.315	391	0	0	0	0
Current Year + 16	0.292	379	0	0	0	0
Current Year + 17	0.270	367	0	0	0	0
Current Year + 18	0.250	533	0	0	0	0
Current Year + 19	0.232	516	0	0	0	0
Current Year + 20	0.215	498	0	0	0	0
Current Year + 21	0.199	481	0	0	0	0
Current Year + 22	0.184	654	0	0	0	0
Current Year + 23	0.170	631	0	0	0	0
Current Year + 24	0.158	609	0	0	0	0
Current Year + 25	0.146	586	0	0	0	0
Current Year + 26	0.135	766	0	0	0	0
Current Year + 27	0.125	738	0	0	0	0
Current Year + 28	0.116	710	0	0	0	0
Current Year + 29	0.107	681	0	0	0	0
Total CPVRR =		$2,656	$0	$0	$0	$0

As mentioned earlier, because this is a Supply Only resource plan, there are no new DSM resources added. Consequently, the various DSM costs categories in Columns (7) through (10) have zero costs. Therefore, the CPVRR cost for the DSM Costs is zero as indicated at the bottom of Column (10). And, after accounting for both Fixed and DSM costs, the CPVRR cost for Supply Only Resource Plan 1 (CC) with CC Unit A remains at $2,656 million (= $2,656 million (Fixed) + $0 million (DSM)). However, we have not yet accounted for Variable Costs associated with this resource plan.

We now account for the Variable Costs in Table 5.13 that appears on the next page. In this table, we carry over the total Fixed Costs (Column (6)) and the total DSM Costs (Column (10)).

TABLE 5.13

Variable Cost Calculations for Supply Only Resource Plan 1 (CC)

	(1)	(6) = Sum of Cols. (2) thru (5)	(10) = Col (7) + Col (8) − Col (9)	(11)	(12)	(13)	(14) = Sum of Cols. (11) thru (13)
Year	Annual Discount Factor 8.00%	Total Genera- tion Fixed Costs (Millions)	DSM Net Costs (Millions)	Genera- tion Variable O & M (Millions)	System Net Fuel (Millions)	System Environ- mental Compliance (Millions)	Total Variable Costs (Millions)
Current Year	1.000	0	0	0	1,445	175	1,620
Current Year + 1	0.926	0	0	0	1,501	280	1,781
Current Year + 2	0.857	0	0	0	1,557	388	1,945
Current Year + 3	0.794	0	0	0	1,616	497	2,112
Current Year + 4	0.735	0	0	0	1,676	607	2,283
Current Year + 5	0.681	141	0	0	1,702	712	2,415
Current Year + 6	0.630	138	0	0	1,765	825	2,590
Current Year + 7	0.583	134	0	0	1,830	939	2,769
Current Year + 8	0.540	130	0	0	1,896	1,054	2,951
Current Year + 9	0.500	282	0	1	1,926	1,161	3,088
Current Year + 10	0.463	274	0	1	1,996	1,278	3,275
Current Year + 11	0.429	266	0	1	2,068	1,397	3,466
Current Year + 12	0.397	259	0	1	2,142	1,518	3,660
Current Year + 13	0.368	416	0	1	2,176	1,625	3,803
Current Year + 14	0.340	404	0	1	2,253	1,748	4,003
Current Year + 15	0.315	391	0	1	2,333	1,872	4,206
Current Year + 16	0.292	379	0	1	2,415	1,998	4,414
Current Year + 17	0.270	367	0	1	2,499	2,125	4,625
Current Year + 18	0.250	533	0	2	2,540	2,234	4,776
Current Year + 19	0.232	516	0	2	2,628	2,363	4,993
Current Year + 20	0.215	498	0	2	2,719	2,494	5,214
Current Year + 21	0.199	481	0	2	2,812	2,626	5,440
Current Year + 22	0.184	654	0	2	2,858	2,737	5,597
Current Year + 23	0.170	631	0	3	2,955	2,871	5,829
Current Year + 24	0.158	609	0	3	3,056	3,007	6,065
Current Year + 25	0.146	586	0	3	3,159	3,144	6,306
Current Year + 26	0.135	766	0	3	3,211	3,256	6,470
Current Year + 27	0.125	738	0	3	3,319	3,396	6,718
Current Year + 28	0.116	710	0	3	3,430	3,536	6,970
Current Year + 29	0.107	681	0	3	3,544	3,679	7,226
Total CPVRR =		$2,656	$0	$9	$24,342	$14,803	$39,154

The Variable Costs grouping represents the following three types of costs: generation variable O&M costs in Column (11), total utility system net fuel costs in Column (12), and total utility system environmental compliance costs in Column (13).[22] The CPVRR Variable Cost value for this resource plan is $39,154 million as presented at the bottom of Column (14).

As suggested by the values in this table, the Variable Cost value for a resource plan ($39,154 million CPVRR for Supply Only Resource Plan 1(CC)) is typically much larger than the Fixed Cost value ($2,656 million CPVRR for this same resource plan) for several reasons. First, the two largest (by far) Variable Cost components, system fuel costs and system environmental compliance costs, address costs for *all* of a utility's generating units, existing and new, while the Fixed Costs typically address only the new generating units that are added to the system. There are (usually) many more existing units on a utility system than the number of new units that will be added in any analysis period.[23] For this reason, a utility's annual fuel and environmental compliance costs in any given year will typically be much greater than new Fixed Costs for the same year.

Furthermore, these annual Variable Costs escalate over time due to rising fuel and environmental compliance costs while the capital cost for an individual generating unit declines from year to year due to depreciation of the generating unit's capital cost.

Table 5.14 further condenses the tabular format to now summarize the three groups of cost components for Supply Only Resource Plan 1 (CC). This table carries over the total Fixed Costs (Column (6)), the total DSM Costs (Column (10)), and the total Variable Costs (Column (14)). The sum of these three cost components results in a total CPVRR cost for Supply Only Resource Plan 1 (CC) of $41,810 million CPVRR. This is indicated in each of the new Columns (15), (16), and (17) which, respectively, present annual nominal costs, annual net present value (NPV) costs, and cumulative NPV (or CPVRR) costs.

The total costs for Supply Only Resource Plan 1 (CC) have now been presented in detail in Tables 5.11 through 5.13. In addition, a summary version of those total costs is now presented in Table 5.14 that is presented on the next page.

---

[22] In this table, the variable O&M costs in Column (11) show zero values until Current Year + 5 when the first new Supply Unit begins operation. Thus, we are accounting for variable O&M costs for new generating unit additions only. Although existing generating units do have variable O&M costs, some utilities account for these costs as fixed costs in their budgeting process. This can make it difficult to accurately calculate a projection of variable O&M costs for existing generating units. For that reason we do not attempt this projection, but we will account for variable O&M costs for existing generating units later when we discuss how levelized system average electric rates are calculated.

[23] As we shall see in Part II of this book, this will not be true if our utility decides to pursue a zero-carbon goal.

TABLE 5.14

Total Cost Calculations for Supply Only Resource Plan 1 (CC)

	(1)	(6) = Sum of Cols. (2) thru (5)	(10) = Col (7) + Col (8) − Col (9)	(14) = Sum of Cols. (11) thru (13)	(15) = (6) + (10) + (14)	(16) = (1) × (15)	(17)
Year	Annual Discount Factor 8.00%	Total Generation Fixed Costs (Millions)	DSM Net Costs (Millions)	Total Variable Costs (Millions)	Total Annual Costs (Millions)	Total Annual NPV Costs (Millions)	Cumulative Total NPV Costs (Millions)
Current Year	1.000	0	0	1,620	1,620	1,620	1,620
Current Year + 1	0.926	0	0	1,781	1,781	1,649	3,269
Current Year + 2	0.857	0	0	1,945	1,945	1,668	4,937
Current Year + 3	0.794	0	0	2,112	2,112	1,677	6,614
Current Year + 4	0.735	0	0	2,283	2,283	1,678	8,292
Current Year + 5	0.681	141	0	2,415	2,556	1,739	10,031
Current Year + 6	0.630	138	0	2,590	2,728	1,719	11,750
Current Year + 7	0.583	134	0	2,769	2,903	1,694	13,444
Current Year + 8	0.540	130	0	2,951	3,081	1,665	15,108
Current Year + 9	0.500	282	0	3,088	3,370	1,686	16,794
Current Year + 10	0.463	274	0	3,275	3,549	1,644	18,438
Current Year + 11	0.429	266	0	3,466	3,732	1,601	20,039
Current Year + 12	0.397	259	0	3,660	3,919	1,556	21,595
Current Year + 13	0.368	416	0	3,803	4,218	1,551	23,146
Current Year + 14	0.340	404	0	4,003	4,406	1,500	24,646
Current Year + 15	0.315	391	0	4,206	4,598	1,449	26,096
Current Year + 16	0.292	379	0	4,414	4,793	1,399	27,495
Current Year + 17	0.270	367	0	4,625	4,992	1,349	28,844
Current Year + 18	0.250	533	0	4,776	5,308	1,328	30,172
Current Year + 19	0.232	516	0	4,993	5,509	1,276	31,449
Current Year + 20	0.215	498	0	5,214	5,713	1,226	32,674
Current Year + 21	0.199	481	0	5,440	5,921	1,176	33,851
Current Year + 22	0.184	654	0	5,597	6,251	1,150	35,000
Current Year + 23	0.170	631	0	5,829	6,460	1,100	36,101
Current Year + 24	0.158	609	0	6,065	6,674	1,052	37,153
Current Year + 25	0.146	586	0	6,306	6,892	1,006	38,159
Current Year + 26	0.135	766	0	6,470	7,236	978	39,138
Current Year + 27	0.125	738	0	6,718	7,455	933	40,071
Current Year + 28	0.116	710	0	6,970	7,679	890	40,961
Current Year + 29	0.107	681	0	7,226	7,908	849	41,810
Total CPVRR =		$2,656	$0	$39,154	$41,810	$41,810	

The resource plan-specific CPVRR total costs for the other two Supply Only resource plans are calculated in a similar way. We now provide the corresponding values for the other two Supply Only resource plans in condensed summary form in Table 5.15.

TABLE 5.15
Economic Evaluation Results of Supply Only Resource Plans: CPVRR Costs

	(1)	(2)	(3)	(4) = Sum of Cols. (1) through (3)	(5)
Resource Plan	Fixed Costs (Millions, CPVRR)	DSM Net Costs (Millions, CPVRR)	Variable Costs (Millions, CPVRR)	Total Costs (Millions, CPVRR)	Difference from Lowest Cost Supply Only Plan (Millions, CPVRR)
Supply Only Resource Plan 1 (CC)	$2,656	$0	$39,154	$41,810	$0
Supply Only Resource Plan 2 (CT)	$2,564	$0	$39,319	$41,883	$73
Supply Only Resource Plan 3 (PV)	$2,843	$0	$38,986	$41,829	$19

The values for the three types of cost components presented earlier for Supply Only Resource Plan 1 (CC) appear in the first row of the table in Columns (1) through (3). The total CPVRR cost for this resource plan is provided in Column (4). The corresponding values for the other two Supply Only resource plans, featuring the CT and PV options are then provided on the second and third rows, respectively.

A quick glance at the CPVRR values in Column (4) shows that Supply Only Resource Plan 1 (CC) is projected to have the lowest CPVRR costs of any of the three resource plans. In order to more clearly see the CPVRR cost difference between the resource plans, Column (5) has been added which presents the CPVRR cost difference between the resource plan with the lowest CPVRR cost, Supply Only Resource Plan 1 (CC), and the other two Supply Only resource plans.

The table shows us several things regarding these Supply Only resource plans. First, the lowest CPVRR cost plan is Supply Only Resource Plan 1 (CC). Its cost of $41,810 million CPVRR is $73 million CPVRR lower than Supply Only Resource Plan 2 (CT), and $19 million CPVRR lower than Supply Only Resource Plan 3 (PV). Therefore, from a CPVRR perspective, the CC unit option is the least expensive Supply option to add to our utility system. But why is this so? A further examination of the table provides the answer.

The second thing one can glean from the table is why one resource plan is better from a CPVRR perspective than another. We will start by comparing just Supply Only Resource Plan 1 (CC) and Supply Only Resource Plan 2 (CT). We see that in regard to the Fixed Costs category, Supply Only Resource Plan 2 (CT) has a cost advantage of $92 million CPVRR over the CC option. (In other words, $2,656 million CPVRR for Supply Only Resource Plan 1 (CC) − $2,564 million CPVRR for Supply Only Resource Plan 2 (CT) = $92 million CPVRR.)

One might expect this by remembering several things, including (i) the CT option's cost/kW is $650/kW versus $950/kW for CC Unit A; (ii) the CT option's cost per kW is multiplied by 160 MW, while the CC Unit A's cost per kW is multiplied by 500 MW; and (iii) the amount of firm gas required by the CT option (10,000 mmBTU/day) is much less than the amount of firm gas needed by the CC unit (50,000 mmBTU/day).

Each of these factors favor the CT option in Supply Only Resource Plan 2 in regard to Fixed Costs. However, we also know from our work in creating the resource plans that these advantages are offset to some degree by the fact that the 160 MW of the CT option, compared to the 500 MW of the CC option, results in our utility having to add one more filler unit, and to add filler units earlier, than if the larger CC unit had been chosen. However, as shown by the results of the system economic analyses, we find that the Fixed Cost perspective still favors Supply Only Resource Plan 2 (CT).

However, the Variable Costs perspective provides a significantly different result. A comparison of the Variable Costs for these two resource plans shows that Supply Only Resource Plan 1 (CC) has an economic advantage of $165 million CPVRR over Supply Only Resource Plan 2 (CT) as shown by the calculation: $39,319 million CPVRR for Supply Only Resource Plan 2 (CT) − $39,154 million CPVRR for Supply Only Resource Plan 1 (CC) = $165 million CPVRR.

One might also expect a Variable Costs advantage for the CC unit featured in Supply Only Resource Plan 1 for two reasons. First, the CC unit is much more fuel-efficient (a heat rate of 6,600 BTU/kWh) than the CT option (10,400 BTU/kWh). Second, the more fuel-efficient CC unit will be operated much more (a projected capacity factor of 80%) than the CT option (a projected capacity factor of 5%).

These two factors will result not only in reduced system fuel costs, but also in reduced system environmental compliance costs. This is because the operation of the CC unit will displace a much greater amount of energy that would otherwise have been produced by the utility's existing marginal generating units which we recall, from Chapter 2, are more expensive to operate (and which typically have higher emission rates) than a new CC unit.

As the values in Table 5.15 show, these advantages for the CC unit result in the significant Variable Cost advantage of $165 million CPVRR compared to the CT option. When the comparative results for Fixed Costs and Variable Costs are combined, the net result is an economic advantage of $73 million CPVRR for Supply Only Resource Plan 1 (CC) as shown by the calculation: $165 million CPVRR advantage in Variable Costs − $92 million CPVRR disadvantage in Fixed Costs = $73 million CPVRR net cost advantage for the CC option.

Now, what about the PV option featured in Supply Only Resource Plan 3 that was the second best Supply Only resource plan in terms of CPVRR cost? A similar comparison to Supply Only Resource Plan 1 (CC) shows that Supply Only Resource Plan 1 (CC) unit has an economic advantage compared to Supply Only Resource Plan 3 (PV) in regard to Fixed Costs, but the situation is reversed in regard to Variable Costs.

In regard to Fixed Costs, the resource plan featuring the CC unit has an advantage of $187 million CPVRR as shown by the calculation: $2,843 million CPVRR for Supply Only Resource Plan 3 (PV) − $2,656 million CPVRR for Supply Only Resource Plan 1 (CC) = $187 million CPVRR. As you recall, the capital cost of the PV option was $1,500/kW compared to $950/kW for CC Unit A. Combining this installed cost differential with the facts that the resource plan featuring the PV option will (i) need two 120 MW PV options to meet the reserve margin need in Current Year + 5, and (ii) will result in the utility adding more filler units (and adding them earlier) than will be required with the resource plan featuring the 500 MW CC unit, results in a significant Fixed Costs disadvantage for the PV option.

However, the Variable Cost total shows an advantage for the resource plan featuring the PV option. Supply Only Resource Plan 3 (PV) has a $168 million CPVRR lower Variable Cost than the Supply Only Resource Plan (CC). This is shown by the calculation: $39,154 million CPVRR for Supply Only Resource Plan 1 (CC) − $38,986 million CPVRR for Supply Only Resource Plan 3 (PV) = $168 million CPVRR.

However, in terms of total costs, the Variable Cost advantage of Supply Only Resource Plan 3 (PV) is not enough to overcome its Fixed Cost disadvantage. Consequently, the Supply Only Resource Plan 1 (CC) is projected to be $19 million CPVRR less expensive overall than Supply Only Resource Plan 3 (PV).

We have now established, based on the assumptions our utility is using, that Supply Only Resource Plan 1 (CC) is the best economic choice among the Supply Only resource plans from a total CPVRR cost perspective. We now turn our attention to examining an electric rate perspective of these three resource plans.

ELECTRIC RATE PERSPECTIVE (LEVELIZED SYSTEM AVERAGE ELECTRIC RATE) FOR THE SUPPLY ONLY RESOURCE PLANS

In Chapter 3, we discussed that when conducting economic evaluations of Supply options only, once we knew what Supply option resulted in the lowest total cost (the CPVRR perspective), this Supply option would also be the Supply option that resulted in the lowest electric rates. We will now see whether this is true. (Of course it is or I wouldn't have told you earlier that it was.)

Table 5.16 presents a revised and expanded version of previously presented Table 5.14. That table provided the annual and total CPVRR revenue requirements for Supply Only Resource Plan 1 (CC). Table 5.16, that is presented on the next two pages, provides the electric rate impact perspective for this same resource plan.[24]

---

[24] Note that the calculation method presented in Table 5.16 is one of several ways in which levelized system average electric rates for utilities might be calculated. The calculation method is shown is the one that I have always used.

TABLE 5.16
Levelized System Average Electric Rate Calculation for Supply Only Resource Plan 1 (CC)

	(1)	(15)	(18)	(19)	(20)	(21)	(22)	(23)	(24)	(25)	(26)
		= Col(6) + Col10) + Col(14)		= Col(15) + Col(18)			= Col(20) − Col(21)	= (Col(19) × 100) / Col (22)	= Col(1) × Col(23)		= Col(1) × Col(25)
Year	Annual Discount Factor 8.00%	Total Annual Costs (Millions)	Other System Costs Not Affected by Plan (Millions)	Total Utility Costs (Millions)	Forecasted Utility Sales/NEL (GWh)	DSM Energy Reduction (GWh)	Net Utility Sales/NEL (GWh)	System Avg. Electric Rate Nominal (Cents/ kwh)	System Avg. Electric Rate NPV (Cents/kwh)	Levelized System Avg Electric Rate (Cents/ kwh)	System Avg. Electric Rate Nominal (Cents/ kwh)
Current Year	1.000	1,620	2,700	4,320	50,496	0	50,496	8.5553	8.555281	12.053	12.053399
Current Year + 1	0.926	1,781	2,754	4,535	51,022	0	51,022	8.8884	8.229985	12.053	11.160555
Current Year + 2	0.857	1,945	2,809	4,754	51,548	0	51,548	9.2229	7.907118	12.053	10.333847
Current Year + 3	0.794	2,112	2,865	4,978	52,074	0	52,074	9.5588	7.588083	12.053	9.568377
Current Year + 4	0.735	2,283	2,923	5,205	52,600	0	52,600	9.8963	7.274059	12.053	8.859608
Current Year + 5	0.681	2,556	2,981	5,537	53,126	0	53,126	10.4222	7.093180	12.053	8.203341
Current Year + 6	0.630	2,728	3,041	5,768	53,652	0	53,652	10.7513	6.775135	12.053	7.595686
Current Year + 7	0.583	2,903	3,101	6,004	54,178	0	54,178	11.0824	6.466451	12.053	7.033043
Current Year + 8	0.540	3,081	3,163	6,245	54,704	0	54,704	11.4155	6.167439	12.053	6.512077
Current Year + 9	0.500	3,370	3,227	6,596	55,230	0	55,230	11.9434	5.974691	12.053	6.029700
Current Year + 10	0.463	3,549	3,291	6,840	55,756	0	55,756	12.2686	5.682719	12.053	5.583056
Current Year + 11	0.429	3,732	3,357	7,089	56,282	0	56,282	12.5962	5.402293	12.053	5.169496
Current Year + 12	0.397	3,919	3,424	7,343	56,808	0	56,808	12.9264	5.133252	12.053	4.786571

Current Year + 13	0.368	4,218	3,493	7,711	57,334	0	57,334	13.4494	4.945328	12.053	4.432010
Current Year + 14	0.340	4,406	3,563	7,969	57,860	0	57,860	13.7724	4.688963	12.053	4.103713
Current Year + 15	0.315	4,598	3,634	8,231	58,386	0	58,386	14.0983	4.444386	12.053	3.799734
Current Year + 16	0.292	4,793	3,707	8,499	58,912	0	58,912	14.4273	4.211206	12.053	3.518272
Current Year + 17	0.270	4,992	3,781	8,773	59,438	0	59,438	14.7595	3.989029	12.053	3.257660
Current Year + 18	0.250	5,308	3,856	9,165	59,964	0	59,964	15.2834	3.824656	12.053	3.016351
Current Year + 19	0.232	5,509	3,933	9,442	60,490	0	60,490	15.6090	3.616800	12.053	2.792918
Current Year + 20	0.215	5,713	4,012	9,725	61,016	0	61,016	15.9382	3.419510	12.053	2.586035
Current Year + 21	0.199	5,921	4,092	10,013	61,542	0	61,542	16.2710	3.232318	12.053	2.394477
Current Year + 22	0.184	6,251	4,174	10,425	62,068	0	62,068	16.7959	3.089438	12.053	2.217108
Current Year + 23	0.170	6,460	4,258	10,718	62,594	0	62,594	17.1227	2.916262	12.053	2.052878
Current Year + 24	0.158	6,674	4,343	11,017	63,120	0	63,120	17.4536	2.752420	12.053	1.900813
Current Year + 25	0.146	6,892	4,430	11,322	63,646	0	63,646	17.7885	2.597445	12.053	1.760012
Current Year + 26	0.135	7,236	4,518	11,754	64,172	0	64,172	18.3170	2.476487	12.053	1.629641
Current Year + 27	0.125	7,455	4,609	12,064	64,698	0	64,698	18.6466	2.334305	12.053	1.508927
Current Year + 28	0.116	7,679	4,701	12,380	65,224	0	65,224	18.9806	2.200117	12.053	1.397154
Current Year + 29	0.107	7,908	4,795	12,702	65,750	0	65,750	19.3193	2.073488	12.053	1.293661
Total CPVRR =		$41,810						Sum =	148.857062		148.857062

This table appears complicated at first glance, but we have seen some of it before and the remainder is straightforward. The table is constructed as follows:

- Table 5.14 is the starting point for Table 5.16. However, in the interests of space, we have hidden all columns except the unnumbered column on the left-hand side that shows the year, Column (1) that shows the annual discount rate factors, and Column (15) that shows the annual costs (and the total CPVRR cost at the bottom).
- To this revised version of Table 5.14, we have added nine new columns and have labeled the new columns as Columns (18) through (26).
- Column (18) provides a projection of "Other System Costs" that are not affected by the selection of the resource plan. As previously discussed, these costs include costs of existing utility power plants, transmission and distribution lines, buildings, and utility staff. These "Other System Costs" are identical for each of the resource plans being analyzed in this chapter and in Chapter 6.[25]
- Column (19) then adds the values in Columns (15) and (18) to derive a total utility cost per year. (These annual total costs are the total revenue requirements that need to be recovered from the utility's customers each year. The utility's electric rates are designed to recover these revenue requirements.)
- Columns (20) through (22) then present, respectively, the forecasted total annual sales (or Net Energy for Load values used by some utilities), the projected reduced energy usage from new DSM options (if any), and the resulting net energy usage by our utility's customers that the utility must serve. (Note that in this Supply Only resource plan there is no reduction in energy sales from additional DSM as shown by the zero values in Column (21)).
- Column (23) then calculates (for each year) an annual system average electric rate by dividing the total utility cost per year shown in Column (19) by the net energy usage for that year shown in Column (22).
- Column (24) calculates the present value of each annual system average electric rate and then sums these annual values at the bottom of the page.
- Column (25) presents the constant (or levelized) system average electric rate which, when present valued and summed in Column (26), results in an identical present valued sum as that shown at the bottom of Column (24).

This levelized system average electric rate value presented in Column (25) is the electric rate perspective value for the resource plan in question.[26]

As we see from the table, the levelized electric rate for Supply Only Resource Plan 1 (CC) is 12.053 cents/kWh. This value does not represent the annual electric rate for any one year or for any one class of customer (*i.e.*, for residential and commercial

---

[25] The previously mentioned variable O&M for the utility's existing generating units are also accounted for in the cost values in Column (18).

[26] The levelized electric rate is also referred to by the term "levelized system average electric rate." Although this term is more descriptive of what the value actually represents, it is a mouthful to say/read. Therefore, we will often use the shorter terminology of "levelized electric rate" from this point forward.

customers). As previously stated, this one value represents a system average electric rate (cents/kWh) value over the life of the time period covered in the analysis.[27]

Similar calculations have been performed for the remaining two Supply Only resource plans. We show these results in Table 5.17 which expands the previously presented Table 5.15 to include a new Column (6) that provides the corresponding levelized electric rate for each of the three Supply Only resource plans.

TABLE 5.17

Economic Evaluation Results of Supply Only Resource Plans: CPVRR Costs and Levelized System Average Electric Rates

	(1)	(2)	(3)	(4) = Sum of Cols. (1) thru (3)	(5)	(6)
Resource Plan	Fixed Costs (Millions, CPVRR)	DSM Net Costs (Millions, CPVRR)	Variable Costs (Millions, CPVRR)	Total Costs (Millions, CPVRR)	Difference from Lowest Cost Supply Only Plan (Millions, CPVRR)	Levelized System Average Electric (Cents / kWh)
Supply Only Resource Plan 1 (CC)	$2,656	$0	39,154	41,810	**0**	**12.053**
Supply Only Resource Plan 2 (CT)	$2,564	$0	39,319	41,883	**73**	**12.064**
Supply Only Resource Plan 3 (PV)	$2,843	$0	38,986	41,829	**19**	**12.056**

As shown in this table, a ranking of the three Supply Only resource plans based on the CPVRR values in Column (4), and a ranking based on the levelized electric rates in Column (6), would be identical. In either case, Supply Only Resource Plan 1 (CC) is the most economical Supply Only resource plan from either the CPVRR or electric rate perspective. This resource plan is followed, in economic ranking order, by Supply Only Resource Plan 3 (PV), then by Supply Only Resource Plan 2 (CT).

We walk away from the economic evaluation of the Supply Only resource plans considered by our utility system with (at least) four pieces of information.

---

[27] Electric utilities typically use multi-part electric rates for certain types of customers. An example is the use of separate demand ($/kW) and energy (cents/kWh) charges for larger commercial and industrial customers. The levelized electric rate is presented as a single "cents per kWh" rate value for ease in comparing the levelized electric rates for different resource plans. This is also part of our ongoing effort to simplify the narrative.

- First, we now see why a utility may choose to stop its economic evaluation of Supply options once it has determined which Supply option results in the lowest CPVRR total cost. This is because the Supply option that has the lowest CPVRR total costs will also be the Supply option that results in the lowest levelized electric rate. Although taking the next step of calculating the levelized electric rate is absolutely necessary when Supply options are compared to DSM options, it is not necessary when comparing only Supply options.

- Second, by examining the differences between the three Supply Only resource plans from both the CPVRR cost and levelized electric rate perspectives, we see that even large differences in resource plan costs (such as a $73 million CPVRR difference between Supply Only Resource Plan 1 (CC) and Supply Only Resource Plan 2 (CT)) will typically result in only seemingly small differences in levelized electric rate values (0.011 cents/kWh) for these same two resource plans). The important point is that even seemingly small differences in levelized electric rate values are equivalent to very large CPVRR cost differences. This becomes especially important to remember when examining DSM options for which the electric rate perspective is a necessity.

- Third, we have seen how even relatively small dissimilarities in just one of the four key characteristics of resource options, capacity (MW), can result in "downstream" changes in the number and timing of future resource additions (*i.e.*, the filler units in our analyses). The capital and fixed cost impacts associated with such changes in resource plans are simply not captured in a screening curve analytical approach. In addition, we have seen that dissimilarities in both capacity (MW) and capacity factor will also result in changes in system fuel costs and system environmental costs. These cost impacts are also not captured if one attempts to only use a screening curve analytical approach.

  We can also now understand why the other two key characteristics (the percentage of the resource option's capacity (MW) that can be considered firm capacity at the utility's system peak hours and the projected life of the resource option) are important. Differences in either of these two characteristics between two otherwise identical resource options would also result in differences in the multi-year resource plans for the two options in regard to the number and timing of filler unit additions. In turn, these resource plan differences would drive differences in both Fixed and Variable Costs between the resource plans over the analysis period. Consequently, we now see why if even one of the four key characteristics of two resource options is dissimilar, the use of a screening curve analytical approach is fundamentally flawed.

- Fourth, our hypothetical utility system has now identified, from an economic perspective, its best Supply option: the CC unit. Our utility is now ready to turn its attention to examining the other type of resource option, DSM, which it is considering. We do so in the next chapter.

6 Resource Option Analyses for Our Utility System: DSM Options

In Chapter 5, our utility performed economic analyses of four Supply options. In this chapter, our utility will perform economic analyses of two demand side management (DSM) options it will also consider to meet its projected resource need 5 years in the future (Current Year + 5). Just as we noted in our discussion of Supply options, in practice a utility would likely consider many more DSM options than the two DSM options that we will be examining here. However, in order to simplify the discussion, while still illustrating the important points regarding economic analyses of DSM options, we have condensed this list to two different DSM options that can reduce our utility's customers' demand for electricity.

TYPES OF DSM RESOURCE OPTIONS UNDER CONSIDERATION

All DSM options that would be considered for meeting a utility's resource needs by lowering electrical demand at the utility's peak hour will have three main characteristics. We will first briefly introduce these characteristics and then discuss them in more detail.

First, the DSM options will require certain types of expenditures. These expenditures typically include administrative costs and usually include incentive payment costs. The administrative costs include expenditures to market, advertise, operate, and monitor the DSM program on an ongoing basis. Incentive costs are typically either a one-time payment to a participating customer to pay a portion of the cost that the participating customer will have to pay to install the DSM measure (such as higher levels of ceiling insulation or a higher efficiency air conditioner), or recurring (typically monthly) credits on the electric bill in exchange for the customer's continuing participation in a DSM program. In addition, for certain types of DSM options, there will also be capital costs for utility-owned equipment, usually equipment that the utility places at the customer's premises and/or at the utility's electrical substation.

Second, DSM options that reduce electrical usage will typically reduce demand at the utility's peak hour. This is commonly referred to as the kilowatt reduction at the peak hour (kW reduction) aspect of the DSM option or program.

The third characteristic that DSM options that reduce electrical usage have in common is that these DSM options will typically reduce the annual electricity

DOI: 10.1201/9781003301509-7

consumption of participating customers (and may also shift the timing during the day of when the electricity is used). This is referred to as the annual kilowatt-hour reduction (kWh reduction) aspect of the DSM program.

Now that these three characteristics of DSM options have been introduced, we will discuss them in more detail. This information should help in understanding the presentation of the economic and non-economic evaluations of DSM options that follow in the remainder of this chapter and then again in Chapter 7 where we take a combined look at the Supply option results from Chapter 5 and the DSM option results from this chapter.

The first characteristic, DSM expenditures, is straightforward. The administrative costs and incentive payments directly associated with a specific DSM option are those expenditures required to implement and operate a DSM program. These costs are typically presented in terms of dollars per participating customer ($/participant) or dollars per peak hour kW reduction ($/kW). The administrative costs are primarily a one-time cost that is incurred when a participating customer is signed up for the DSM program. As mentioned earlier, the incentive payment can either be a one-time payment when the customer signs up for the program (or when the customer pays for the DSM measure), or an ongoing payment such as a monthly credit on the participating customer's monthly bill.

These $/participant, or $/kW, costs are typically small in comparison to most of the $/kW cost values we discussed in regard to the Supply options in Chapter 5. However, a DSM option will usually only provide a relatively small amount of kW reduction for each participating customer compared to the MW output of Supply options. For example, the kW reduction may be 1 kW or less for any single residential or small commercial participating customer. Therefore, a very large number of participating customers must be signed up and retained for the DSM option to be able to compete with Supply options that will typically contribute capacity in terms of (at least) a hundred or more MW (in which 1 MW = 1,000 kW).

For example, recall that our utility system has a resource need in 5 years of either 120 MW of new generating capacity or a peak load reduction of 100 MW. If a DSM option is projected to offer 1 kW of demand reduction per participating customer, the utility would need to sign up 100,000 participating customers in order to meet this 100 MW peak load reduction objective (100,000 participating customers × 1 kW per customer = 100,000 kW or 100 MW.)

Therefore, although the DSM option's $/kW administrative costs and incentive payments are small relative to the $/kW values we have discussed for new Supply options, these "per kW" DSM costs will be multiplied by very large numbers of participating customers.

In regard to the second characteristic of DSM options, kW reduction at the utility's peak hour, this is the aspect of DSM options that drives most of the categories of benefits provided by DSM options. It is the kW reduction characteristic of DSM that results in avoiding the new generating unit (Supply option) that otherwise would have been built absent the DSM option. In addition, the kW reduction characteristic will be the sole driver of a number of additional types of utility system cost savings, many of which would be the result of avoiding the construction and operation of a new Supply option if a DSM option is chosen instead.

The following list contains ten DSM-related utility cost savings categories that are solely driven by the kW reduction characteristic of DSM options:

1. The avoided capital cost of the new generating unit;
2. The avoided transmission capital costs of interconnecting the new generating unit to the existing transmission system (often referred to as transmission interconnection costs) and of any modifications to the existing transmission system to handle the new amount of power that will be supplied from the location of the new generating unit (often referred to as transmission integration costs);
3. The avoided fixed operating and maintenance (O&M) costs for the new generating unit;
4. The avoided capital replacement costs for the new generating unit;
5. The avoided firm natural gas transportation cost (if applicable) associated with the new generating unit;
6. The avoided variable O&M costs for the new generating unit;
7. The avoided capital costs of other transmission facilities throughout the utility system that otherwise would have been built if the peak load had not been lowered.
8. The avoided O&M costs for these other system transmission facilities or modifications that otherwise would have been built if the peak load had not been lowered by the DSM option;
9. The avoided capital costs of system distribution facilities that otherwise would have been built if the peak load had not been lowered by the DSM option; and
10. The avoided O&M costs for system distribution facilities that otherwise would have been built if the peak load had not been lowered by the DSM option.

In addition, the kW reduction characteristic of DSM options will also drive two of three calculations that, in total, comprise the net changes in the utility system's fuel costs and environmental compliance costs that are the result of the addition of the DSM option. In order to see this, we will examine the three specific types of changes in system fuel costs that will result from the introduction of a DSM option.

These three types of changes can be described as follows:

1. If the new generating unit is not built due to selection of the DSM option, the fuel that would have been burned in the new generating unit is not burned. This, by itself, will result in lower system fuel costs. (This calculation is driven by DSM's kW reduction characteristic because the kW reduction characteristic is what avoids the need for the new generating unit.)
2. However, if the new generating unit is not built and, therefore, does not operate, the amount of energy the new generating unit would have supplied to the utility system will now have to be provided by the existing generating units on the utility system. Because these existing generating units are typically not as fuel-efficient as most new generating units (recall that new

generating units would typically not be operated unless they produced electricity at a lower cost than the existing generating units), the operation of the utility's existing generating units to supply this same amount of energy will result in higher annual fuel costs for the utility system. We will refer to this as the "fuel penalty" for avoiding the new generating unit. (Because this fuel penalty is a result of avoiding the new generating unit, and the avoidance of the new generating unit is driven by DSM's kW reduction, the fuel penalty is also driven by DSM's kW reduction.)

3. The third calculation involves how much fuel is saved by the utility system not having to serve as much annual energy (kWh) during a year's 8,760 hours as it otherwise would if the DSM option is not selected. The reduction in the amount of system energy that the utility must serve is driven by the kWh reduction characteristic of the DSM option.

Therefore, the net impact of DSM on the utility system's fuel cost can be viewed as a three-part calculation involving both reduced and increased fuel costs[1]:

Net system fuel cost impact from DSM = (i) fuel cost savings from fuel not being burned in the avoided new generating unit (driven by kW reduction), *minus* (ii) higher system fuel costs ("fuel penalty") from existing generating units now supplying energy that would have been supplied by the new unit (driven by kW reduction), *plus* (iii) system fuel cost savings from reduced annual energy that must be served by the utility (driven by kWh reduction).[2]

In a similar fashion, the net environmental compliance (or emission) cost impact to the utility system from avoiding a new generating unit with DSM can be derived in three calculations:

1. By avoiding the new generating unit, the air emissions that would have occurred from burning fuel in this new generating unit are also avoided. This results in emission cost savings for the utility system. This impact is solely driven by DSM's kW reduction because it is the kW reduction that results in avoiding the need for the new generating unit.

2. Because the existing generating units on the utility system must now supply the same amount of energy the avoided new generating unit would have supplied, this increased output from the existing generating units typically results in increased emissions from the existing generating units. Because these existing generating units are typically less fuel-efficient than the avoided unit, the result is typically higher system emissions for this amount of energy than would have been the case if the avoided unit had been built and had supplied the energy. This "emission penalty" is also solely driven

---

[1] In reality, these three impacts generally occur simultaneously. We have discussed these calculations separately to clarify how the overall net impact of DSM on system fuel is calculated.

[2] The result of this calculation can be either a net savings, or a net cost, in system fuel costs. This is dependent upon the individual utility system's existing generating units and the types of Supply and DSM options being considered. Note that the First Principle of Electric Utility Resource Planning has shown its face again.

by DSM's kW reduction because it is the kW reduction characteristic, that avoids the need for the new generating unit.

3. The lower amount of annual energy (kWh) that the utility system must supply over each year's 8,760 hours due to the selection of the DSM option will lower the utility system's emissions. This emission saving is solely driven by DSM's kWh reduction.

Therefore, the net impact of DSM on the utility system's environmental compliance costs can be viewed, similar to that for system fuel cost impacts discussed earlier, as a three-part calculation involving both reduced and increased environmental compliance costs:

Net system environmental compliance cost impact from DSM = (i) environmental compliance cost savings from fuel not being burned in the avoided new generating unit (driven by kW reduction), *minus* (ii) higher system environmental compliance costs ("environmental compliance cost penalty") from existing generating units now supplying energy that would have been supplied by the new unit (driven by kW reduction), *plus* (iii) system environmental compliance cost savings from reduced annual energy that must be served by the utility (driven by kWh reduction).[3]

In regard to the third characteristic of DSM, kWh reduction, we have seen that kWh reduction is responsible for only one of the three DSM-based impacts regarding either system fuel costs or environmental compliance costs. In other words, the kWh reduction characteristic, of DSM options drives only two DSM-based cost savings categories. By comparison, the kW reduction characteristic of DSM options drives a total of 14 DSM cost savings categories: the original list of ten cost savings categories we previously discussed, plus the two fuel cost impact categories, and the two environmental compliance cost impact categories, that were just discussed.

Therefore, we see that the second characteristic of DSM options, kW reduction, is responsible for driving many more categories (14 categories) of DSM-related utility benefits than does the third characteristic, kWh reduction (2 categories). However, the two cost savings categories that are driven by the kWh reduction characteristic usually involve relatively large cost savings values.

The key point is that there are 16 utility cost impact, or net benefit, categories and that neither the kW reduction characteristic, nor the kWh reduction characteristic, of DSM options should be overlooked in determining the true impact of DSM options.[4]

However, we are not quite through with the utility impacts resulting from the DSM kW reduction and kWh reduction characteristics. Although we have examined 16 categories of utility impacts that can result from selecting a DSM option, there is one more impact of DSM that must be considered.

---

[3] The result of this calculation can also be either a net savings, or a net cost, in system environmental compliance costs. This is again dependent upon the individual utility system's existing generating units and the types of Supply and DSM options being considered.

[4] In the years leading up to the writing of this book, there has been a trend by certain non-utility parties to focus solely on the kWh reduction characteristic of DSM options. Such a focus is a mistake because, as discussed above, there are many more categories of cost savings that are driven by kW reduction than are driven by kWh reduction. Therefore, a focus solely on kWh reduction does not provide a complete picture of the impacts, both positive and negative, of DSM options.

That impact is the reduction in monies (or revenues) the utility receives from its customers due to the introduction of a DSM option. This is due to the DSM option's kW and kWh reduction characteristics. This impact is often referred to as "lost revenues" but is more correctly referred to as "revenue requirements not received" or, more succinctly, as "unrecovered revenue requirements." In other words, the result of a DSM option is that there is an amount of the utility's costs for which planned revenues to cover those costs are not received due to lower sales. These unrecovered costs or revenue requirements equate to a cost imposed by the DSM option.

Without going into too much detail into how electric rates are set for a utility by its regulatory authority, here is how electric rates are basically set. A projection of the total demand (kW) and total energy (kWh) the utility will be expected to serve over a given time period is developed. The utility's projected fixed and variable operating costs over this same time period are also developed. The regulatory authority then develops an allowed rate of return (simplistically, think "return on the investments made in the utility by investors."). These factors lead to a determination of the total amount of money per year the utility must take in to meet both its operating costs and its allowed rate of return.

This amount of money per year that the utility must take in is referred to as the utility's annual "revenue requirements." Electric rates are then set on both a cents/kWh basis and a \$/kW basis[5] that will allow the recovery of this amount of revenue requirements assuming the projected amounts of kW and kWh are served.

For example, using a very simplistic example, let's assume that some (again, very small) utility's projected total revenue requirements are \$1 million/year and that its projected amount of energy it must serve (its sales) are 10 million kWh. Therefore, if this utility's electric rates are set solely on a cents/kWh basis, its electric rate will be set as follows:

$$\text{Electric rate (cents/kWh)} = \left(\$1,000,000 \times 100 \text{ cents/\$1}\right)/10,000,000 \text{ kWh}$$
$$= 10.0 \text{ cents/kWh}$$

This electric rate of 10.0 cents/kWh will allow the utility to recover the desired \$1 million of revenue requirements if its sales are, as projected, 10,000,000 kWh. But what happens if the utility's projected sales drop by 500,000 kWh to 9,500,000 kWh due to the addition of a particular DSM option that only reduces costs by \$25,000? The utility will recover only \$950,000 (= 9,500,000 kWh × 10.0 cents/kWh) in revenue requirements, but the utility's cost is now \$975,000 (=\$1,000,000−\$25,000). In order to recover the revised revenue requirement amount of \$975,000, the utility's electric rates would have to be raised to approximately 10.26 cents/kWh (= \$975,000 in revenue requirements divided by 9,500,000 kWh).

---

[5] All customers served by a traditional regulated utility typically pay a cents per kWh charge for each kWh they use. And there are usually different cents per kWh rates for different groups of customers (*i.e.*, "rate classes") who fall into different monthly kWh usage ranges. In addition, non-residential customers (*i.e.*, commercial and industrial customers) whose highest monthly demand is above a given threshold (for example, 20 kW), also typically pay a \$ per kW charge based on their highest monthly demand. For the sake of simplicity, we are discussing only a single average cents per kWh for all of the utility's customers.

Now, switching our perspective from that of the utility system as a whole to the perspective of an individual residential customer, let's see what happens. Let's assume you are that customer and you have decided not to participate (or were ineligible to participate) in this particular DSM option. If your electric usage was 1,000 kWh/month, you originally would be paying $100/month with the electric rate at 10.0 cents/kWh. But if the electric rate increased to 10.26 cents/kWh, and your usage had not changed, you would be paying $102.60/month. (You might not be pleased with this outcome, especially if the DSM option was available only to non-residential customers making you ineligible to participate in the DSM option.)

We note that with this particular DSM option, the percentage decrease in utility costs (revenue requirements) of 2.5% (= $25,000/$1,000,000) was less than the percentage decrease in kWh sales of 5% (= 500,000 kWh/10,000,000 kWh). This resulted in an increase in the electric rate. We could have assumed another DSM option that had identical percentage decreases in both revenue requirements and kWh sales. In such a case, there would have been no change in electric rates. Or we could have assumed yet another DSM option that had a greater percentage decrease in revenue requirements than the percentage decrease in kWh sales. In this case, there would have been a decrease in the utility's electric rate.

The purpose of this simple example is to show that DSM options which result in kW and kWh reductions will have impacts on both the numerator (the projected revenue requirements or costs of the utility due to the costs and benefits of the DSM option) and also on the denominator (the projected number of sales in kWh). Both of these impacts affect the utility's electric rates and also affect customers' bills.[6]

We have discussed which categories of utility costs (revenue requirements) are either incurred when offering a DSM option (such as administrative costs, incentive payments, and unrecovered revenue requirements) or are avoided by the DSM option (such as avoided generation, and fuel savings). Thus, it is clear that DSM impacts the numerator value in an electric rate calculation. However, because DSM impacts the number of kWh that the utility system will serve, the denominator in an electric rate calculation is also impacted by DSM. We will return later in this chapter to examine the electric rate impact on our hypothetical utility system of the two DSM options it will be evaluating.

With this discussion of the three basic characteristics of DSM options, we will now introduce the two DSM options our utility is considering. Our assumptions for the key inputs for these two DSM options are presented in Table 6.1. The values shown for these inputs are assumed to be average values for each participating customer.

---

[6] Obviously, the amount of electricity that a customer uses also affects the customer's bill. However, some customers (such as those with special medical conditions and/or limited income) may not be able to reduce their usage enough to offset any increase in electric rates resulting from the selection of certain DSM options (or of too many DSM options). Therefore, a utility should be very careful in its selection of DSM options.

TABLE 6.1

Key Inputs and Descriptors for Economic Evaluation: DSM Options

Input	Units of Measurement	DSM Option 1	DSM Option 2
(1) Demand reduction (peak hour)	kW reduction per participant	1.00	0.50
(2) Energy reduction (annual)	kWh reduction per participant	1,500	3,000
(3) Administrative cost	$ per participant	$100	$100
(4) Incentive payment	$ per participant	$200	$200
(5) Participant equipment cost	$ per participant	$800	$800
(6) Life of DSM measure	Years	15	5
(7) Energy-to-demand reduction ratio	Equivalent annual hours	1,500	6,000
(8) Equivalent capacity factor	% of annual hours	17%	68%

As shown in Row (1) of this table, the two DSM options differ in regard to the projected demand (kW) reduction per participant value. DSM Option 1 reduces the peak hour demand of each participating customer by 1.00 kW, while DSM Option 2 reduces the peak hour demand of each participating customer by 0.50 kW.

Based on our discussion in Chapter 5 of the PV option having a firm capacity value of 50%, compared to the 100% firm capacity values of the CC and CT options, the alert reader may quickly realize that, because DSM Option 2 will provide only 0.50 kW per participating customer compared to 1.00 kW per participating customer for DSM Option 1, twice as many participating customers (200,000), will need to be signed up for DSM Option 2 compared to DSM Option 1 (100,000). Only in that way can both DSM options achieve the same total desired demand reduction level (100 MW) to meet the projected resource needs of our utility system in Current Year + 5 if the resource need in that year is to be met with a DSM option.

There are other differences between the two DSM options as can be seen from Rows (2) through (6) in the table, which present other key characteristics of the DSM options. In Row (2), the annual energy (kWh) reduction per participant is presented. DSM Option 1 will result in 1,500 kWh of energy use being reduced by each participant and DSM Option 2 will result in a 3,000 kWh reduction per participant.

Rows (3) and (4) present, respectively, the "per participant" administrative cost and incentive payment. These are identical for the two DSM options: an administrative cost of $100 per participant (row 3) and an incentive cost of $200 per participant (row 4).

However, a note regarding the total incentive payment costs for the two DSM options is in order. Despite the fact that the per participant administrative and incentive costs are identical for the two DSM options, the total administrative and incentive costs that will be incurred for DSM Option 2 will be (at least) twice as high as these costs for DSM Option 1 in meeting the 100 MW objective for DSM. This is because the kW reduction per participant for DSM Option 2 (0.50 kW) is only half that of DSM Option 1 (1.00 kW). Therefore, twice as many total participants will need to be signed up for DSM Option 2 than for DSM Option 1 in order to achieve the same level of demand reduction (100 MW). As a consequence, the total administrative and incentive payment costs for DSM Option 2 will be (at least) twice as high as for DSM Option 1. (You may ask why these costs for DSM Option 2 are not exactly twice as high as for DSM Option 1. The answer is the "life expectancy" of the equipment installed as part of either DSM option. We will get to that factor momentarily when we discuss Row (6).)

Row (5) presents the incremental equipment cost that will be paid by the DSM participant for each option: $800. For example, if a DSM option were a much more efficient air conditioning unit, a customer might have to pay $800 more, even after factoring in the utility's incentive payment, to install this higher efficiency air conditioner instead of a more standard efficiency air conditioner. (Note that this is not a utility cost that will be borne by all of the utility's customers.)

Row (6) then presents the life expectancy of the DSM equipment before it needs to be replaced: 15 years for DSM Option 1 and 5 years for DSM Option 2. The life expectancy of the DSM equipment installed for DSM Option 2 is only 1/3 as long as the life expectancy is for the different equipment installed for DSM Option 1. Therefore, it will be necessary to replace that DSM equipment much earlier for DSM Option 2 than for DSM Option 1.

For purposes of this discussion, we will assume that we perform this replacement for either DSM option by signing up new customers and paying an incentive to the new customers. Because we are doing this every 5 years for DSM Option 2, or three times in the same 15-year period that constitutes the life expectancy of DSM Option 1, the *total* administrative and incentive costs over the life of the analyses are further increased for DSM Option 2 in comparison to those costs for DSM Option 1. This explains why the total costs over the analysis period will actually be more than twice as high for DSM Option 2 as for DSM Option 1.

Returning to Table 6.1, Rows (1) through (6) present all of the key characteristics of the two DSM options in the manner they are normally discussed. However, I have found that an examination of the information presented on these 6 rows often fails to provide a clear picture of how DSM options really compare with each other, especially in regard to how they will "operate" once they are implemented on a utility system. Consequently, Rows (7) and (8) have been added.

Rows (7) and (8) are not actually inputs for the DSM options themselves but are "descriptors" of DSM options that I have found useful in helping to explain differences between DSM options and the impacts that these different DSM options have on utility systems. Row (7) calculates an energy-to-demand reduction ratio by dividing the annual kWh reduction value in Row (2) by the kW reduction value in Row (1). The resulting value represents the equivalent number of hours the DSM option will impact the utility system assuming its kW reduction value remains constant during all of these hours (which is a simplistic assumption). When two DSM options have different kW reduction values and/or different kWh reduction values, this descriptor allows one to better understand the differences between the DSM options and how the utility system will be impacted by the different DSM options.

For example, Row (7) helps shed more light on how much energy reduction will actually result from each of the two DSM options. A glance at the information presented in Row (2) shows that, on a per participant basis, DSM Option 1 will reduce 1,500 kWh and DSM Option 2 will reduce 3,000 kWh. However, recall that DSM options are typically utilized as a resource option to meet a pre-determined resource need (100 MW for our utility if DSM is selected to meet its resource needs 5 years in the future).

Also recall that the per participant kW reduction for DSM Option 2 is 0.50 kW compared to 1.00 kW for DSM Option 1, which leads to the fact that twice as many participants will need to be signed up for DSM Option 2 than for DSM Option 1 to meet the pre-determined resource need of 100 MW.

Therefore, Row (7) shows that DSM Option 2 will actually reduce 6,000 kWh for every 1 kW reduction (= 3,000 kWh/0.5 kW) versus a 1,500 kWh reduction per kW for DSM Option 1. We have termed the value of 6,000 as the "equivalent annual hours" that DSM option will reduce for each 1 kW of demand reduction. Similarly, DSM Option 1 has an equivalent annual hour value of 1,500. Thus, implementing the desired 100 MW of either DSM option will result in DSM Option 2, reducing four times more energy than DSM Option 1 (= 6,000/1,500).

Row (8) then calculates an "equivalent capacity factor" by dividing the equivalent annual hour values in Row (7) by 8,760 hours in a year. The resulting ratio can be thought of as the DSM equivalent of the capacity factor values for generating units such as the four Supply options previously evaluated in Chapter 5. This "equivalent capacity factor" information is also useful in understanding and discussing how the DSM options will impact the utility system in comparison to the Supply options that DSM will be competing with.

Recall that our four Supply options had capacity factors, in ascending order, of 5% for the CT unit, 20% for the PV option, and 80% for the two CC units. By comparing these capacity factors with the equivalent annual capacity factors for the two DSM options of approximately 17% (= 1,500/8,760) for DSM Option 1 and approximately 68% (= 6,000/8,760) for DSM Option 2, we see that DSM Option 2's equivalent capacity factor is higher than the capacity factors for the CT and PV options, but lower than the capacity factors for the CC options. The equivalent annual capacity factor for DSM Option 1 is higher than the capacity factor for the CT option, but lower than the capacity factors for the PV and CC options.

Thus, these DSM options can be seen as impacting the utility system—or "operating"—in a range that would be similar to how a low-level intermediate generating unit (DSM Option 1) to a high-level intermediate generating unit (DSM Option 2) would operate on our utility system.[7] This information is important not only in regard to the economic impacts the resource options will have, but also how the resource options will affect system fuel use and system emissions (*i.e.*, environmental compliance costs).

This completes the introduction of the DSM option-specific inputs that will be used in the economic evaluation of these options. Before we leave this section, let's look back (just as we did for the Supply options) and see what appears to be obvious, and perhaps more importantly, what is not obvious, from looking at these inputs.

First, from looking at the inputs, the following statements about these two DSM options appear to be obvious:

- Because twice as many participating customers will have to be signed up for DSM Option 2 than for DSM Option 1 in order to reach the same 100 MW objective, and because the life expectancy of the DSM equipment for DSM

---

[7] The majority of DSM options will "operate" on a utility system in a range similar to a peaking-to-intermediate level generating unit. DSM options that operate similar to baseload generating units are less common.

Option 2 is only 1/3 as long as for DSM Option 1, DSM Option 2 will have significantly higher total administrative and incentive costs over the period of the analysis than will DSM Option 1.

- Assuming that the same 100 MW level is implemented with either DSM option, all of the kW reduction-driven impacts to the utility system will be identical for DSM Option 1 and DSM Option 2.

- Conversely, the kWh reduction-driven impacts to the utility system regarding fuel usage, air emissions, and unrecovered revenue requirements will be significantly greater for DSM Option 2 than for DSM Option 1.

However, two outcomes are not clear. First, it is unclear which of the two combinations: DSM Option 1's lower total administrative and incentive costs, plus lower total kWh reduction impacts, or DSM Option 2's higher total administrative and incentive costs, plus higher total kWh reduction impacts, will result in a lower total cost (CPVRR).

Second, in regard to the much more important electric rate perspective, this CPVRR value will represent only one-half of the picture, *i.e.*, the numerator in the electric rate calculation. While the previously mentioned determination of CPVRR costs will identify the numerator of an electric rate calculation, the denominator (the amount of total energy to be served by the utility if a particular DSM option is selected) must also be accounted for. It is only after both the numerator and denominator are determined that we will know which DSM option will result in the lower electric rate for our utility's customers. And it is only by calculating the electric rate impact that we will have a full picture of the economic impacts of both of the DSM options.

We will see how all of this actually plays out in the next sections of this chapter. We will start by performing preliminary economic screening evaluations of the two DSM options.

PRELIMINARY ECONOMIC SCREENING EVALUATION OF DSM OPTIONS: UNDERSTANDING THE COST-EFFECTIVENESS SCREENING TESTS

As mentioned in Chapter 3, preliminary economic screening evaluations of DSM options typically utilize specific "cost-effectiveness screening tests" that compare the DSM option with a comparably sized Supply option.[8] Therefore, we will next perform a preliminary economic screening evaluation of these two DSM options using the Participant test, RIM test, and TRC test. The Supply option to which the

---

[8] The use of DSM cost-effectiveness tests can be a meaningful way to perform preliminary economic screening of DSM options. Conversely, using a screening curve approach does not work with DSM options. In the previous chapter, we discussed the inherent problems with using a screening curve analysis approach in which even one of four key characteristics are dissimilar for Supply options. We now see that using such an approach for DSM options (such as the two options discussed here) is again flawed due to significant differences in the DSM options in regard to at least two key characteristics: capacity factor and life of the measure. Furthermore, when comparing DSM options to Supply options, the DSM options are dissimilar in all four key characteristics.

two DSM options will be compared will be the most economic Supply option for our hypothetical utility system: CC Unit A which was featured in Supply Only Resource Plan 1 (CC).

We will begin by reviewing these preliminary economic screening tests. These three tests are comparisons of DSM-driven benefits and costs from the perspective of either a potential participating customer (as the Participant test is designed to do) or all of the utility's customers (as the RIM and TRC tests are purported to do). The test results are typically presented in a benefit-to-cost ratio format in which the present value of the benefits derived from DSM is divided by the present value of the DSM costs.

A benefits-divided-by-costs resulting value of 1.00 means that the present value of the benefits is exactly equal to the present value of the costs. A ratio greater than 1.00 indicates that the benefits exceed the costs. In such a case, the DSM option is said to have passed that particular cost-effectiveness screening test. Conversely, a ratio of less than 1.00 indicates that benefits are less than the costs. In this case, the DSM option is said to have failed that particular cost-effectiveness test.

In order to better understand the results of these cost-effectiveness screening tests when our two DSM options are evaluated, we need to better understand the specific cost-effectiveness tests. We will start with the Participant test and Table 6.2 which looks only at the projected benefits of a DSM option to a prospective participating customer.

TABLE 6.2

Economic Elements Included in the Participant Test: Benefits Only

Economic Elements	Participant Incurred Economic Impacts	Included in the Participant Test?
Benefits of DSM		
(= avoided costs and/or direct benefits)		
Bill savings by participants	Yes	Yes
Incentives received by participants	Yes	Yes
Tax credits received by participants	Yes	Yes

Table 6.2 first indicates, in the shaded column, the types of economic impacts, or benefits, that a participating customer would actually receive from a DSM option or program. These include the bill savings from the DSM-induced reduction in energy usage by the participant, the incentive payments received from the utility, and tax credits that may be available for some DSM measures. The unshaded column on the right-hand side of the table then indicates whether the Participant test accounts for these categories of benefits in its calculations. As shown in this table, the Participant test does account for all of these categories of benefits that a participating customer would actually receive from a DSM program.

Now we turn our attention to the costs that a participating customer would incur from his/her participation in the DSM program. This information is presented in Table 6.3 which is an expansion of the previous table.

TABLE 6.3
Economic Elements Included in the Participant Test: Benefits and Costs

Economic Elements	Participant Incurred Economic Impacts	Included in the Participant Test?
Benefits of DSM (= avoided costs and/or direct benefits)		
Bill savings by participants	Yes	Yes
Incentives received by participants	Yes	Yes
Tax credits received by participants	Yes	Yes
Costs of DSM (= incurred costs)		
Participants' capital and O&M costs	Yes	Yes

As shown in the shaded column of Table 6.3, a participating customer would expect to incur initial purchase (*i.e.*, capital) costs (and, perhaps, some ongoing operation and/or maintenance costs). As the unshaded column on the right-hand side of the table also shows, the Participant test does account for these DSM-related costs in its calculation.

Therefore, these two tables show that the Participant test correctly includes all of the DSM-related benefits and costs that a potential participating customer should consider when deciding whether to participate in a DSM program offered by a utility. For that reason, the Participant test is a meaningful test with which to perform preliminary economic screening of DSM options.

We now turn our attention to the RIM and TRC tests. We again begin by looking only at the benefits side of the calculation for these two tests in Table 6.4.

TABLE 6.4
Economic Elements Included in the RIM and TRC Tests: Benefits Only

Economic Elements	Utility-Incurred Economic Impacts	Included in the RIM Test?	Included in the TRC Test?
Benefits of DSM (=avoided costs and/or cost impacts)			
Generation capital and O&M	Yes	Yes	Yes
Transmission capital and O&M	Yes	Yes	Yes
Distribution capital and O&M	Yes	Yes	Yes
Net system fuel impacts	Yes	Yes	Yes
Net system environmental compliance impacts	Yes	Yes	Yes

This table shows that a listing of the benefit categories of DSM includes the following costs that will be avoided by DSM's kW and kWh reduction characteristics: (i) generation capital and O&M, (ii) transmission capital and O&M, (iii) distribution capital and O&M, (iv) the net fuel impacts to the utility system, and (v) the net environmental compliance impacts to the utility system.[9]

And, similar to what we observed with the Participant test, the unshaded portion of the table shows that all of the benefit categories that will result from the implementation of DSM options are accounted for in the benefits calculation of both the RIM and TRC tests. In fact, assuming all else equal, the RIM and TRC tests will calculate identical benefit values for a given DSM program.

Two logical questions remain. First, as was the case with the Participant test, are all of the relevant costs of DSM included in the RIM and TRC tests? Second, in regard to the RIM and TRC tests calculating identical benefits from DSM, do the RIM and TRC tests also calculate identical DSM costs? The answer is "no" to both of these questions. This is seen from Table 6.5 which is an expanded version of Table 6.4.

TABLE 6.5

Economic Elements Included in the RIM and TRC Tests: Benefits and Costs

Economic Elements	Utility-Incurred Economic Impacts	Included in the RIM Test?	Included in the TRC Test?
Benefits of DSM			
(=avoided costs and/or cost impacts)			
Generation capital and O&M	Yes	Yes	Yes
Transmission capital and O&M	Yes	Yes	Yes
Distribution capital and O&M	Yes	Yes	Yes
Net system fuel impacts	Yes	Yes	Yes
Net system environmental compliance impacts	Yes	Yes	Yes
Costs of DSM			
(=incurred costs)			
Utility equipment & administration	Yes	Yes	Yes
Incentives paid to participants	Yes	Yes	No
Unrecovered revenue requirements	Yes	Yes	No
Participants' capital and O&M	No	No	Yes

This table shows that we have listed four types of DSM-related costs: (i) utility equipment and administration costs, (ii) incentives paid to participating customers, (iii) unrecovered revenue requirements, and (iv) capital and O&M costs that

---

[9] These five categories of DSM benefits is a "collapsed" version of the 16 types of DSM cost impacts that were discussed earlier. This collapsed set of types of impacts is used here solely to make the RIM and TRC tests easier to understand. The use of the word "net" in regard to system fuel and environmental compliance impacts denotes that DSM options can have both positive and negative impacts in these areas as previously discussed.

participating customers pay themselves. As shown by the bottom portion of the shaded column in Table 6.5, the rows marked with a "Yes" indicate that there are three types of DSM-related costs that are incurred by the utility and which are passed on to, or otherwise impact, all of a utility's customers through the utility's electric rates. These three types of DSM costs are: (i) utility equipment and administration, (ii) incentives paid to participants, and (iii) unrecovered revenue requirements.

As the unshaded columns of Table 6.5 show, both the RIM and TRC tests include the DSM-related costs of utility equipment and administration. However, at this point, the RIM and TRC tests significantly diverge in regard to the accounting of DSM costs.

The RIM test includes the remaining two DSM costs that will be incurred by the utility and which will be passed on to, or otherwise impact, all of the utility's customers through electric rates: incentives paid to participants and unrecovered revenue requirements. Conversely, the TRC test does not include either of these two utility-incurred DSM cost impacts. Furthermore, the TRC test includes the participants' capital and O&M costs even though these costs are not incurred by the utility and are not passed on to all of its customers through electric rates. (Recall that these costs, which are incurred only by the participant, are already properly accounted for in the Participant test.)

Therefore, the RIM test correctly accounts for all of the DSM-related benefits and costs that will be incurred by the utility and passed on to, or otherwise impact, all of the utility's customers through electric rates. Therefore, the RIM test is a meaningful test with which to perform preliminary economic screening of DSM options.

Conversely, although the TRC test does account for all of the DSM-related benefits, it falls far short of accounting for all of the DSM-related costs that will impact all of the utility's customers through the utility's electric rates. If one's objective is to perform preliminary economic screening analyses of DSM options that account for all relevant cost impacts for all of a utility's customers, then the TRC test falls far short. Therefore, the TRC test is a fundamentally flawed test.

Because of the fundamental flaws in the TRC test, the TRC test will provide—in almost every case—different benefit-to-cost ratios from those provided by the RIM test when evaluating the same DSM option.[10] This shall be evident in the next section as we present the actual results of the RIM and TRC tests for DSM Options 1 and 2.[11]

PRELIMINARY ECONOMIC SCREENING ANALYSES OF DSM OPTIONS: RESULTS

The benefit-to-cost ratio results of the three DSM cost-effectiveness screening tests for our hypothetical utility, when comparing DSM Options 1 and 2 to our utility's most economic Supply option (CC Unit A), are presented in Table 6.6.

---

[10] As one might expect from the fact that the TRC test does not account for all relevant cost impacts, the resulting benefit-to-cost values from the TRC test are generally higher than (and often considerably higher than) the benefit-to-cost values resulting from the RIM test that does account for all relevant cost impacts.

[11] A further discussion of the RIM and TRC tests is presented in Appendix E which includes a real-life example of using the two tests on an actual DSM program.

TABLE 6.6

Preliminary Economic Screening Analyses of DSM Options: Benefit-to-Cost Ratios

Cost-Effectiveness Test	for DSM Option 1	for DSM Option 2
Participant test	2.04	3.60
RIM test	1.18	0.68
TRC test	2.27	2.10

We will start with the Participant test results shown in Row (1). Both of the DSM options are projected to have benefit-to-cost ratios significantly higher than 1.00. Therefore, the projected CPVRR benefits values for potential participants exceed the projected CPVRR cost values for potential participants. Therefore, both of the two DSM options are projected to be cost-effective from the perspective of a potential participant and our utility's customers should find either of these two DSM options attractive.

The question now becomes: should our utility attempt to sign up its customers for either of these two DSM options? In other words, should our utility offer either of these DSM options?

In an attempt to answer that question, we turn to the RIM and TRC test results that are shown, respectively, on Rows (2) and (3). The RIM test results show that DSM Option 1 has a benefit-to-cost ratio of 1.18. Therefore, DSM Option 1 is projected to be cost-effective from the perspective of the RIM test. However, DSM Option 2 has a RIM test benefit-to-cost ratio of 0.68, i.e., less than 1.00. Therefore, DSM Option 2 is *not* projected to be cost-effective from the perspective of the RIM test.

However, the TRC test results show that DSM Option 1 has a TRC benefit-to-cost ratio of 2.27 and DSM Option 2 has a TRC benefit-to-cost ratio of 2.10. Both DSM options are projected to be cost-effective from a TRC test perspective (and, as previously mentioned, the benefits-to-costs ratio for the TRC test is significantly higher than the corresponding RIM test values).

Our utility now has a dilemma based on the results of its preliminary economic screening analyses. Both DSM options are projected to be cost-effective from a participant's perspective (the Participant test) and both DSM options are projected to be cost-effective under at least one of the two cost-effectiveness tests that are intended to provide a perspective of the value of the DSM options to the utility system as a whole.

Unless a utility has previously determined which of the so-called utility-perspective cost-effectiveness screening tests (RIM or TRC) it believes provides the more meaningful perspective, it will need to perform a full economic analysis of its utility system to determine which DSM option is best for it to add, and whether either DSM option is better than the most economic Supply option (CC Unit A). Although the answer to the question of *"Which DSM option should be carried forward?"* is probably obvious to the astute reader based on the test results, for purposes of discussion, our utility system has graciously (again) offered to include both DSM options in its ongoing analyses.

In order for the utility to perform this analysis, it is necessary to create two DSM-based resource plans in a manner similar to what was done when the competing Supply options were evaluated. Therefore, we now turn our attention to creating two "With DSM" resource plans.

CREATING THE COMPETING "With DSM" RESOURCE PLANS

As discussed in Chapter 3, an integrated resource planning (IRP) approach looks at long-term resource plans in order to ensure that all of the impacts that a proposed resource option addition will have on the utility system over an extended period of time are captured. In Chapter 5, we first created and then analyzed, three different resource plans that each featured a different Supply option added in Current Year + 5. We are now at a point where we will create two resource plans to address the two DSM options. Then, in the next section of this chapter, these resource plans will be analyzed.

Recall that with the Supply options we examined, each Supply option had a different amount of capacity (MW) that would be added in the decision year (Current Year + 5). These different amounts of capacity that would be added in the decision year by each Supply option had a different impact on the timing and magnitude of resource additions in years after the decision year. Therefore, each Supply option required the development of a separate resource plan to address this impact.

This is not the case with our two DSM options. That is because we are assuming that either DSM Option 1 or DSM Option 2 will be implemented to achieve the same amount of MW reduction, 100 MW, by the decision year (Current Year + 5).[12] Therefore, in regard to the timing and magnitude of resource additions in the years following the decision year, both of the DSM-based resource plans will have identical resource needs. Therefore, for purposes of our discussion, we need only look at a single DSM-based resource plan to examine the impact that 100 MW of DSM will have on the timing and number of the filler units. We will refer to this DSM-based resource plan as the starting point "With DSM" resource plan.

However, recall that the two DSM options have different costs and also have different values for the total amount of energy that will be reduced. These different impacts will be addressed by actually evaluating two separate "With DSM" resource plans in which the underlying DSM cost and energy reduction values for each DSM option are accounted for. This evaluation will also be discussed in the next section.

We now look at the impact of either of our DSM options—both of which will achieve 100 MW of demand reduction by the decision year—on future resource needs using the starting point resource plan that has the same format as that previously used in Chapter 5. That format is presented in Table 6.7.

---

[12] We are assuming that our utility has projected that it is possible to sign up at least 100,000 and 200,000 participating customers by the decision year for DSM Options 1 and 2, respectively. In utility parlance, the "achievable potential" for the DSM options are at least 100,000 and 200,000 participants, respectively, in the desired time frame.

TABLE 6.7
Long-Term Projection of Reserve Margin for the Hypothetical Utility: With New DSM Added in "Current Year + 5"

	(1a)	(1b)	(1c)	(1d)	(1e)	(1f)	(2a)	(2b)	(2c)	(3)	(4)	(5)
						= (See Formula Below)			= (2a) - (2b)	= (1f) - (2c)	= (3)/(2c)	= ((2c) × 1.20) - (1f)
	Previously Projected Generating Capacity (MW)	CC Unit A (No. Units)	CT (No. Units)	PV (No. Units)	Filler Units (No. Units)	Total Projected Generating Capacity (MW)	Previously Forecasted Electrical Demand (MW)	Newly Projected DSM (MW)	Firm Peak Demand (MW)	Reserves (MW)	Reserve Margin (%)	Generation Only MW Needed to Meet Reserve Margin (MW)
		Cumulative No. of New Unit Additions										
Year												
Current Year	11,600	0	0	0	0	11,600	9,600	0	9,600	2,000	20.8	(80)
Current Year + 1	12,000	0	0	0	0	12,000	9,700	20	9,680	2,320	24.0	(384)
Current Year + 2	12,000	0	0	0	0	12,000	9,800	40	9,760	2,240	23.0	(288)
Current Year + 3	12,000	0	0	0	0	12,000	9,900	60	9,840	2,160	22.0	(192)
Current Year + 4	12,000	0	0	0	0	12,000	10,000	80	9,920	2,080	21.0	(96)
Current Year + 5	12,000	0	0	0	0	12,000	10,100	100	10,000	2,000	20.0	0
Current Year + 6	12,000	0	0	0	0	12,000	10,200	100	10,100	1,900	18.8	120
Current Year + 7	12,000	0	0	0	0	12,000	10,300	100	10,200	1,800	17.6	240
Current Year + 8	12,000	0	0	0	0	12,000	10,400	100	10,300	1,700	16.5	360
Current Year + 9	12,000	0	0	0	0	12,000	10,500	100	10,400	1,600	15.4	480
Current Year + 10	12,000	0	0	0	0	12,000	10,600	100	10,500	1,500	14.3	600
Current Year + 11	12,000	0	0	0	0	12,000	10,700	100	10,600	1,400	13.2	720
Current Year + 12	12,000	0	0	0	0	12,000	10,800	100	10,700	1,300	12.1	840
Current Year + 13	12,000	0	0	0	0	12,000	10,900	100	10,800	1,200	11.1	960
Current Year + 14	12,000	0	0	0	0	12,000	11,000	100	10,900	1,100	10.1	1,080

Current Year + 15	12,000	0	0	12,000	11,100	100	11,000	1,000	9.1	1,200
Current Year + 16	12,000	0	0	12,000	11,200	100	11,100	900	8.1	1,320
Current Year + 17	12,000	0	0	12,000	11,300	100	11,200	800	7.1	1,440
Current Year + 18	12,000	0	0	12,000	11,400	100	11,300	700	6.2	1,560
Current Year + 19	12,000	0	0	12,000	11,500	100	11,400	600	5.3	1,680
Current Year + 20	12,000	0	0	12,000	11,600	100	11,500	500	4.3	1,800
Current Year + 21	12,000	0	0	12,000	11,700	100	11,600	400	3.4	1,920
Current Year + 22	12,000	0	0	12,000	11,800	100	11,700	300	2.6	2,040
Current Year + 23	12,000	0	0	12,000	11,900	100	11,800	200	1.7	2,160
Current Year + 24	12,000	0	0	12,000	12,000	100	11,900	100	0.8	2,280
Current Year + 25	12,000	0	0	12,000	12,100	100	12,000	0	0.0	2,400
Current Year + 26	12,000	0	0	12,000	12,200	100	12,100	(100)	(0.8)	2,520
Current Year + 27	12,000	0	0	12,000	12,300	100	12,200	(200)	(1.6)	2,640
Current Year + 28	12,000	0	0	12,000	12,400	100	12,300	(300)	(2.4)	2,760
Current Year + 29	12,000	0	0	12,000	12,500	100	12,400	(400)	(3.2)	2,880

CC Unit A =	500	MW
CT =	160	MW
PV =	60	MW (firm)
Filler Units =	500	MW

Formula: $(1f) = (1a) + ((1b) \times 500) + ((1c) \times 160) + ((1d) \times 60) + ((1e) \times 500)$

Table 6.7 is almost identical to the format presented in Chapter 5 in Tables 5.4 through 5.9. The only change is that two new columns have now been added in order to address the DSM options.

In the earlier Chapter 5 version of this table, Column (2) presented the "Peak Electrical Demand (MW)" values which represent the forecasted system peak loads assuming no incremental DSM is implemented. We have now labeled the former Column (2) as Column (2a). Then two new columns are added immediately after Column (2a). Column (2b) presents the annual MW reduction values at the system peak hour from the incremental DSM added if either DSM Option 1 or 2 is selected.

As shown by the values in Column (2b), we have assumed that either DSM option will reduce peak load by 20 MW more each year for the five years from Current Year + 1 through Current Year + 5 to achieve the desired 100 MW of incremental demand reduction by Current Year + 5. This 100 MW demand reduction is then assumed to be maintained for all subsequent years.

Then Column (2c) computes what we earlier termed the "firm" peak electrical demand that results when subtracting the incremental DSM in Column (2b) from the forecasted peak demand in Column (2a). This firm peak demand value is what the utility will actually have to serve at its peak hour. This value is also used to calculate the utility's reserve margins and resource needs.

As shown in Table 6.7, the incremental 100 MW of incremental demand reduction by Current Year + 5 will allow our hypothetical utility to achieve a 20% reserve margin without adding any additional new generating units in that year. However, the table also shows that additional resources will be needed starting with the next year, Current Year + 6, because our utility's reserve margin is projected to drop to 18.8% in that year. This fact is highlighted by shading the Current Year + 6 row. Additional resources will also be needed in subsequent years as well to continue to meet the 20% reserve margin criterion as the demand for electricity continues to increase.

These additional resource needs will be addressed in the same manner as was the case for the Supply options: through the addition of 500 MW CC filler units starting in Current Year + 6. The number and timing of these filler units is presented in Table 6.8 in Column (1e).

Table 6.8 is presented on pages 156 and 157.

(This page/space is intentionally left blank.)

TABLE 6.8

Long-Term Projection of Reserve Margin for the Hypothetical Utility: With New DSM Added in "Current Year + 5" Plus Filler Units

	(1a)	(1b)	(1c)	(1d)	(1e)	(1f)	(2a)	(2b)	(2c)	(3)	(4)	(5)
						= (See Formula Below)			= (2a) − (2b)	= (1f) − (2c)	= (3) / (2c)	= ((2c) × 1.20) − (1f)
	Previously Projected Generating Capacity (MW)	CC Unit A (No. Units)	CT (No. Units)	PV (No. Units)	Filler Units (No. Units)	Total Projected Generating Capacity (MW)	Previously Forecasted Electrical Demand (MW)	Newly Projected DSM (MW)	Firm Peak Demand (MW)	Reserves (MW)	Reserve Margin (%)	Generation Only MW Needed to Meet Reserve Margin (MW)
		Cumulative No. of New Unit Additions										
Year												
Current Year	11,600	0	0	0	0	11,600	9,600	0	9,600	2,000	20.8	(80)
Current Year + 1	12,000	0	0	0	0	12,000	9,700	20	9,680	2,320	24.0	(384)
Current Year + 2	12,000	0	0	0	0	12,000	9,800	40	9,760	2,240	23.0	(288)
Current Year + 3	12,000	0	0	0	0	12,000	9,900	60	9,840	2,160	22.0	(192)
Current Year + 4	12,000	0	0	0	0	12,000	10,000	80	9,920	2,080	21.0	(96)
Current Year + 5	12,000	0	0	0	0	12,000	10,100	100	10,000	2,000	20.0	0
Current Year + 6	12,000	0	0	0	1	12,500	10,200	100	10,100	2,400	23.8	(380)
Current Year + 7	12,000	0	0	0	1	12,500	10,300	100	10,200	2,300	22.5	(260)
Current Year + 8	12,000	0	0	0	1	12,500	10,400	100	10,300	2,200	21.4	(140)
Current Year + 9	12,000	0	0	0	1	12,500	10,500	100	10,400	2,100	20.2	(20)
Current Year + 10	12,000	0	0	0	2	13,000	10,600	100	10,500	2,500	23.8	(400)
Current Year + 11	12,000	0	0	0	2	13,000	10,700	100	10,600	2,400	22.6	(280)

Current Year + 12	12,000	0	0	2	13,000	10,800	100	10,700	2,300	21.5	(160)
Current Year + 13	12,000	0	0	2	13,000	10,900	100	10,800	2,200	20.4	(40)
Current Year + 14	12,000	0	0	3	13,500	11,000	100	10,900	2,600	23.9	(420)
Current Year + 15	12,000	0	0	3	13,500	11,100	100	11,000	2,500	22.7	(300)
Current Year + 16	12,000	0	0	3	13,500	11,200	100	11,100	2,400	21.6	(180)
Current Year + 17	12,000	0	0	3	13,500	11,300	100	11,200	2,300	20.5	(60)
Current Year + 18	12,000	0	0	3	13,500	11,400	100	11,300	2,200	19.5	60
Current Year + 19	12,000	0	0	4	14,000	11,500	100	11,400	2,600	22.8	(320)
Current Year + 20	12,000	0	0	4	14,000	11,600	100	11,500	2,500	21.7	(200)
Current Year + 21	12,000	0	0	4	14,000	11,700	100	11,600	2,400	20.7	(80)
Current Year + 22	12,000	0	0	4	14,000	11,800	100	11,700	2,300	19.7	40
Current Year + 23	12,000	0	0	5	14,500	11,900	100	11,800	2,700	22.9	(340)
Current Year + 24	12,000	0	0	5	14,500	12,000	100	11,900	2,600	21.8	(220)
Current Year + 25	12,000	0	0	5	14,500	12,100	100	12,000	2,500	20.8	(100)
Current Year + 26	12,000	0	0	5	14,500	12,200	100	12,100	2,400	19.8	20
Current Year + 27	12,000	0	0	6	15,000	12,300	100	12,200	2,800	23.0	(360)
Current Year + 28	12,000	0	0	6	15,000	12,400	100	12,300	2,700	22.0	(240)
Current Year + 29	12,000	0	0	6	15,000	12,500	100	12,400	2,600	21.0	(120)

CC Unit A =	500	MW
CT =	160	MW
PV =	60	MW (firm)
Filler Units =	500	MW

Formula: (1f) = (1a) + ((1b) × 500) + ((1c) × 160) + ((1d) × 60) + ((1e) × 500)

As shown in Column (1e), this general "With DSM" resource plan for either DSM option will have 6 filler units added after the decision year. This is the same number of filler units as in Supply Only Resource Plan 3 (PV) that added 120 MW of capacity.

In fact, the timing and number of filler units in both Supply Only Resource Plan 3 (PV) presented in Chapter 5, and the With DSM resource plan just presented, are identical. One would expect this to be the case because the 120 MW of capacity added in Supply Only Resource Plan 3 (PV) and the 100 MW of demand reduction added in the With DSM resource plan will have an identical impact on the utility's reserve margin. Consequently, for all years after the decision year, the utility's resource needs will be the same regardless of whether 120 MW of new generating unit capacity, or 100 MW of demand reduction from DSM is added, in Current Year + 5.

Now that we know what the timing and magnitude of filler units will be for either of the DSM options, we will move forward to an economic analysis of a resource plan that features 100 MW of DSM Option 1 and a resource plan that features 100 MW of DSM Option 2. We will (again, showing great imagination) name these two resource plans the With DSM Resource Plan 1 (which features DSM Option 1), and the With DSM Resource Plan 2 (which features DSM Option 2), respectively. With these two resource plans, we will account for the differences in the two DSM options in regard to DSM total administration and incentive costs, the life of the respective DSM options, the amount of energy (kWh) reduced, etc.

FINAL (OR SYSTEM) ECONOMIC ANALYSIS OF DSM OPTIONS

OVERVIEW

In Chapter 5, our discussion of the Supply options ultimately focused on two tables that summarized the final economic analyses for the Supply Only options. Table 5.16 presented both the CPVRR total cost values and levelized system average electric rate for one of the Supply Only Resource Plans in detail: Supply Only Resource Plan 1 (CC). Then Table 5.17 summarized the CPVRR and levelized system average electric rates for all three of the Supply Only resource plans.

We will take the same approach in presenting the results for the With DSM resource plans. And, just as we did in the discussion of the analyses of the Supply options, we will also focus in detail on only one DSM option: DSM Option 1 which is featured in the With DSM Resource Plan 1. As we did in Chapter 5, we will present three different cost components for this resource plan: Fixed Costs, DSM Costs, and Variable Costs. We will then present the total costs for this plan. Once these cost values have been presented, we will discuss how the costs for the With DSM Resource Plan 1 differ from the costs for the Supply Only resource plans.

Finally, the levelized system average electric rate calculation for the With DSM Resource Plan 1 will be presented in detail, followed by a summary of the CPVRR costs and levelized system average electric rate values for both With DSM resource plans.

The calculation methodology leading to the results shown is identical to those presented in detail in Chapter 5 for Supply Only Resource Plan 1 (CC). However, as we shall see, the DSM Costs columns, and the column showing the DSM energy reductions for the levelized system average rate calculation, will no longer show zero values as they did in Chapter 5 for Supply Only Resource Plan 1 (CC).

RESULTS FOR THE WITH DSM RESOURCE PLAN 1

We begin by looking at the Fixed Costs for the With DSM Resource Plan 1. These
values are presented in Table 6.9.

TABLE 6.9
Fixed Cost Calculations for With DSM Resource Plan 1

Year	Annual Discount Factor 8.00% (1)	Generation Capital (Millions) (2)	Generation Fixed O&M (Millions) (3)	Generation Capital Replacement (Millions) (4)	Firm Gas Transportation (Millions) (5)	Total Generation Fixed Costs (Millions) (6) = Sum of Cols. (2) thru (5)
Current Year	1.000	0	0	0	0	0
Current Year + 1	0.926	0	0	0	0	0
Current Year + 2	0.857	0	0	0	0	0
Current Year + 3	0.794	0	0	0	0	0
Current Year + 4	0.735	0	0	0	0	0
Current Year + 5	0.681	0	0	0	0	0
Current Year + 6	0.630	102	3	5	38	148
Current Year + 7	0.583	98	3	5	38	145
Current Year + 8	0.540	94	3	5	38	141
Current Year + 9	0.500	90	3	5	38	137
Current Year + 10	0.463	196	7	11	77	290
Current Year + 11	0.429	188	7	11	77	282
Current Year + 12	0.397	179	7	11	77	274
Current Year + 13	0.368	171	7	12	77	266
Current Year + 14	0.340	282	11	18	115	425
Current Year + 15	0.315	268	11	18	115	413
Current Year + 16	0.292	255	11	19	115	400
Current Year + 17	0.270	242	11	19	115	387
Current Year + 18	0.250	229	12	19	115	375
Current Year + 19	0.232	347	16	26	153	543
Current Year + 20	0.215	329	16	27	153	525
Current Year + 21	0.199	310	16	27	153	507
Current Year + 22	0.184	292	17	28	153	490
Current Year + 23	0.170	416	21	36	192	665
Current Year + 24	0.158	392	22	36	192	641
Current Year + 25	0.146	367	22	37	192	618
Current Year + 26	0.135	343	23	38	192	595
Current Year + 27	0.125	473	28	46	230	777
Current Year + 28	0.116	443	28	47	230	748
Current Year + 29	0.107	412	29	48	230	720
Total CPVRR =		$1,553	$72	$119	$706	$2,450

At this point, we will only briefly pause to mention that the total Fixed Costs value for With DSM Resource Plan 1 of $2,450 million CPVRR shown in Table 6.9 is lower than the total Fixed Cost value for Supply Only Resource Plan 1 (CC) of $2,656 million CPVRR shown previously in Table 5.11. We will return to see why this is the case after we have first introduced the DSM and Variable Costs for this With DSM resource plan.

Now, as we did in Chapter 5, we carry over Column (6) from this table which provides the total Fixed Costs and then add the DSM Costs. This information is presented in Table 6.10 that is presented on the next page.

TABLE 6.10
DSM Cost Calculations for With DSM Resource Plan 1

Year	(1) Annual Discount Factor 8.00%	(6) = Sum of Cols. (2) thru (5) Total Generation Fixed Costs (Millions)	(7) DSM Administrative Costs (Millions)	(8) DSM Incentive Payments (Millions)	(9) T & D Costs Avoided by DSM (Millions)	(10) = Col (7) + Col (8) − Col (9) DSM Net Costs (Millions)
Current Year	1.000	0	0	0	0	0
Current Year + 1	0.926	0	2	4	0	6
Current Year + 2	0.857	0	2	4	1	5
Current Year + 3	0.794	0	2	4	3	3
Current Year + 4	0.735	0	2	4	4	2
Current Year + 5	0.681	0	2	4	5	1
Current Year + 6	0.630	148	0	0	7	(7)
Current Year + 7	0.583	145	0	0	6	(6)
Current Year + 8	0.540	141	0	0	6	(6)
Current Year + 9	0.500	137	0	0	6	(6)
Current Year + 10	0.463	290	0	0	6	(6)
Current Year + 11	0.429	282	0	0	5	(5)
Current Year + 12	0.397	274	0	0	5	(5)
Current Year + 13	0.368	266	0	0	5	(5)
Current Year + 14	0.340	425	0	0	5	(5)
Current Year + 15	0.315	413	0	0	4	(4)
Current Year + 16	0.292	400	3	4	4	3
Current Year + 17	0.270	387	3	4	4	3
Current Year + 18	0.250	375	3	4	4	3
Current Year + 19	0.232	543	3	4	3	4
Current Year + 20	0.215	525	3	4	3	4
Current Year + 21	0.199	507	0	0	3	(3)
Current Year + 22	0.184	490	0	0	3	(3)
Current Year + 23	0.170	665	0	0	2	(2)
Current Year + 24	0.158	641	0	0	2	(2)
Current Year + 25	0.146	618	0	0	2	(2)
Current Year + 26	0.135	595	0	0	2	(2)
Current Year + 27	0.125	777	0	0	1	(1)
Current Year + 28	0.116	748	0	0	1	(1)
Current Year + 29	0.107	720	0	0	1	(1)
Total CPVRR =		$2,450	$12	$21	$43	($10)

We again pause only briefly to note two aspects of the DSM Costs. First, they are no longer zero as was the case with the Supply Only resource plans. Second, the net DSM cost for this resource plan featuring DSM Option 1 is actually a negative net cost value of ($10) million CPVRR due to the reduced electrical load on the utility system from the DSM option resulting in avoided transmission and distribution costs, *i.e.*, a cost savings. We will soon return to discuss these DSM net costs. However, we now carry over the total Fixed Costs (Column [6]) and total DSM Costs (Column [10]) and then add in the Variable Costs. This information is presented in Table 6.11 that is presented on the next page.

TABLE 6.11
Variable Cost Calculations for With DSM Resource Plan 1

	(1)	(6) = Sum of Cols. (2) through (5)	(10) = Col (7) + Col (8) − Col (9)	(11)	(12)	(13)	(14) = Sum of Cols. (11) through (13)
Year	Annual Discount Factor 8.00%	Total Generation Fixed Costs (Millions)	DSM Net Costs (Millions)	Genera- tion Variable O&M (Millions)	System Net Fuel (Millions)	System Environ- mental Compliance (Millions)	Total Variable Costs (Millions)
Current Year	1.000	0	0	0	1,445	175	1,620
Current Year + 1	0.926	0	6	0	1,500	280	1,780
Current Year + 2	0.857	0	5	0	1,555	388	1,943
Current Year + 3	0.794	0	3	0	1,612	496	2,108
Current Year + 4	0.735	0	2	0	1,670	606	2,276
Current Year + 5	0.681	0	1	0	1,731	717	2,448
Current Year + 6	0.630	148	(7)	0	1,757	823	2,580
Current Year + 7	0.583	145	(6)	0	1,821	937	2,758
Current Year + 8	0.540	141	(6)	0	1,888	1,052	2,940
Current Year + 9	0.500	137	(6)	0	1,956	1,169	3,125
Current Year + 10	0.463	290	(6)	1	1,987	1,276	3,263
Current Year + 11	0.429	282	(5)	1	2,059	1,394	3,454
Current Year + 12	0.397	274	(5)	1	2,133	1,515	3,648
Current Year + 13	0.368	266	(5)	1	2,209	1,636	3,846
Current Year + 14	0.340	425	(5)	1	2,244	1,745	3,990
Current Year + 15	0.315	413	(4)	1	2,323	1,868	4,193
Current Year + 16	0.292	400	3	1	2,405	1,994	4,400
Current Year + 17	0.270	387	3	1	2,489	2,121	4,611
Current Year + 18	0.250	375	3	1	2,576	2,249	4,826
Current Year + 19	0.232	543	4	2	2,617	2,359	4,978
Current Year + 20	0.215	525	4	2	2,708	2,489	5,199
Current Year + 21	0.199	507	(3)	2	2,801	2,621	5,424
Current Year + 22	0.184	490	(3)	2	2,897	2,755	5,653
Current Year + 23	0.170	665	(2)	3	2,944	2,866	5,812
Current Year + 24	0.158	641	(2)	3	3,044	3,001	6,048
Current Year + 25	0.146	618	(2)	3	3,147	3,138	6,288
Current Year + 26	0.135	595	(2)	3	3,253	3,277	6,532
Current Year + 27	0.125	777	(1)	3	3,306	3,389	6,699
Current Year + 28	0.116	748	(1)	3	3,417	3,530	6,950
Current Year + 29	0.107	720	(1)	3	3,531	3,672	7,207
Total CPVRR =		$2,450	($10)	$8	$24,345	$14,804	$39,157

Finally, we summarize the total costs for the With DSM Resource Plan 1 in Table 6.12.

TABLE 6.12
Total Cost Calculations for With DSM Resource Plan 1

	(1)	(6) = Sum of Cols. (2) thru (5)	(10) = Col (7) + Col (8) − Col (9)	(14) = Sum of Cols. (11) thru (13)	(15) = (6) + (10) + (14)	(16) = (1) × (15)	(17)
Year	Annual Discount Factor 8.00%	Total Generation Fixed Costs (Millions)	DSM Net Costs (Millions)	Total Variable Costs (Millions)	Total Annual Costs (Millions)	Total Annual NPV Costs (Millions)	Cumulative Total NPV Costs (Millions)
Current Year	1.000	0	0	1,620	1,620	1,620	1,620
Current Year + 1	0.926	0	6	1,780	1,786	1,654	3,274
Current Year + 2	0.857	0	5	1,943	1,947	1,670	4,944
Current Year + 3	0.794	0	3	2,108	2,111	1,676	6,620
Current Year + 4	0.735	0	2	2,276	2,279	1,675	8,295
Current Year + 5	0.681	0	1	2,448	2,449	1,667	9,961
Current Year + 6	0.630	148	(7)	2,580	2,722	1,715	11,677
Current Year + 7	0.583	145	(6)	2,758	2,897	1,690	13,367
Current Year + 8	0.540	141	(6)	2,940	3,075	1,661	15,028
Current Year + 9	0.500	137	(6)	3,125	3,256	1,629	16,657
Current Year + 10	0.463	290	(6)	3,263	3,548	1,644	18,300
Current Year + 11	0.429	282	(5)	3,454	3,731	1,600	19,901
Current Year + 12	0.397	274	(5)	3,648	3,917	1,556	21,456
Current Year + 13	0.368	266	(5)	3,846	4,107	1,510	22,966
Current Year + 14	0.340	425	(5)	3,990	4,410	1,502	24,468
Current Year + 15	0.315	413	(4)	4,193	4,601	1,450	25,918
Current Year + 16	0.292	400	3	4,400	4,803	1,402	27,320
Current Year + 17	0.270	387	3	4,611	5,001	1,352	28,672
Current Year + 18	0.250	375	3	4,826	5,204	1,302	29,974
Current Year + 19	0.232	543	4	4,978	5,524	1,280	31,254
Current Year + 20	0.215	525	4	5,199	5,728	1,229	32,483
Current Year + 21	0.199	507	(3)	5,424	5,928	1,178	33,661
Current Year + 22	0.184	490	(3)	5,653	6,140	1,129	34,790
Current Year + 23	0.170	665	(2)	5,812	6,474	1,103	35,893
Current Year + 24	0.158	641	(2)	6,048	6,687	1,055	36,947
Current Year + 25	0.146	618	(2)	6,288	6,904	1,008	37,955
Current Year + 26	0.135	595	(2)	6,532	7,126	963	38,919
Current Year + 27	0.125	777	(1)	6,699	7,475	936	39,855
Current Year + 28	0.116	748	(1)	6,950	7,698	892	40,747
Current Year + 29	0.107	720	(1)	7,207	7,925	851	41,597
Total CPVRR =		$2,450	($10)	$39,157	$41,597	$41,597	

It is now time to examine more closely the costs for the With DSM Resource Plan 1 to see what the effects are of the introduction of DSM Option 1, and the accompanying avoidance of the most economic Supply option (CC Unit A) in Current Year + 5. By looking at the CPVRR cost information on the bottom row of Table 6.12, we immediately see that the With DSM Resource Plan 1 differs from the most economical Supply option resource plan, Supply Only Resource Plan 1 (CC), in two ways.

First, as previously mentioned, the DSM Cost values presented in Column (10) are no longer zero as was the case with any of the Supply Only resource plans. Second, the total CPVRR cost for the With DSM Resource Plan 1, $41,597 million is significantly lower than the CPVRR cost value for Supply Only Resource Plan 1 (CC) of $41,810 million shown previously in Table 5.14.

The lower CPVRR total cost for With DSM Resource Plan 1 can best be explained by looking at differences in the Fixed Costs, DSM Costs, and Variable Costs sections of the table in comparison to the previously presented (Table 5.14) values for the most economic Supply Only resource plan: Supply Only Resource Plan 1 (CC).

In regard to Fixed Costs, the With DSM Resource Plan 1 has lower costs ($2,450 million CPVRR) than the Supply Only Resource Plan 1 (CC) ($2,656 million CPVRR). This $206 million CPVRR cost savings in regard to Fixed Costs is primarily due to the fact that the 100 MW DSM option avoids the need to build a 500 MW CC unit in the decision year. Avoiding the 500 MW CC unit saves a significant amount of Fixed Costs in that year.

However, recall that the With DSM Resource Plan 1 does require the addition of six CC filler units in later years compared to only five filler units with Supply Only Resource Plan 1. This lowers the Fixed Cost savings value from what it would have been if there had been no difference between the two resource plans in the number of filler units. However, the resulting net Fixed Cost savings value of $206 million CPVRR still represents a significant total fixed cost savings for the With DSM Resource Plan 1.

We next turn our attention to DSM Costs. We have seen in Table 6.10 that the sum of the DSM administration costs ($12 million CPVRR) and incentive costs ($21 million CPVRR). The sum of these two DSM Costs equals $33 million CPVRR. This cost value represents the direct costs of implementing the DSM option. However, we have also seen that the reduced electrical load that our utility system must meet will also result in a combined capital and fixed O&M savings of $43 million CPVRR in transmission and distribution facilities that would otherwise have been needed. Consequently, the *net* DSM Costs are actually a net *savings* of $10 million (= $33 million CPVRR for administration and incentive costs minus $43 million CPVRR in transmission and distribution savings).[13]

---

[13] The combining of direct DSM administrative and incentive costs with savings from avoided transmission and distribution (T&D) facilities is not a universal practice. However, the final result in terms of total utility costs for resource plans is unaffected regardless of whether these costs are combined. (Combining these costs is one way to address the fact that certain computer models do not directly address T&D costs that would be avoided by DSM. One approach for dealing with this is to combine these avoided T&D costs with the direct DSM costs to ensure that the T&D avoided costs are accounted for.)

In regard to Variable Costs, we see that the With DSM Resource Plan 1 results in slightly higher Variable Costs of $3 million CPVRR (= $39,157 million CPVRR for the With DSM Resource Plan 1 − $39,154 million CPVRR for the Supply Only Resource Plan 1 [CC]). There are two reasons for this result. First, the With DSM Resource Plan 1 will result in the utility system serving less energy each year due to the kWh reduction from DSM.

This lowers fuel and environmental compliance costs for our utility system. If there were no other differences in the two resource plans, this impact would have resulted in the With DSM Resource Plan 1 having lower Variable Costs than the Supply Only Resource Plan 1 (CC). However, there are other differences between the resource plans.

Second, the With DSM Resource Plan 1, even though it ends up with the same number (6) of new, fuel-efficient CC units (*i.e.*, CC units in the decision year and subsequent CC filler units) as in the Supply Only Resource Plan 1 (CC), will have a number of years in which the cumulative number of these new fuel-efficient CC units is less than the cumulative number of these units in the Supply Only Resource Plan 1. Therefore, the utility system's "fleet" of generation units will be less efficient in those particular years, resulting in higher fuel and emission costs in those years compared to the Supply Only Resource Plan 1 (CC).

These two factors serve to counterbalance each other to a degree for our hypothetical utility system, but the second factor is more important in this particular comparison. The net result is a disadvantage in regard to Variable Costs of $3 million CPVRR for the With DSM Resource Plan 1.

Consequently, we see that the total CPVRR cost of the With DSM Resource Plan 1 ($41,597 million CPVRR) is $213 million CPVRR less expensive than the cost of the Supply Only Resource Plan 1 (CC) ($41,810 million CPVRR). As just discussed, virtually all of these savings come from the Fixed Cost category ($206 million CPVRR). The other two main cost categories, DSM Costs resulting in a net savings of $10 million CPVRR and a Variable Cost increase of $3 million CPVRR, are of lesser impact.

However, before we are tempted to announce that DSM Option 1 is a better choice for our utility system than the best Supply option (CC Unit A), we need to remember that we have not yet completed the final step of an economic evaluation involving both DSM and Supply options. Namely, we have not performed a calculation of the electric rates that will result from the With DSM resource plan.

In fact, to help ensure that we are not tempted to forget this, we will introduce my third Fundamental Principle of Electric Utility Resource Planning:

Fundamental Principle #3 of Electric Utility Resource Planning: "Electric Rate Impacts Are the Most Important Consideration When Analyzing DSM and Supply Options; Total Cost Impacts Are Less Important"

In regard to economic analyses, projections of total utility system costs for competing Supply options can be used to correctly select the most economic Supply resource option. As previously discussed, this is because the Supply option which results in the lowest total costs will also result in the lowest electric rates. However, when evaluating DSM options versus Supply options, economic analyses must be carried out one step further. Analyses of DSM versus Supply options must account for the fact that DSM options reduce the number of kWh of sales over which a utility's costs (revenue requirements) are recovered. This unique characteristic of DSM options makes it necessary to conduct an electric rate calculation in order to really determine which option, DSM or Supply, is the best economic choice from a customer's perspective. The importance of total costs is lessened in these calculations because total costs are merely one input into the calculation of electric rates.

One almost always sees that the total CPVRR costs for a utility system are lower if a DSM option that reduces electrical load, rather than a Supply option, is selected. However, the important issue is how the utility's electric rates compare for resource plans with the two types of resource options. In our example, we need to see how the previously calculated levelized system average electric rate of 12.053 cents/kWh for the Supply Only Resource Plan 1 (CC) compares with the levelized system average electric rate for the With DSM Resource Plan 1.

The levelized electric rate calculation for the With DSM Resource Plan 1, including DSM-driven costs, benefits, and kWh reductions, is presented in Table 6.13 that appears on the next two pages.

TABLE 6.13
Levelized System Average Electric Rate Calculation for With DSM Resource Plan 1

Year	(1) Annual Discount Factor 8.00%	(15) = Col (6) + Col 10 + Col(14) Total Annual Costs (Millions)	(18) Other System Costs Not Affected by Plan (Millions)	(19) = Col(15) + Col(18) Total Utility Costs (Millions)	(20) Forecasted Utility Sales/NEL (GWh)	(21) DSM Energy Reduction (GWh)	(22) = Col(20) − Col(21) Net Utility Sales/NEL (GWh)	(23) = (Col(19) × 100) / Col (22) System Avg. Electric Rate Nominal (Cents/kwh)	(24) = Col(1) × Col(23) System Avg. Electric Rate NPV (Cents/kwh)	(25) Levelized System Avg Electric Rate (Cents/kwh)	(26) = Col(1) × Col(25) System Avg. Electric Rate Nominal (Cents/kwh)
Current Year	1.000	1,620	2,700	4,320	50,496	0	50,496	8.5553	8.555281	12.048	12.048371
Current Year + 1	0.926	1,786	2,754	4,540	51,022	15	51,007	8.9013	8.241907	12.048	11.155899
Current Year + 2	0.857	1,947	2,809	4,756	51,548	45	51,503	9.2352	7.917691	12.048	10.329536
Current Year + 3	0.794	2,111	2,865	4,977	52,074	75	51,999	9.5707	7.597567	12.048	9.564385
Current Year + 4	0.735	2,279	2,923	5,201	52,600	105	52,495	9.9080	7.282669	12.048	8.855912
Current Year + 5	0.681	2,449	2,981	5,430	53,126	135	52,991	10.2470	6.973940	12.048	8.199919
Current Year + 6	0.630	2,722	3,041	5,763	53,652	150	53,502	10.7709	6.787463	12.048	7.592517
Current Year + 7	0.583	2,897	3,101	5,998	54,178	150	54,028	11.1019	6.477846	12.048	7.030109
Current Year + 8	0.540	3,075	3,163	6,238	54,704	150	54,554	11.4350	6.177973	12.048	6.509360
Current Year + 9	0.500	3,256	3,227	6,483	55,230	150	55,080	11.7702	5.888053	12.048	6.027185
Current Year + 10	0.463	3,548	3,291	6,840	55,756	150	55,606	12.3002	5.697355	12.048	5.580727
Current Year + 11	0.429	3,731	3,357	7,088	56,282	150	56,132	12.6275	5.415701	12.048	5.167340
Current Year + 12	0.397	3,917	3,424	7,341	56,808	150	56,658	12.9573	5.145537	12.048	4.784574

Current Year + 13	0.368	4,107	3,493	7,600	57,334	150	57,184	13.2899	4.886655	12.048	4.430161
Current Year + 14	0.340	4,410	3,563	7,973	57,860	150	57,710	13.8153	4.703584	12.048	4.102001
Current Year + 15	0.315	4,601	3,634	8,235	58,386	150	58,236	14.1407	4.457729	12.048	3.798149
Current Year + 16	0.292	4,803	3,707	8,509	58,912	150	58,762	14.4806	4.226735	12.048	3.516805
Current Year + 17	0.270	5,001	3,781	8,782	59,438	150	59,288	14.8121	4.003246	12.048	3.256301
Current Year + 18	0.250	5,204	3,856	9,060	59,964	150	59,814	15.1468	3.790476	12.048	3.015093
Current Year + 19	0.232	5,524	3,933	9,458	60,490	150	60,340	15.6738	3.631800	12.048	2.791753
Current Year + 20	0.215	5,728	4,012	9,740	61,016	150	60,866	16.0020	3.433210	12.048	2.584956
Current Year + 21	0.199	5,928	4,092	10,021	61,542	150	61,392	16.3225	3.242557	12.048	2.393478
Current Year + 22	0.184	6,140	4,174	10,314	62,068	150	61,918	16.6581	3.064095	12.048	2.216183
Current Year + 23	0.170	6,474	4,258	10,732	62,594	150	62,444	17.1865	2.927118	12.048	2.052022
Current Year + 24	0.158	6,687	4,343	11,030	63,120	150	62,970	17.5162	2.762293	12.048	1.900020
Current Year + 25	0.146	6,904	4,430	11,334	63,646	150	63,496	17.8500	2.606423	12.048	1.759278
Current Year + 26	0.135	7,126	4,518	11,644	64,172	150	64,022	18.1880	2.459050	12.048	1.628961
Current Year + 27	0.125	7,475	4,609	12,084	64,698	150	64,548	18.7204	2.343545	12.048	1.508297
Current Year + 28	0.116	7,698	4,701	12,399	65,224	150	65,074	19.0530	2.208503	12.048	1.396571
Current Year + 29	0.107	7,925	4,795	12,720	65,750	150	65,600	19.3901	2.081089	12.048	1.293122

Total CPVRR =	$41,597								Sum = 148.794961		148.794961

From Column (25) in this table, we see that the levelized system average electric rate for the With DSM Resource Plan 1 is 12.048 cents/kWh. Therefore, this DSM-based resource plan featuring DSM Option 1 will result in lower electric rates than the most economical resource plan featuring a Supply Option: Supply Only Resource Plan 1 (CC) which had a levelized system average electric rate of 12.053 cents/kWh.

From an economic perspective, DSM Option 1 is clearly superior to the best Supply option. This is what we would expect by examining the RIM preliminary screening test analysis results comparing DSM Option 1 versus the CC unit. DSM Option 1 passed the RIM screening test which meant that it was potentially more cost-effective than the CC unit for our utility system. And, very importantly, this outcome was essentially guaranteed by not adding more of DSM Option 1 than was called for by our utility's resource needs of 100 MW.[14]

But what about DSM Option 2 versus the same CC unit? DSM Option 2 passed the TRC test but failed the RIM test. What will the final (or system) economic evaluation show for this DSM option? We will now find out.

RESULTS FOR THE WITH DSM RESOURCE PLAN 2

We now add in the economic information for the resource plan that features DSM Option 2. A summary of both the cost and electric rate information for both With DSM resource plans is presented in Table 6.14.

TABLE 6.14

Economic Evaluation Results of Both With DSM Resource Plans: CPVRR Costs and Levelized System Average Electric Rates

	(1)	(2)	(3)	(4) = Sum of Cols. (1) thru (3)	(5)	(6)
Resource Plan	Fixed Costs (Millions, CPVRR)	DSM Net Costs (Millions, CPVRR)	Variable Costs (Millions, CPVRR)	Total Costs (Millions, CPVRR)	Difference from Lowest Cost With DSM Plan (Millions, CPVRR)	Levelized System Average Electric (cents/kWh)
With DSM Resource Plan 1	$2,450	($10)	$39,157	$41,597	$221	**12.048**
With DSM Resource Plan 2	$2,450	$102	$38,824	$41,376	$0	**12.093**

---

[14] Adding more DSM than is needed to meet projected near-term resource needs will increase the present value costs of DSM and can result in an otherwise economically preferred DSM option (or portfolio of DSM options) no longer being the economic choice.

Table 6.14 shows us several things about the two With DSM resource plans. First, the Fixed Cost value of $2,450 million CPVRR is identical for the two resource plans. This is to be expected because both resource plans have added an identical amount of DSM (100 MW) that avoided a new generating unit in the decision year and they both result in the identical number, and timing, of filler units that are added after the decision year. Therefore, the Fixed Costs for our utility system will be identical with either DSM Option 1 or 2.

Second, the two plans differ considerably in regard to net DSM Costs. The difference is $112 million CPVRR between the two plans, *i.e.*, a $10 million CPVRR net *savings* for the resource plan featuring DSM Option 1 versus a $102 million CPVRR net *cost* for the resource plan featuring DSM Option 2. A logical question is: "*Why does this cost difference exist?*"

We just discussed how the various components of DSM Costs were calculated for DSM Option 1. In addition, we previously discussed how the administration and incentive costs would be more than twice as high for DSM Option 2 compared to DSM Option 1. We will now examine how these net DSM Costs actually worked out.

The previously discussed CPVRR administration and incentive costs for DSM Option 1, $12 million and $21 million, respectively, add up to an administrative and incentive cost total of $33 million. For DSM Option 2, this total increases initially to $66 million CPVRR just due to the fact that twice as many participants need to be signed up for DSM Option 2 as for DSM Option 1 (because each DSM Option 2 participant only provides 0.5 kW reduction compared to 1.0 kW reduction for DSM Option 1 participants).

The $43 million CPVRR savings in avoided T&D expenditures we saw previously for DSM Option 1 are driven solely by the peak demand (kW) reductions achieved by DSM Option 1. Because the identical peak demand reduction, both annually and in total, is also achieved for DSM Option 2, the Resource Plan with DSM Option 2 will also realize the same $43 million CPVRR savings from avoided T&D expenditures. In other words, there is no difference between the two DSM Options in regard to avoided T&D expenditures.

Consequently, when considering just the doubling effect on the administration and incentive costs, plus the identical T&D savings, we realize that the net DSM Costs have (so far) increased from a net *savings* of $10 million CPVRR for DSM Option 1 to a net *cost* of $23 million CPVRR for DSM Option 2 (= $66 million CPVRR in administration and incentive costs − $43 million CPVRR in T&D savings).

However, as we see from the results of Table 6.14, the total net cost for DSM Option 2 is not $23 million CPVRR, but $102 million CPVRR. Where does this additional cost for DSM Option 2 come from? As suggested earlier in this chapter, the answer lies in the difference in the life expectancy of the equipment that is installed for DSM Option 2 versus DSM Option 1.

We previously discussed how the fact that the much shorter (5-year) "life expectancy" for the DSM equipment installed with DSM Option 2, compared to the 15-year life expectancy for DSM Option 1's equipment, would further result in higher DSM administrative and incentive costs for DSM Option 2. This is because the utility must continue to replenish the kW (and accompanying kWh) savings from DSM Option

2 that would otherwise begin to vanish 5 years after a customer is signed up to participate in DSM Option 2. (This replenishment is also needed for DSM Option 1's equipment, but less frequently due to the 15-year life expectancy for that equipment.)

By looking at the final total net DSM Cost for DSM Option 2 of $102 million CPVRR, we see that the life expectancy of a DSM measure can be a truly significant factor in the total DSM CPVRR cost. In our example, the difference in the life expectancy between DSM Option 2 and DSM Option 1 increased the net DSM Costs for DSM Option 2 by $79 million CPVRR (= $102 million CPVRR total net cost − the previously calculated net cost of $23 million CPVRR prior to accounting for life expectancy).

As a result, the total CPVRR DSM Cost for DSM Option 2 is $102 million (= $66 million CPVRR in administration and incentive costs due to the need to sign up twice as many participating customers than with DSM Option 1 − $43 million CPVRR in T&D savings + $79 million CPVRR in additional administrative and incentive costs due to the need to replenish DSM Option 2 more frequently due to its shorter life expectancy).

Finally, we examine the Variable Costs for our utility system. From the table, we see that the Variable Costs for DSM Option 2 ($38,824 million CPVRR) are significantly lower than for DSM Option 1 ($39,157 million CPVRR) as expected. This difference of $333 million CPVRR is driven by the fact that DSM Option 2 will result in four times the energy savings on the utility system compared to DSM Option 1. (Recall that from Row [7] of Table 6.1, the "Energy-to-Demand Reduction Ratio," or MWh-to-MW reduction ratio, was 6,000 for DSM Option 2 versus 1,500 for DSM Option 1.)

When all three cost categories are combined, the net result is a CPVRR cost for the With DSM Resource Plan 2 of $41,376 million CPVRR, or $221 million CPVRR less than for the With DSM Resource Plan 1. However, as we just reminded ourselves, DSM Option 2 will also result in significantly fewer kWh sales over which our utility's costs must be recovered. Therefore, we now turn our attention to the more important issue of what effect the With DSM Resource Plan 2 will have on the utility's electric rates.

From the last column in Table 6.14, we see that, despite the lower CPVRR costs for the Resource Plan with DSM Option 2, this resource plan will result in a higher levelized electric rate (12.093 cents/kWh) than will result with the With DSM Resource Plan 1 (12.048 cents/kWh). Therefore, DSM Option 2 is a worse choice than DSM Option 1 because of the higher electric rates that our utility's customers will be charged with DSM Option 2.

Therefore, DSM Option 1 (which passed the RIM test) is a better economic choice for our utility's customers than DSM Option 2 (which failed the RIM test) because it will result in lower electric rates for all customers.

We can now close this chapter by summing up the results of both the previous and current chapters. Our utility has now completed the final (or system) economic analyses of all three Supply options: the CC option, the CT option, and the PV option. The utility has also completed the final (or system) economic analyses for both of the DSM Options: DSM Option 1 and DSM Option 2.

Therefore, from an economic perspective, our utility should have the information it needs to determine which one of these five resource options is the best economic choice for its customers. Our utility is also at a point at which it can begin to examine these five resource options from a non-economic perspective. Then, with information from both the economic and non-economic perspectives, a final decision can be made regarding which of the five resource options is the best overall choice of resource option with which to meet its resource needs 5 years in the future.

Our utility takes these next steps in Chapter 7.

7 Final Resource Option Analyses for Our Utility System

ECONOMIC COMPARISON OF THE RESOURCE PLANS

Our hypothetical utility system has now completed its economic analyses of five resource plans, each resource plan containing one of the five resource options it is considering in order to meet its next resource need that occurs in Current Year + 5. The numeric results of these analyses are summarized in Table 7.1. The resource plans are listed in the order they were discussed in previous chapters.

TABLE 7.1

Economic Evaluation Results for All Five Resource Plans: CPVRR Costs and Levelized System Average Electric Rates

Resource Plan	Total Costs (Millions, CPVRR)	Difference from Lowest Cost Resource Plan (Millions, CPVRR)	Levelized System Average Electric Rate (cents/kWh)	Difference from Lowest Levelized Average Electric Rate Resource Plan (cents/kWh)
Supply Only Plan 1 (CC)	41,810	434	12.053	0.005
Supply Only Plan 2 (CT)	41,883	507	12.064	0.016
Supply Only Plan 3 (PV)	41,829	453	12.056	0.008
With DSM Plan 1	41,597	221	12.048	0
With DSM Plan 2	41,376	0	12.093	0.045

This table presents the economic results from two perspectives: (i) total costs (presented in the first and second columns), and (ii) electric rates (presented in the third and fourth columns). The table summarizes what we have previously discussed: (i) the resource plan with the lowest electric rate is the With DSM Resource Plan 1, and (ii) the resource plan with the lowest CPVRR cost is With DSM Resource Plan 2.

We will now examine the results from each of these two perspectives separately by rearranging the information to show the rankings of the five resource plans from each perspective. This should allow us to understand the results more clearly.

We shall start with a ranking of resource plans from a total cost (CPVRR) perspective. This information is presented in Table 7.2.

DOI: 10.1201/9781003301509-8

TABLE 7.2

Ranking of Resource Plans in Regard to CPVRR Costs

Resource Plan	Total Costs (Millions, CPVRR)	Difference from Lowest Cost Resource Plan (Millions, CPVRR)	Relative Ranking
With DSM Plan 2	41,376	0	1st
With DSM Plan 1	41,597	221	2nd
Supply Only Plan 1 (CC)	41,810	434	3rd
Supply Only Plan 3 (PV)	41,829	453	4th
Supply Only Plan 2 (CT)	41,883	507	5th

This table shows that—from a total cost perspective—the With DSM Resource Plan 2 (that featured DSM Option 2 that failed the RIM test, but which passed the TRC test) is projected to result in the lowest total CPVRR costs over the analysis period. The With DSM Resource Plan 1 (that featured DSM Option 1 that passed both the RIM and TRC tests) is projected to result in the second lowest total CPVRR costs. (Note that both of the resource plans featuring the DSM options are projected to result in lower total CPVRR costs than any of the three Supply Only resource plans. This is a common outcome in analyses of DSM and Supply resource options.)

From a total cost perspective, Supply Only Resource Plan 1 (CC) is projected as the most economic Supply Only resource plan and as the resource plan that has the third lowest total CPVRR costs. Supply Only Resource Plan 3 (PV) is projected as the resource plan with the fourth lowest total CPVRR costs and Supply Only Resource Plan 2 (CT) is projected as having the highest total CPVRR costs.

However, as we have discussed previously, the total cost perspective ignores a very important economic impact that occurs when DSM resource options are considered: the upward pressure that will be placed on electric rates for all of the utility's customers because the utility's total costs will be recovered over fewer kWh of sales due to the kWh reduction characteristic of DSM options. For that reason, a total cost perspective when evaluating both Supply and DSM options can only provide an incomplete economic picture. No final decision regarding resource options should ever be made based solely on projections of total costs when one or more DSM options are being evaluated.

Therefore, we now turn our attention to examining a ranking of the five resource plans from the more important electric rate perspective. Unlike a total cost perspective, the electric rate perspective *does* provide a complete economic picture because it *does* account for the kWh reduction characteristic of DSM options. Table 7.3 provides the electric rate perspective.

TABLE 7.3

Ranking of Resource Plans in Regard to Electric Rates

Resource Plan	Levelized System Average Electric Rate (cents/kWh)	Difference from Lowest Levelized Average Electric Rate Resource Plan (cents/kWh)	Relative Ranking
With DSM Plan 1	12.048	0	1st
Supply Only Plan 1 (CC)	12.053	0.005	2nd
Supply Only Plan 3 (PV)	12.056	0.008	3rd
Supply Only Plan 2 (CT)	12.064	0.016	4th
With DSM Plan 2	12.093	0.045	5th

From the more complete, and meaningful, perspective of electric rates, we see that the ranking of the five resource plans has significantly changed. The With DSM Resource Plan 1 (featuring DSM Option 1 that passed both the RIM and TRC tests) has moved up to first place in the ranking. This resource plan not only remains better than all three Supply Only resource plans, but is also now seen as significantly better than the With DSM Resource Plan 2 (featuring DSM Option 2 that failed the RIM test but passed the TRC test). The With DSM Resource Plan 2 is now seen as the worst of the five resource plans from an electric rate perspective.

In regard to the three Supply Only resource plans, the relative ranking of these three plans remains the same. As previously discussed, this is expected because Supply options do not result in changes in the number of kWh over which the utility system's total costs are spread. Therefore, the best Supply Only resource plan from a total cost perspective will also be the best Supply Only resource plan from an electric rate perspective. Consequently, Supply Only Resource Plan 1 (CC) remains more economical from an electric rate perspective than Supply Only Resource Plan 3 (PV). And both of these resource plans remain more economical than Supply Only Resource Plan 2 (CT).

With this information in hand, our utility is able to determine which resource plan is its best plan in regard to economics. Recalling Fundamental Principle #3 (which states that an electric rate perspective is the meaningful economic perspective to take when evaluating both DSM and Supply options), our utility selects With DSM Resource Plan 1, which features DSM Option 1 (*i.e.*, the DSM option that passed both the RIM and TRC tests), as the best resource plan and resource option from an economic perspective.

However, our utility is not quite finished in its analyses. It decides that it wants to see how the five resource plans compare in regard to three non-economic considerations: the amount of time until the winning economic resource plan becomes the best economic resource plan (*i.e.*, the cross-over time), system fuel use, and system emissions. We will examine these considerations next.

NON-ECONOMIC ANALYSES OF THE RESOURCE PLANS

"CROSS-OVER" TIME TO BEING THE MOST ECONOMIC RESOURCE PLAN

The first of the non-economic considerations we will examine is the determination of how long it takes until one resource plan emerges as clearly the most economic resource plan. For this determination, we will use the tabular format for cross-over time that was introduced in Chapter 3. Table 7.4 presents the cumulative present value

TABLE 7.4

"Cross-Over" Table for All Five Resource Plans Electric Rate of Resource Plans by Year

Year	Supply Only Resource Plan 1 (CC)	Supply Only Resource Plan 2 (CT)	Supply Only Resource Plan 3 (PV)	With DSM Resource Plan 1	With DSM Resource Plan 2
Current Year	1st	1st	1st	1st	1st
Current Year + 1	1st	1st	1st	4th	5th
Current Year + 2	1st	1st	1st	4th	5th
Current Year + 3	1st	1st	1st	4th	5th
Current Year + 4	1st	1st	1st	4th	5th
Current Year + 5	4th	1st	3rd	2nd	5th
Current Year + 6	4th	2nd	3rd	1st	5th
Current Year + 7	2nd	3rd	4th	1st	5th
Current Year + 8	2nd	3rd	4th	1st	5th
Current Year + 9	3rd	2nd	4th	1st	5th
Current Year + 10	2nd	3rd	4th	1st	5th
Current Year + 11	2nd	3rd	4th	1st	5th
Current Year + 12	2nd	3rd	4th	1st	5th
Current Year + 13	2nd	3rd	4th	1st	5th
Current Year + 14	2nd	3rd	4th	1st	5th
Current Year + 15	3rd	2nd	4th	1st	5th
Current Year + 16	3rd	2nd	4th	1st	5th
Current Year + 17	2nd	3rd	4th	1st	5th
Current Year + 18	2nd	3rd	4th	1st	5th
Current Year + 19	2nd	3rd	4th	1st	5th
Current Year + 20	2nd	3rd	4th	1st	5th
Current Year + 21	2nd	3rd	4th	1st	5th
Current Year + 22	2nd	3rd	4th	1st	5th
Current Year + 23	2nd	3rd	4th	1st	5th
Current Year + 24	2nd	4th	3rd	1st	5th
Current Year + 25	2nd	4th	3rd	1st	5th
Current Year + 26	2nd	4th	3rd	1st	5th
Current Year + 27	2nd	4th	3rd	1st	5th
Current Year + 28	2nd	4th	3rd	1st	5th
Current Year + 29	2nd	4th	3rd	1st	5th

Note: 1st = lowest electric rate, etc.

levelized system average electric rate information for each year for all of the five resource plans that our utility has analyzed. This information is presented as a ranking in which "1st" represents the resource plan with the lowest cumulative present value (CPV) levelized electric rates through that year, "2nd" represents the resource plan with the second lowest CPV levelized electric rates through that year, etc.

As shown in this table, there is no difference in the electric rates in the Current Year because no new resources intended to meet the Current Year + 5 resource need have yet been added. However, this is not the case during the next four years for the two With DSM resource plans. Due to the need to add either 100,000 participants (for DSM Option 1), or 200,000 participants (for DSM Option 2), by Current Year + 5, our utility begins signing up DSM programs participants in Current Year + 1. Administrative and incentive costs for these DSM programs are incurred, utility kWh sales are reduced, variable costs are reduced, and T&D system costs are reduced. All of these impacts have an effect on electric rates. The net effect is an increase in electric rates from Current Year + 1 through Current Year + 4 for the two With DSM resource plans. The electric rate increase is higher, and therefore worse for customers, for the With DSM Resource Plan 2.

Consequently, the three Supply Only plans have the lowest (and the same) electric rates through Current Year + 4 (because, typically, no costs for construction of generating options are passed on to customers until the new generating unit goes in-service which will be Current Year + 5). Then, in Current Year + 5, as denoted by the solid line across the page after the Current Year + 4 row, and the shading of the Current Year + 5 values, the economic ranking of these five resource plans begins to change. The most noticeable changes are that the resource plan with the lowest installed cost Supply option, the CT, moves into the first position and the resource plan with DSM Option 1 moves into the second position. The resource plans with the next lowest installed costs (the PV and CC options, respectively) move into the third and fourth positions. The resource plan with DSM Option 2 remains in the fifth (last) position.

In Current Year + 6, the With DSM Resource Plan 1 emerges as the resource plan with the lowest electric rate (*i.e.*, moves into first position). The With DSM Resource Plan 2 remains in the last (fifth) position. These two resource plans then maintain those positions for the remainder of the analysis years. Only the positions of the three Supply Only resource plans jockey for the second, third, and fourth positions in the subsequent years.

Starting in Current Year + 17, as indicated by the shading, the Supply Only Resource Plan 1 (CC) takes (and maintains) the second place ranking. Then, in Current Year + 24, as indicated by the shading, the Supply Only Resource Plan 3 (PV) takes (and maintains) the third place ranking and the Supply Only Resource Plan 2 (CT) takes (and maintains) the fourth place ranking.

Recalling that our utility's resource need was in Current Year + 5, and that the best resource plan from an electric rate perspective has been virtually established within only 1 year of the resource need date, the generational equity consideration that we discussed in Chapter 3 is not an issue for our utility if the With DSM Resource Plan 1 is selected.

Before we leave this subject, there is one interesting aspect of the two types of resource options, Supply and DSM options, which shows up in this table. That aspect is that DSM options, no matter how economic they are over the full term of the analysis (as is the case with the With DSM Resource Plan 1), typically result in higher electric rates in the years prior to the year of resource need (Current Year + 5) and perhaps for a short time thereafter.

The reason for this is that the DSM options require expenditures (which are usually recovered almost immediately from the utility's customers) prior to the year of resource need in order to sign up participating customers.[1] In those early years, these DSM costs are generally not completely offset by avoiding utility costs for system fuel, system environmental compliance, transmission, and distribution. For truly cost-effective DSM options (such as DSM Option 1 that passes the RIM cost-effectiveness test and that results in the lowest system electric rates), the additional benefits derived from avoiding the generating unit that would otherwise be needed are enough to result in this type of DSM option beginning to deliver lower electric rates to the utility's customers either in the year of resource need or shortly thereafter.

A further note of caution is also warranted. We have just seen how the selection of a DSM option will typically result in increased electric rates in the near-term compared to the selection of a Supply option. We have also seen that selection of a DSM option that fails the RIM test (such as DSM Option 2) will result in increased electric rates in both the near- and long-term compared to a DSM option that passes the RIM test (DSM Option 1). In addition, there is yet one other way in which electric rates can be further increased if the resource option selected is a DSM option.

This occurs if a *greater amount* of DSM is implemented than is needed to meet the projected resource need. This drives up electric rates for two reasons. First, DSM expenditures are incurred earlier than they need to be to meet any subsequent resource need (*i.e.*, this would be a resource need after Current Year + 5 in our example). This drives up the present value of these additional DSM costs compared to what the present value of those costs would have been if the additional DSM had been signed up later in time (*i.e.*, closer to the time when the additional DSM could have met a subsequent resource need). Second, the additional kWh reduction resulting from the additional DSM further reduces the number of kWh, which is the denominator in electric rate calculations.

Therefore, the implementation of more DSM than is needed to meet the next projected resource need both increases the numerator (costs) and decreases the denominator (kWh sales), in an electric rate calculation. This "double whammy" will definitely increase electric rates. Thus, it is simply not a good idea from an electric rate perspective to "push" for more DSM than is needed to meet a utility's next resource need because customers' electric rates are increased needlessly.

---

[1] The construction of new Supply options (*i.e.*, new generating units) also requires expenditures prior to the resource need year when the new generating unit will begin to operate. However, a utility will not typically recover these construction expenditures from its customers before the unit goes into operation. Exceptions do exist, particularly when the construction cost of the new generating unit is very high (as, for example, with a new nuclear unit).

SYSTEM FUEL USE

The next non-economic consideration that our utility takes a look at is the impact of the resource plans on its system fuel use. We start by examining our utility's total system fuel use over all of the years of the analysis period.[2] Our utility system's fuel use will be expressed in terms of million mmBTU.[3] The projected total amount of fuel that would be used by our utility over the analysis period from each of the five resource plans is presented in Table 7.5. The order of the resource plans in this table is the same order in which the resource plans were originally discussed earlier in the book.

TABLE 7.5

Comparison of Total System Fuel Use by Resource Plan

Resource Plan	System Total Fuel Use (Million mmBTU)
Supply Only Plan 1 (CC)	15,736
Supply Only Plan 2 (CT)	15,784
Supply Only Plan 3 (PV)	15,683
With DSM Plan 1	15,737
With DSM Plan 2	15,639

The projected total fuel use results are not too surprising. One would expect that, all else equal, the resource plans featuring the DSM options, and the PV option, might result in lower system total fuel usage. This is because the DSM options themselves not only consume no fuel but also reduce the number of kWh that our utility must serve. The PV option also consumes no fuel (but it does not reduce the number of kWh the utility must serve). Conversely, the two resource plans that feature the CC and CT options are based on resource options that do burn fossil fuel. One might expect that the total system fuel usage for these two resource plans would be higher than the other three resource plans.

Therefore, it is not surprising that the With DSM Resource Plan 2 (that features the DSM option with the greatest kWh reduction) results in the lowest total fuel usage of any of the five resource plans for our utility system.

The resource plan that results in the second lowest total fuel usage is the Supply Only Resource Plan 3 (PV). This result was "helped" by the fact that the PV option's 50% firm capacity value required the utility to sign up 240 MW of nameplate PV in order to meet its 120 MW resource need in Current Year + 5.

---

[2] The use of all of the years addressed in the analysis when examining total system fuel use allows us to include the longer term system impacts of differences between the resource options that could be chosen for the decision year. The primary differences are the capacity (MW) of these options (which, as we have seen, affects the timing and number of the filler units and their fuel use) and the options' respective capacity factors or, in the case of the DSM options, equivalent capacity factors.

[3] This value, a million mmBTU, represents a lot of energy. To put this value in perspective, one million mmBTU represents the equivalent energy content that could be supplied by roughly 170,000 barrels of oil.

In a virtual tie are the With DSM Resource Plan 1 and the Supply Only Resource Plan 1 (CC). The Supply Only Resource Plan 2 (CT) is projected to have the highest system fuel usage. In addition, if one were to compare the differences in the fuel totals for all five resource plans on a percentage basis, one would see that there is less than a 1% difference between any two resource plans.[4]

At this point, the astute reader is likely to ask the following question:

This total fuel use information in terms of mmBTU might be interesting, but how important is it really? After all, the cost of the fuel for each resource plan has already been accounted for in the economic analyses and the percentage differences between the resource plans is very small. Is there another reason that it is important to look at system fuel use information?

This is a very good question. (You really are becoming quite proficient at this.) It is true that the actual fuel costs have been accounted for in the economic analyses. Furthermore, utilities do not often look at their system fuel use in terms of total mmBTU of fuel consumed. The reason they do not often do so is actually evident if one takes another look at Table 7.5 to see what it does not show. Such a look tells us nothing about the *types of fuels* that are being consumed with each resource plan.

As mentioned in the overview discussion at the end of Chapter 3, utilities frequently examine their projected system fuel usage to ensure that they are not becoming overly dependent upon any one type of fuel. There are at least three reasons for this concern.

First, utilities want to see if a resource option makes them overly dependent upon a particular type of fuel in regard to the risk of sudden cost increases in that type of fuel. The costs of certain fuels are prone to significant cost changes in a relatively short time. This has historically been the case for natural gas and oil at various times. For this reason, utilities strive over the long-term to avoid becoming too reliant on such fuels because rapid fuel cost increases must be paid by the utility's customers and can lead to unexpected shocks in utility bills. In most regulatory jurisdictions, fuel costs can be recovered in a relatively short time without the lag time inherent in a formal rate case. Therefore, customers can feel the impact of fuel cost increases relatively quickly after the increases occur (if the utility chooses to request cost recovery of fuel cost amounts that are higher than what they originally projected as soon as it is possible to do so).

Second, too much reliance on any one type of fuel can become a system reliability issue. If the available supply of any type of fuel is reduced, even temporarily, the utility's ability to produce electricity from generating units that rely on that type of fuel will be diminished. This can also result in system fuel cost increases that must be passed on to customers if the utility has to switch to higher cost (but more available) fuels. Furthermore, if the availability of the fuel supply is either significantly reduced and/or reduced for a long period of time, the reliability of the utility system as a whole (*i.e.*, the ability of the utility to meet its customers' demand for electricity) may come into question.

Taken together, these two considerations are often referred to under a general heading as the "fuel diversity" profile of a utility system. And, before we introduce a third reason why there one might be interested in examining utility system fuel

---

[4] The differences in the levelized average electric rates among the five resource plans are also less than 1%.

usage, we will take a look at how our utility system's fuel diversity profile is affected by each of the five resource plans.

Table 7.6 presents this information. It shows the percentage of the total energy served (GWh) by our utility system that is generated by the following types of fuel/energy: nuclear, coal, natural gas, and solar. This information is first shown for the Current Year (*i.e.*, the year that the utility will be making its decision as to how to address its resource needs 5 years in the future).

TABLE 7.6

Fuel Diversity Profile of the Five Resource Plans (% of System GWh Served by Type of Fuel)

Year	Resource Plan	Nuclear (%)	Coal (%)	Natural Gas (%)	Solar (%)	Total (%)
Current Year	Not applicable	14.7	48.6	36.7	0.0	100.0
Current Year + 5	Supply Only Plan 1 (CC)	14.0	46.2	39.8	0.0	100.0
Current Year + 5	Supply Only Plan 2 (CT)	14.0	46.2	39.8	0.0	100.0
Current Year + 5	Supply Only Plan 3 (PV)	14.0	46.2	39.0	0.8	100.0
Current Year + 5	With DSM Plan 1	14.1	46.3	39.6	0.0	100.0
Current Year + 5	With DSM Plan 2	14.2	46.6	39.2	0.0	100.0

The table then shows the projected values for Current Year + 5 (*i.e.*, the decision year). The values for the Current Year provide a point of comparison that enables our utility to see how the selection of any of the five resource plans will affect or alter the utility's fuel use profile in 5 years. (Subsequent decisions about resource options that would be added in later years would impact our utility's fuel use profile in those years. However, because our utility is not currently making a resource decision for any year after Current Year + 5, projections of our utility's fuel mix for years after Current Year + 5 are of relatively little importance to our utility at this time.)

The first row of Table 7.6 shows us that the energy generated by our utility system's existing generating units to serve its customers in the Current Year is derived 14.7% from nuclear fuel, 48.6% from coal, 36.7% from natural gas, and 0% from solar.

The remaining rows of the table then show us the projected percentages of total energy served from each of these fuel types 5 years in the future (Current Year + 5) for each of the resource plans under consideration.

A review of this information reveals several things. Most importantly, for the two considerations we are discussing under the "fuel diversity" heading, there is not a significant difference in these percentage values among the five resource plans. All five resource plans will generally "move" our utility system's fuel diversity profile in the same direction: slightly lower percentage contributions from nuclear and coal (the utility's non-marginal fuels), and slightly higher percentage contributions from natural gas (the utility's primary marginal fuel).

This result that the fuel diversity profiles vary relatively little among the five resource plans, is important information for our utility to have. As a result, our utility will not reject the results of its economic analyses due solely to fuel diversity considerations.

But before we leave this table, let us further our education a bit more and look at two more items that are revealed by the table's results. First, we note that the percentage values 5 years in the future for nuclear and coal decline for each resource plan. However, because these are not our utility's marginal fuels, nuclear and coal units are projected to provide just as much energy (GWh) in 5 years as they do in the Current Year.

However, the number of total GWh that will be served by the utility has increased due to our assumption of continuing load growth for our utility. Because no new nuclear or coal capacity is being added by our utility, the percentage of GWh served by nuclear and coal, compared to the total GWh served, will shrink. (Note that this "shrinkage" is slightly less for the two With DSM resource plans. This is because the DSM options will result in somewhat less total GWh being served by our utility.) In addition, the percentage values for natural gas have increased in each resource plan due to the combination of the increase in GWh to be served and the fact that natural gas is the marginal fuel for our utility system.

Finally, we now return briefly to discuss the third reason that utilities may wish to examine their projected fuel use patterns. An analysis of a utility's fuel use patterns can, indirectly, shed considerable light on the utility system's projected air emissions. However, for the sake of expediency, we will move directly to examine our utility's system air emissions. We turn to that non-economic consideration next.

SYSTEM AIR EMISSIONS

As previously mentioned, we will be examining three types of air emissions for the entire utility system: sulfur dioxide (SO_2), nitrogen oxides (NO_x), and carbon dioxide (CO_2). Although there are a number of other emissions that utilities are concerned with (mercury, particulates, etc.), our discussion of air emissions will focus on these three types of air emissions. This will enable us to simplify our discussion while, at the same time, bring out certain key aspects of how resource options may impact system air emissions for a given utility system.

In addition, it allows us to look at two types of air emissions (SO_2 and NO_x) that are currently regulated due to the direct effects of these emissions on human health. It also allows us to look at one type of air emission (CO_2) that, at the time the second edition of this book is written, is not currently regulated in the United States in the same way SO_2 and NO_x are regulated. However, CO_2 emissions are of interest largely due to concerns not directly tied to human health in the same sense SO_2 and NO_x emissions are.[5]

We will start our discussion by presenting Table 7.7 that appears on the next page. This table provides a comparison of total system emissions for our utility system for all of the years addressed in our analysis. The total emissions (presented in terms of total tons of emissions for the analysis period) for each of the three types of emissions are presented separately.

---

[5] Just as a point of interest, electric utilities typically have very accurate data regarding air emissions through the use of equipment that continually monitors air emissions. This data is recorded and reported regularly to various governmental agencies.

TABLE 7.7
Comparison of System Air Emissions by Resource Plan

Resource Plan	SO$_2$ (Million tons)	NO$_x$ (Million tons)	CO$_2$ (Million tons)
Supply Only Plan 1 (CC)	1.011	0.252	1,119
Supply Only Plan 2 (CT)	1.011	0.255	1,122
Supply Only Plan 3 (PV)	1.011	0.253	1,116
With DSM Plan 1	1.011	0.254	1,119
With DSM Plan 2	1.011	0.252	1,113

An initial glance at this table shows that, regardless of which resource plan one may be looking at, there are significant differences in the magnitude of the values among the SO$_2$, NO$_x$, and CO$_2$ columns. The CO$_2$ values are three orders of magnitude (*i.e.*, about 1,000 times) greater than the SO$_2$ values, and the SO$_2$ values are approximately four times larger than the NO$_x$ values.

Now we turn our attention to a result in the table that is perhaps even more interesting. A closer examination of the total emissions presented in Table 7.7 reveals two results that may be counterintuitive at first glance. First, the differences among the five resource plans in regard to any of the three air emissions are very small. Second, the resource plans that end up in a tie regarding the lowest system emissions for NO$_x$ are Supply Only Resource Plan 1 (CC) and With DSM Resource Plan 2. From their descriptions, these resource plans would appear to be at opposite ends of the spectrum. One resource plan features a new generating unit (a CC unit) that will operate on fossil fuel 80% of the hours per year, while the other resource plan results in the largest reduction in energy served by our utility system.[6]

In order to examine these results more clearly, we turn our attention to Table 7.8 that transforms the numerical emission ton values in Table 7.7 into a ranking of total system emissions by type of air emission. The designation of "1st" indicates the resource plan with the lowest system emissions for that type of emission, continuing to the designation of "5th" that indicates the resource plan with the highest system emissions.

TABLE 7.8
Ranking of Resource Plans by System Air Emissions

Resource Plan	SO$_2$ (Million tons)	NO$_x$ (Million tons)	CO$_2$ (Million tons)
Supply Only Plan 1 (CC)	1st-tied	1st-tied	3rd-tied
Supply Only Plan 2 (CT)	1st-tied	5th	5th
Supply Only Plan 3 (PV)	1st-tied	3rd	2nd
With DSM Plan 1	1st-tied	4th	3rd-tied
With DSM Plan 2	1st-tied	1st-tied	1st

Note: 1st = lowest emissions; 5th = highest emissions

---

[6] This result points out that the addition of a new Supply option which is much more fuel-efficient than the existing generating units currently on a utility system can also significantly improve a utility's air emission profile.

Some readers may be puzzled by these results. This is not an uncommon reaction. The idea that a resource plan (Supply Only Resource Plan 1 (CC)) that features a large new generating unit that operates on fossil fuel, and that runs most of the hours in a year can result in lower system air emissions for <u>any</u> type of air emission (*i.e.*, NO_x) compared to a resource plan that features either a renewable energy generator (Supply Only Resource Plan 3 (PV)) or DSM (in the With DSM Resource Plan 1), is a result that may seem counterintuitive to some readers.

In fact, this possibility may never have occurred to those readers when they read about energy efficiency and/or renewable energy options being "clean" or "green" resource options. The conclusion that one is tempted to leap to from reading some of the literature about resource options is that anything touted as "clean" or "green" must result in a lowering of all types of air emissions. As we can see from the analyses of our utility system example, this conclusion may well be wrong.

From experiences gained in previous "real -life" discussions regarding this topic, I believe that some readers will find these results counterintuitive, while other readers will readily understand how the results presented in Table 7.8 could have occurred after accounting for a number of factors such as: (i) the greater efficiency of the new generating units compared to our utility's existing generation fleet (resulting in these more efficient generating units being dispatched more than the less efficient existing generators), and (ii) the lower emission rates of the new generating units compared to the emission rates of the utility's existing generating units.

(For those readers who do understand why these results make sense, no explanation is needed. For the other group of readers, an explanation is necessary. In either case, an explanation of why results such as these may occur on a utility system takes a bit of time. Rather than provide that explanation here and take a chance of disrupting the flow of the narrative, I refer readers who need/wish to read an explanation to refer to Appendix G. This appendix provides an explanation of why results such as these can occur using a completely different type of "system" that also uses fossil fuel.)

Our examination of air emissions for our utility system has revealed insights into how resource options will impact a utility system's total air emissions. However, the importance of those insights goes far beyond the issue of air emissions which is being discussed at the moment. The importance of those insights is summarized in my Fundamental Principle #4, which applies to *all* aspects of resource planning, not just in regard to system emissions:

Fundamental Principle #4 of Electric Utility Resource Planning: "Always Ask Yourself: 'Compared to What?'"

In all aspects of resource planning, one must always ask the question "compared to what" when one is analyzing a particular resource option.

For example, the fact that Resource Option A is projected to either lower the number of kWh served by a utility (such as our utility's DSM options), or to produce kWh without burning fossil fuel (such as our utility's PV option), is no guarantee that Resource Option A will actually result in lower total <u>system</u> air emissions for a specific utility. The key point is *"What is resource option A being compared to?"*

Although Resource Option A may reduce air emissions from what the emissions otherwise would be *if no other resource option were added or considered*, there may be a Resource Option B that would result in even lower system air emissions.[7] (In such a case, the selection of Resource Option A would actually *increase* system emissions compared to the selection of Resource Option B.) Therefore, in order to know which resource option will really result in lower system air emissions, all applicable resource options should be evaluated. As we have seen in the non-economic analyses of air emissions for our hypothetical utility system, the answer may vary from one type of air emission to another.

Our utility system has performed just such an analysis that examined all resource options available to it. The results from the analyses of system air emissions, combined with the economic analysis results and the results of the analysis of economic cross-over times and system fuel diversity profiles, provide comprehensive information that our utility system can now use to make its final resource decision.

SUMMARY OF RESULTS FROM THE RESOURCE OPTION ANALYSES FOR OUR UTILITY SYSTEM

In its decision-making process, our utility first summarizes the results of the analyses it has conducted. Its summary consists of the following points:

1. From a system reliability perspective, all five resource plans will enable the utility system to meet its reliability (*i.e.*, reserve margin) criterion in the decision year (as well as in subsequent years through the addition of filler units). Consequently, all five resource plans are judged to be acceptable from a system reliability perspective.
2. In regard to economics, the With DSM Resource Plan 1 is projected to result in the lowest electric rates for our utility's customers. By comparison, all other resource plans will result in electric rates higher than would be experienced with the With DSM Resource Plan 1 (*i.e.*, these other four resource plans would result in electric rate increases for the utility's customers compared to the selection of the With DSM Resource Plan 1).
3. In addition, the "cross-over" time to when the With DSM Resource Plan 1 becomes the best economic plan is very short: Current Year + 6, which is only 1 year after our utility needs to add a new resource option. Consequently, the cross-over time for With DSM Resource Plan 1 is acceptable.
4. From a system fuel perspective, two points are evident. First, the costs of system fuel use have already been accounted for in the economic analyses. Second, the fuel use profile analysis for each of the five resource plans does not show significant differences in the relative percentages of energy by fuel type. In addition, none of the resource plans will result in an overdependence

---

[7] In addition, there are many circumstances in which "doing nothing," *i.e.*, considering no other resource option, is not a viable option. An example of such a circumstance is our hypothetical utility's need to add new resources in Current Year + 5 to maintain system reliability. Our utility did not have the option of "doing nothing."

on one type of fuel. Therefore, there are no significant differences among the five resource plans in regard to system fuel-based issues regarding the volatility of fuel costs or the reliability of fuel supply. Consequently, all five resource plans are acceptable in this regard.

5. Two points are also evident in regard to a system air emission's (SO_2, NO_X, and CO_2) perspective for our utility. First, the projected compliance costs of system air emissions have already been accounted for in the economic analyses. Second, no one resource plan emerged as the single resource plan with the lowest emissions for all three types of emissions. All five resource plans had the same SO_2 emission totals. The Supply Only Resource Plan 1 (CC) and With DSM Resource Plan 2 are projected to be the resource plan with the lowest NO_X system emissions. And the With DSM Resource Plan 2 is projected to be the resource plan with the lowest CO_2 system emissions. However, there was less than a 1% difference in the system emission totals among any of the five resource plans regarding SO_2 and CO_2 emissions. (In regard to NO_X emission, the largest differential between any two resource plans was less than 1.2%).

Our utility further summarizes economic and non-economic information for the five resource plans in tabular form in Table 7.9.

TABLE 7.9
Summary of Rankings of Resource Plans

Resource Plan	System Reliability (Meeting Reserve Margin Criterion by the Decision Year)	System Economics (Levelized Electric Rate)	System Fuel Diversity (Fuel Use Profile)	System Air Emissions		
				SO_2	NO_X	CO_2
Supply Only Plan 1 (CC)	Acceptable	2nd	Acceptable	1st-tied	1st-tied	3rd-tied
Supply Only Plan 2 (CT)	Acceptable	4th	Acceptable	1st-tied	5th	5th
Supply Only Plan 3 (PV)	Acceptable	3rd	Acceptable	1st-tied	3rd	2nd
With DSM Plan 1	Acceptable	1st*	Acceptable	1st-tied	4th	3rd-tied
With DSM Plan 2	Acceptable	5th	Acceptable	1st-tied	1st-tied	1st

Note: 1st = best in category; 5th = worst in category
* Time for the With DSM Plan 1 to become the most economic plan is acceptable

Based on this information, our utility system selects the With DSM Resource Plan 1 as the best resource option with which it will meet its resource needs 5 years in the future. This resource plan is acceptable in regard to both system reliability and system fuel use profile, places the utility in a reasonable position (*i.e.*, system air emission rankings of 1st-tied, 4th-tied, and 3rd-tied) in case environmental regulations were to unexpectedly tighten, and, most importantly of all, will accomplish all of the aforementioned items while serving our utility's customers with the lowest levelized system average electric rates.

As a backup resource plan, our utility would select the Supply Only Resource Plan 1 (CC). It provides a bit more protection in regard to the possibility of tightening in environmental regulations (*i.e.*, it has rankings of 1st-tied, 1st-tied, and 3rd-tied in regard to system air emissions) and does so with the second lowest levelized system average electric rates for the utility system's customers.[8]

The remaining three resource plans, the Supply Only Resource Plan 2 (CT), Supply Only Resource Plan 3 (PV), and the With DSM Resource Plan 2, are clearly less desirable choices from an economics-only perspective, or from a combined economic and system emission perspective. The utility system clearly has two better overall choices than any of these three remaining resource plans.

Therefore, our utility system has completed its resource planning work. It has conducted a comprehensive integrated resource planning (IRP) analysis and it has made its selection of how it will meet its future resource needs. You may well be asking: "What is left for us to discuss?"

I'm glad you asked. We will next discuss in qualitative terms (yes, that's correct—no more numbers for a while—you are welcome) various issues that could have complicated the resource planning analyses our hypothetical utility system has just completed, and which will complicate the resource planning efforts of many utilities in the future.

---

[8] If such a tightening of environmental regulations were to occur, it is possible that the utility would have to take certain actions (install more pollution control equipment, perform fuel switching, etc.) for any of the five plans. The economic advantage of the With DSM Resource Plan 1, and to a lesser degree the Supply Only Resource Plan 1 (CC), allows our utility more "room" or "cushion" economically to make such changes because it would be starting from the position of the lowest electric rates for its customers.

8 Are We Done Yet?
Other Factors That Can (and Will) Complicate Resource Planning Analyses

CONSTRAINTS ON SOLUTIONS: SIX EXAMPLES

We ended the previous chapter by stating that we would now discuss various issues that could have complicated the resource planning analyses our hypothetical utility system just completed (and which will complicate the resource planning efforts of many utilities in the future). In fact, the issues we will discuss in this chapter are routinely addressed by many utilities today in their resource planning. Not all issues apply to every utility, but at least some of these issues have been faced, to one degree or another, by most electric utilities.

From the perspective of a utility resource planner, these issues can be thought of as "constraints" that must be dealt with as the resource planner attempts to determine through his or her analyses a solution to the question of which resource option(s) is best for his or her specific utility. There are a number of such constraints, or types of constraints, that a particular utility may face. For purposes of our discussion, we will focus on six specific constraints or types of constraints. We shall first place each of these six constraints into one of three general categories.

These three general categories of constraints are ones we label as: (i) "absolute" constraints (*i.e.*, types of constraints that have existed for many years and over which the utility has little/no direct control); (ii) legislative/regulatory-imposed constraints; and (iii) utility-imposed constraints. We shall first list two examples of constraints that could be placed into the three categories. Then we shall examine each type of constraint in more depth.

The six constraints that we shall discuss, and the three categories into which we shall put these constraints, are as follows:

"Absolute" constraints

1. Siting/Geographic constraints
2. Tightening of environmental regulations

Legislative/Regulatory-imposed constraints

3. Imposition of "standards"/Quotas for specific resource options
4. Prohibition of specific resource options

Utility-imposed constraints
   5. System reliability constraints on specific resource options
   6. Load shape constraints

EXAMPLES OF "ABSOLUTE" CONSTRAINTS

SITING/GEOGRAPHIC CONSTRAINTS

This first type of constraint, siting/geographic constraints, almost exclusively affects Supply options. This type of constraint has been a factor in utility resource planning almost since the first day that the electric utility industry emerged and the industry began to build new generating units.

A number of considerations come into play regarding where new generating units of various types could be sited in a utility's service territory. A partial list of the considerations that come into play as a utility examines a potential site for a new generating unit includes the following:

- Proximity of the potential site to existing transmission lines, and
- Proximity of the potential site to the utility's "load center."[1]

The next three considerations are "as applicable" considerations. They typically apply for conventional fossil-fueled generation options (and the first applicable consideration listed below would also apply to the siting of new nuclear units) but do not apply for other generation options such as solar or wind.

- Proximity of the site to adequate water resources for the operation of the generating unit;
- Proximity of the site to existing natural gas pipelines, railroad lines, navigable waters/ports, etc., in regard to fuel transportation; and
- Proximity of the site to non-attainment air emission zones.

The first of these five considerations, the proximity of the site to existing transmission lines, takes into account the fact that the energy produced by any new generating unit must be carried over electrical lines away from the generating unit site before the electricity can be delivered to customers. All else equal, a potential new-generation site is more attractive from an economic perspective if the site is close to existing transmission lines.

For example, assume that a utility has two potential sites. Site A is 10 miles from an existing transmission line and Site B is 25 miles from the same existing transmission line. In either case, a new transmission line will need to be built that connects the new generating site with the existing transmission line. Assuming a rough cost of $1 million to $2 million per mile for building a new transmission line that will connect with the existing transmission line, the utility's costs to build this new

---

[1] The term "load center" refers to a geographical region of a utility's service territory or area in which a large portion of the utility's entire electric load exists. Such a region may also be referred to as a "load pocket."

transmission line will be \$10 million to \$20 million if Site A is chosen or will be \$25 million to \$50 million if Site B is chosen. Assuming all else equal, the utility will likely choose Site A for the new generating unit.

This consideration is one that is common to all types of generating units. However, it is one that can be especially relevant in regard to certain renewable energy generators. Wind energy generators are a good example because the geographic region of higher winds is often remote from the utility's load center (perhaps because people found it less desirable to live in, much less play golf in, areas with frequent high winds). The same may also apply to solar resource options due to the need for large tracts of land needed for large arrays of solar.[2] The location of these large land tracts may be miles from existing transmission lines that can handle the additional flow of the energy from the solar facility.

Thus, new transmission lines that cover a great distance may need to be built in order to access the wind and/or solar resource and bring this resource's electrical energy output to existing transmission lines.

The second of these five considerations, the proximity of the site to the location of the utility's load center, addresses how far the new generation site being considered is from where a significant portion of the utility's customers are located. The greater the distance is between the new generation site and the load center, the further the energy produced by the new generating plant has to travel before it can be delivered to customers.

This distance to the load center may affect the cost to build the new transmission lines (as just discussed), and this distance also will likely affect the magnitude of electrical losses that occur as electricity travels along the electrical lines from the generating plant to the customers.[3] The greater these losses are, the less energy the generating unit is able to actually deliver to the utility's customers.

The third of these five considerations is the proximity of the site to adequate water resources. Sufficient quantity and quality of water must be available for operating a generating unit that requires water for its operation (such as is the case with CC or nuclear units, for example). These requirements may vary considerably depending upon the type of generating unit in question.

The fourth of these five considerations, the proximity of the site to existing natural gas pipelines, railroad lines, navigable waters/ports, etc., addresses the fact that most types of new generating units (except for renewable energy facilities such as solar and wind, or nuclear) will typically require on-going delivery of fuel to the plant site. Depending on the type of generating unit being considered, the fuel may be delivered by a number of delivery methods such as by pipeline, rail, truck, or barge/ship.

---

[2] We will take a look at the amount of land that may be needed for PV for our hypothetical electric utility in Part II of this book.

[3] These losses are typically referred to as "line losses." In a very rough sense, one can think of them as a type of electrical "friction" loss as electricity travels a long distance through electrical wires. The amount of the line losses will vary from one utility to another, but losses in the ballpark of 5% are fairly typical. In general, the longer that electricity must travel, the greater the line losses are. Obviously, the lower the line loss value, the more efficient the utility system is and the lower its costs (assuming all else equal).

The lack of existing delivery systems or avenues near a potential site may eliminate one or more type of generating units for that site because the fuel which that type of plant will require cannot be delivered by existing infrastructure (pipelines, rail lines, etc.), and because the cost to develop new delivery systems to serve the site may be prohibitive. In certain cases, the potential plant site may be completely eliminated because no type of fuel can be delivered at a reasonable cost.

The last of these five considerations, the proximity of the site to non-attainment air emission zones, recognizes that certain federal and/or state regulations have designated certain geographic areas as zones in which air emission standards either are not being met or are in danger of not being met. Consequently, types of new generating units that produce air emissions during its operation, such as CC and CT units, may not be able to be permitted for a potential site if the projected air emission output from that unit's operation will result in a failure to meet air emission standards in that geographic zone. (However, other types of new generating units may still be acceptable for the site.)

How do utility resource planners attempt to address these considerations? For illustrative purposes, think of the general approach as one in which a utility starts with computer-generated maps of their service territory. One map may show the general geography of the area, including rivers and roads. The next map may show major population areas (which can be thought of as the utility's electrical load centers) and show areas of potentially available unoccupied land. A subsequent map may show the location of existing transmission lines. Another map may show the location of existing natural gas pipelines etc. This process continues until maps that present all of the relevant considerations for the utility in regard to siting new generating units have been obtained.

Then these maps are (figuratively) laid on top of each other, or otherwise combined, to develop a single composite map that includes all of this information. From this composite map, the utility can begin to locate promising potential sites for new generating units of various types. Once these potential sites are identified, the utility can begin to develop estimates for each site of the costs to build and operate types of generating units. In addition, the utility will develop estimates of the time it will take to obtain all of the necessary permits that will be needed for the new units at each site.

In this way, the utility identifies the types of new generating units that can be built at specific sites, the costs associated with each site, and the projected time it will take to permit and construct the facility.

POTENTIAL TIGHTENING OF ENVIRONMENTAL REGULATIONS[4]

The next type of "Absolute" constraint we will discuss is a tightening of environmental regulations. At the end of Chapter 7, our utility system considered the projected

---

[4] This type of constraint could have been placed in the grouping that will be discussed next: the "legislative/regulatory-imposed" constraint grouping. I have chosen not to do so because I view the tightening of environmental regulations constraint as being primarily based on attempts to reduce environmental impacts from utility operation of existing generating units. Conversely, the two constraints that will be discussed in the "legislative/regulatory-imposed" grouping are ones that I view as being primarily based on forcing utility resource planning into selecting certain types of new resource options that are "favored" by the legislature or regulators.

levels of system air emissions for each resource plan in the course of making a decision regarding which resource option was the best selection for meeting its resource needs. This examination of system air emissions is, in part, a way to judge the utility's risk level in light of a potential tightening of environmental regulations. Such a tightening of environmental regulations will likely alter the operation of the utility system as a whole. The costs of that altered operation are passed on to the utility's customers through higher electric rates (just as are all other prudently incurred costs of utility system operation).

By "tightening" of environmental regulations, we refer both to the imposition of more restrictive standards for air emissions currently being regulated and to new regulations of emissions (or other environmental concerns such as water usage) not currently being regulated.

Although the possibility of potential tightening of environmental regulations may most directly affect the operation of a utility's existing fleet of generating units, this possibility is also typically considered by utilities when they make a resource option decision. Resource options that the utility is now considering to meet a future resource need will not be placed into service immediately but will become operational in a future year. (Recall that our utility system is placing a resource option in service in Current Year + 5, or 5 years in the future from its Current Year.) In addition, most resource options will have an operating life of many years. For example, a new generating unit will typically have an expected operational life of at least 25 years.

Therefore, a resource decision that is made today may not begin to impact the utility system for a number of years due to the length of time it takes to construct or acquire the resource option, but the decision will certainly impact the utility system for a number of years or decades after the resource is constructed or acquired.

Utilities must make decisions about resource options that will operate in future years, and they must do so recognizing that there is uncertainty about potential environmental regulations that the utility, and its generating units, will face. Utilities are also mindful that the overall trend of environmental regulations is generally one of continually tightening regulations.

Of particular concern to utility resource planning is the potential emergence of new environmental regulations that address an entirely new type of emission that has not been regulated in the past. Relatively speaking, utility resource planners find that an incremental tightening of an air emission (for example, SO_2 or NO_X) that is already being regulated <u>may</u> be relatively easy to address. This is true, in part, because manufacturers of generating equipment, pollution-control equipment, etc. anticipate that such tightening of currently regulated emissions will occur and they guide their development work accordingly.[5]

---

[5] This is almost a "chicken or the egg" situation. Environmental regulations frequently become tighter as soon as it becomes clear that emission control technology has advanced to the point where it is now technologically possible to further lower emissions. Thus, technology advancements and tighter regulations often move in lockstep fashion. One may wonder if decisions on whether environmental regulations should be tightened are based less on analyses of whether the regulations *should* be tightened, than on whether it is technologically *possible* to meet tighter standards regardless of the cost impacts.

However, the emergence of new, or even of potentially new, regulations for air emissions (or other concerns) that are not currently regulated is more problematic in utility resource planning. This situation greatly increases the uncertainty that utility resource planners must consider or address in two ways.

First, there is uncertainty as to whether there will even be regulation of this type and, if so, when the regulations will be in place. Second, there is additional uncertainty regarding the eventual costs of complying with any such regulation. Only when regulations are actually in place will this uncertainty be entirely removed or minimized.

A good example of this is the potential regulation of greenhouse gas (GHG) emissions, especially CO_2. At the time the first edition of this book was published in 2012, there had been no lasting federal regulation of GHG emissions at the utility system level in the United States, despite the fact that the possibility of GHG emission regulation has been seriously discussed for at least three decades.[6] (However, the federal government did pass legislation in 2022 as the second edition of this book was written that seeks to lower methane emissions by placing a fee/tax on certain levels of methane emissions. This 2022 legislation will be discussed in Part II of this book.)

As a result, utility resource planners and utility regulators have wrestled with whether, and how, to address potential GHG emission regulation as decisions are made about resource options. A frequently used approach is to consider various scenarios of compliance costs with GHG emissions, often represented as CO_2 compliance costs, which we used in the economic analyses in Chapters 5, 6, and 7.

Such a scenario approach can certainly utilize a wide range of GHG compliance costs in the analyses of resource options. Typically, one of those scenarios assumes a compliance cost of zero denoting no future regulation of the environmental concern being considered. If the utility finds that one resource option emerges from its analyses as the best resource option under all, or most, of the scenarios, including the zero-cost scenario, there should be little debate regarding the selection of this resource option as the best selection for that utility.

However, a utility may find that Resource Option A is the best solution for a scenario with high GHG compliance cost projections, Resource Option B is the best solution for other scenarios with medium GHG compliance cost projections, and Resource Option C is the best solution for still other scenarios with lower or zero GHG compliance cost projections. In such a situation, the utility is often faced with a dilemma regarding which scenario the utility believes is the "most likely" scenario before the utility can select a resource option as the best solution for its future resource needs.

Then the question becomes whether the GHG compliance cost scenario the utility selects as the most likely scenario is also perceived by the utility's regulators as the

---

[6] To further point out the problems which utility resource planners face regarding uncertainty with potential environmental regulation, during the 1970s there was considerable concern over global *cooling*. (For example, *Time* magazine published an article in 1974 titled "Another Ice Age?" that discussed the potential catastrophic effects of such an occurrence.) The possibility of new environmental regulations to address global cooling began to be discussed and planned increases of CO_2 emissions were considered one possible way to address this potential problem.

"most likely" scenario. The two parties can, and often do, reach different conclusions. In such a case, there will likely be disagreement regarding which resource option is in the best interest of the utility's customers. And, unfortunately, the answer as to whether the best overall resource option was chosen is often not known for many years after the resource option is added to the utility system. These decisions can only be made at the time the decision is needed using the best information available at that time, plus some judgment.

In future years, with the benefit of hindsight, once the question of potential environmental regulation for a specific type of emission is finally settled, a review of that particular resource option decision may show that the utility's customers were harmed by the decision. The customers may be harmed in several ways. Customers can be harmed if a resource option was chosen based on a scenario of passage of a specific environmental regulation, or of a specific projected level of environmental regulation compliance costs, but the regulation never occurred or the actual level of environmental compliance costs is considerably different (higher or lower) than what was projected. Such an outcome, realized only with the benefit of hindsight, could show that the selected resource option was not the best economic choice. As a consequence, the selected resource option is likely to have unnecessarily increased electric rates for the utility's customers.

The potential for tightening of environmental regulations has been problematic for utility resource planners for many years. However, this problem may be minimized or even eliminated by recent developments regarding renewable resource options. This will be discussed in Part II of this book.

EXAMPLES OF LEGISLATIVE/REGULATORY-IMPOSED CONSTRAINTS

"STANDARDS"/(QUOTAS) FOR SPECIFIC TYPES OF RESOURCE OPTIONS

We now turn our attention to what I have labeled as "legislative/regulatory-imposed types of constraints" and discuss the third and fourth constraints. The third constraint is the use of so-called standards for specific types of resource options. These standards are actually quotas for selected types of resource options.[7] At the time this book is written, such "standards"/quotas have been imposed at various places in the United States for one or both of two types of resource options: (i) renewable energy generation options and (ii) demand side management (DSM) options. We will first take a look at "standards"/quotas for renewable energy generation options.

The use of "standards"/quotas for renewable energy options in utility resource planning has been commonly referred to as "renewable portfolio standards" or RPS. The basic rationale behind the introduction of RPS is that, at the time an RPS was introduced, renewable energy-based resource options had not typically been selected in great volume by utilities as they determine what the best resource options are for

---

[7] Because these "standards" are actually quotas, I will refer to them as "standards"/quotas to remind the readers that we are discussing quotas imposed by governmental actions.

their individual utility systems.[8] Proponents of renewable energy resource options offered reasons why these options should be utilized more. Essentially, they argued that greater implementation of renewable energy options will reduce the United States use of fossil fuels to produce electricity as well as reduce air emissions.

As this book is written, roughly half of the states in the United States have some form of RPS regulation in effect (and RPS regulation has also been considered by the federal government at various points in time). A wave of state RPS regulations appeared in the middle of the 2000 to 2010 time period after a national conference of state governors was held that focused on energy issues as a primary topic. Almost immediately, numerous states began to issue RPS regulations.

Many of these RPS regulations have been imposed without the benefit of integrated resource planning (IRP)-type analyses to determine what the full range of projected impacts for the utilities (and, more importantly, their customers) would really be. Therefore, the potential benefits (and, to some extent, costs of complying with the regulations) that accompanied the introduction of RPS regulations were often poorly informed (at best) when the initial RPS regulations were issued.

The individual state RPS regulations largely took a form that suggests that the regulations were designed mostly with a catchy marketing slogan in mind. For example, a number of states issued regulations that touted "15% by 15" (*i.e.*, 15% of the annual energy produced by electric utilities will be produced from renewable energy sources by 2015) or "20% by 20" (*i.e.*, 20% of the annual energy produced by electric utilities will be produced from renewable energy sources by 2020—you get the picture). Such slogans were indeed catchy and looked good in the press.

In addition, the RPS "standards"/quotas frequently have more than one layer of quotas. For example, an RPS quota may have called for 15% of the utility's annual energy production to be produced by renewable energy sources by 2015. In addition, the RPS regulation may also have required that, of the 15% amount, one-third (or 5% of the annual energy production of the utility) is to be produced by a specific type (or combination) of renewable energy source such as wind energy, solar thermal, and/or PV. Therefore, apparently operating on the theory that "more is better," there were often quotas-within-quotas set into RPS regulations.

However, just as with the case regarding whether an RPS regulation should even be implemented, relatively little rigorous analyses were performed to determine the "best" value to set as a quotas-within-quotas for specific renewable energy sources' contribution to annual energy production. Such analyses may have proved problematic in regard to the proponents of these quotas actually justifying setting any quota when viewed from the perspective of a utility's customers and the utility's electric rates.

Again, recall the analyses we conducted for our hypothetical utility system. Our analyses of a specific PV option for our utility system showed that adding this renewable energy option to our utility system was not the best economic option for this

---

[8] The result of the resource option evaluation work for our utility system in earlier chapters provides some insight into why certain renewable energy resource options had not typically been selected by utilities in their resource planning work until the latter part of the 2010–2019 decade. The less-than-100% firm capacity value for most renewable options, combined with the then-much-higher installed costs than what we have assumed for solar in Chapter 6, resulted in relatively few renewable generation options being selected. As we shall discuss in Part II of this book, that situation has certainly changed.

utility. Higher electric rates for our utility's customers would be the outcome if this particular PV option was selected based on the assumptions used in the analyses.

The legislators and regulators across the United States who have imposed RPS regulations in various states recognized this reality of resulting higher electric rates. Consequently, they often sought to protect electric customers, to a degree, by including in the RPS regulation a form of a "safeguard" against electric rates that are "too much higher." This safeguard usually takes the form of a "cap" on annual expenditures for renewable energy generation. Such a cap can take the form of an edict that utilities can (or should) stop spending for renewable energy generation in a given year once these expenditures equal a set percentage (such as 2% or 5%) of the utility's annual revenue requirements.[9]

The theory is that the use of an annual cap on expenditures, with an RPS "standard"/quota, will allow the implementation of renewable energy sources with an "acceptable level" of economic harm to electric utility customers (*i.e.*, electric rates will definitely increase but will not increase beyond a certain level determined by the legislators/regulators). However, the reality of combining an expenditure cap with an RPS "standard"/quota is that one of two things will likely happen.

First, if the expenditure cap does a reasonably adequate job of protecting utility customers from higher electric rates, the annual amount of energy produced from renewable energy sources may be less (perhaps much less) than the amount called for by the RPS. In other words, the RPS "standard"/quota may not be met. Second, the only way that the amount of energy called for by the quota value will be met may be to relax the expenditure cap safeguard. This would result in even greater costs being passed on to utility customers through electric rates that are higher than called for by the cap. Stated another way, the cap is relaxed or ignored.

In either case, at least one of the two objectives of the RPS "standard"/quota and expenditure cap approach may not be met. One either implements less renewable energy than is called for by the RPS "standard"/quota (but the utility's electric rates still increase up to the cap level), or one meets the renewable energy "standard"/ quota and the utility's electric rates increase beyond the amount called for in the original expenditure cap.

When RPS "standards"/quotas are looked at in this way, such "standards"/quotas do seem to be a strange departure from the IRP concept for electric utilities. As previously mentioned in Chapter 3, a key IRP principle is to ensure that all resource options *compete* on a level playing field in order to determine which resource option(s) are best for an individual utility.

By comparison, the use of RPS "standards"/quotas *seeks to avoid competition* and force the selection of one favored type of resource option: renewable energy options. This is clearly a departure from, or an abandonment of, IRP principles.

We earlier mentioned that another type of resource option, DSM, has been the recipient of a "standard"/quota approach imposed by legislators and/or regulators. This approach was first introduced in the early 1990s as a way to force certain

---

[9] Although not often stated in these terms (probably due to public relations concerns), this edict to spend money on renewable energy options up to the expenditure "cap" is an edict to raise electric rates for utility customers by roughly the same percentage as the cap percentage.

utilities to seriously consider (if utilities were not already doing so) DSM in their resource planning work. This approach was "side-lined" to an extent when a number of states and utilities moved away from a traditional, regulated electric utility structure to a competitive, "unregulated" market structure. As part of this move, DSM became less of a factor for utilities that were now in this new market structure.

Other states and utilities retained the traditional, regulated utility market structure. If such a state or utility continued with an IRP approach to utility planning, then DSM was automatically considered on a level playing field with Supply options. Most importantly, if the state adopted DSM "standards"/quotas but set those DSM "standards"/quotas based solely on the outcome of the utility's IRP work, the introduction of DSM "standards"/quotas caused little (if any) economic harm to the utility's customers. In other words, this logical type of DSM "standard"/quota basically mandates the utility to use its IRP process to (i) first determine how much DSM might be needed to fully meet the utility's projected future resource needs and (ii) then determine how much of that DSM amount was actually cost-effective to utilize.

Such an approach ensures that the utility utilizes only the amount of resource options (DSM and/or Supply) that is actually needed to meet future resource needs and is cost-effective. Therefore, the analyses we conducted in previous chapters for our hypothetical utility system would have been the same regardless of whether our utility system was facing a state-imposed DSM "standard"/quota or not, as long as the DSM "standard"/quota was based on the utility's IRP process.

For example, our utility system's projected resource needs were first identified (100 MW by Current Year + 5 if the resource need is met solely by DSM), and then DSM options (along with Supply options) were analyzed to determine how much DSM was cost-effective to add. In the analyses, 100 MW of DSM Option 1 was selected as the most cost-effective resource option for the utility's customers. As long as our utility's regulators set a DSM "standard"/quota based on our utility's IRP analyses, our utility's customers will not be harmed.

However, if another DSM option had been mandated, and/or an amount of DSM larger than 100 MW by Current Year + 5 had been mandated, by a different type of DSM "standard"/quota, then our utility's customers would have been harmed due to the resulting higher electric rates.

At the time this book is written, there have been concerted efforts in a number of states to alter an IRP-based approach to DSM to an approach based on "standards"/ quotas for DSM. These efforts seek to replace this IRP- or competition-based approach to DSM analysis and regulation with a set of largely arbitrary percentage values similar to an RPS "standard"/quota approach. The RPS "standard"/quota approach typically takes the form of trying to establish an arbitrary percentage of the utility's total annual energy production that must be produced by renewable energy sources.

The efforts to change the DSM "standard"/quota approach frequently involve an attempt to move in a somewhat similar, but opposite direction.[10] This type of DSM "standard"/quota approach typically requires that the utility *reduce* its annual energy

---

[10] Such DSM-based quotas are often referred to as DSM "goals" or "energy efficiency resource standards (EERS)."

production by an arbitrary percentage (such as 2%) instead of *producing* an arbitrary percentage of energy as with RPS regulations. This would result in the amount of energy the utility sold to its customers to be lowered, year after year, by an arbitrary amount through the implementation of DSM.

Consequently, the type of DSM options that would be most favored under this type of DSM quota would be those DSM options that have the highest kWh reduction characteristics. In other words, such a DSM "standard"/quota system would move a utility system to implement DSM options that are more similar to DSM Option 2 that was analyzed in Chapters 6 and 7 for our utility system rather than DSM Option 1.

However, as you may recall, DSM Option 2 was demonstrated *not* to be one of the better resource options that our utility system could have selected. In fact, this resource option was the <u>worst</u> resource options in regard to electric rates for our utility's customers.

Furthermore, such a DSM "standard"/quota system could require that a much greater amount of DSM options similar to DSM Option 2 would be implemented *than what is actually needed to meet the utility's projected future resource needs*. This is because this type of DSM "standard"/quota is focused solely on energy (MWh) reduction, not on the demand (MW) reduction aspect of DSM that actually avoids the need for new power plants. This type of DSM "standard"/quota that focuses on energy (MWh) reduction can end up with an associated amount of demand (MW) reduction that is greater than what is needed to meet the utility's resource needs. This greater-than-needed amount of a less desirable type of DSM options results in *even greater increases* in the utility electric rates.

This prescriptive approach to DSM options is, like a prescriptive RPS approach to renewable energy options, also a departure from, or abandonment of, IRP's key principle of requiring all resource options to *compete* with each other to earn a role in the utility's resource plan. Instead, this prescriptive approach ignores this underlying IRP principle and mandates that a certain (usually arbitrarily selected) amount of DSM is to be included in the resource plan.

Later, at the end of this chapter, we will return to these RPS and DSM "standards"/quota constraints, and other constraints that we will soon address, to further discuss the impacts that these constraints typically have on a utility system and its customers. In the meantime, we will now move on to several other constraints that affect utility resource planning.

PROHIBITION OF SPECIFIC RESOURCE OPTIONS

The other type of the "legislative/regulatory-imposed constraints" we will discuss involves either a direct, or de facto, ban on specific types of resource options. State legislators or regulators may impose an outright ban on building a certain type(s) of generating unit. Another form of effective prohibition can occur absent such direct legislation/regulations. In this case, utilities may find that one or more permitting agencies in their state consistently rule against applications to authorize the building of a new generating unit of a particular type.

The types of resource options in question are typically Supply options and, initially, were Supply options that were projected to otherwise operate with high capacity factors (*i.e.*, baseload generating units). The type of generating unit that was initially most affected by some form of a ban is coal units.

The state of Florida's experience in the middle of the 2001 to 2010 decade is a good example of this. Growing concerns over the increasing dependence of the state on natural gas as a utility fuel, caused by the introduction of a number of new natural gas-fueled generating units in the late 1990s and early 2000s, resulted in a clear message being sent from the Florida Public Service Commission (FPSC) to the state's utilities to begin to add new non-gas-fueled generating units. The utilities had also recognized that they were becoming quite dependent upon natural gas and they were already evaluating non-gas-fueled resource options on their own.

New nuclear units were then estimated to take about a decade to permit and build.[11] Facing steady demand growth that resulted in projections of significant future resource needs in much less time than the 10 years or more it would take to add new nuclear units, Florida utilities turned their attention to advanced coal generation technologies. By 2006/2007, several electric utilities in Florida petitioned the FPSC for authorization to build advanced technology coal units.

However, by that time, Florida had a new governor. The new governor was concerned about GHG emissions to the point that he issued an Executive Order that laid out goals for reducing state GHG emissions to certain levels by 2017. Also about that time, a number of analyses emerged from various sources around the country that showed a wide range of compliance costs for GHG emissions that might occur if federal GHG emission regulation were to occur. A projection of high potential compliance costs for GHG emissions would be detrimental to coal units because the CO_2 emission rate for coal units is higher than the CO_2 emission rate for any other type of new Supply option.

Therefore, despite the previous concerns over Florida's high dependence on natural gas, by 2006/2007 the political winds in Florida had changed regarding the desirability of adding new coal-fueled generating units. As a consequence, a couple of Florida utilities' petitions to the FPSC for authorization to build new coal units were denied due, in large part, to uncertainty over potential future GHG compliance costs.[12] In addition, one utility's petition for a new coal unit that had previously been approved by the FPSC was effectively blocked in the environmental permitting process. These actions constituted a de facto ban on new coal units in Florida that continues as this book is written.[13]

---

[11] This time estimate proved to be wildly optimistic. In 2009, FPL applied for approval from the Nuclear Regulatory Commission to build and operate two new nuclear units at the site (the Turkey Point site, south of Miami) where two existing nuclear units were currently operating. The license was eventually granted in 2018, or approximately 9 years from the date FPL applied for approval. (Note that it took about 2 years less time for the United States to announce a goal of putting a man on the moon [1962] and then put a man on the moon [1969], than it took the federal government to simply make a decision on FPL's nuclear license request.)

[12] In fairness to the FPSC's decision, the utility analyses presented scenarios that showed that if GHG regulations were instituted, and if the costs of complying with these regulations were high, then the advanced technology coal units were not projected to be cost-effective.

[13] As a consequence of this, more gas-fueled generating units have been added and the state's dependence upon natural gas has increased. However, due largely to hydraulic fracturing drilling techniques, the cost of natural gas has declined significantly and gas availability has increased significantly.

Examples of direct prohibition of specific types of generating units, including non-coal units, also exist. One example occurred in 1978 via a federal ban on the use of natural gas in all new generating units (other than peaking units). This ban was put in place primarily due to concerns over the projected availability of domestic natural gas supplies. (This federal ban was subsequently lifted as the projections of natural gas availability changed to show that significant quantities of natural gas were available.)

Other examples of direct prohibition of specific types of generating units have focused on new nuclear units. For example, the states of Minnesota and Wisconsin effectively banned serious consideration of new nuclear units in the past. More recently, certain states have effectively banned the construction of new generating units that utilize any type of fossil fuel. This has often been done in conjunction with RPS-type mandates to reach certain percentages, up to 100%, by a certain future year.[14]

We shall return to this prohibition type of constraint at the end of this chapter when we shall further discuss the impacts of various constraints on utility resource planning and utility customers.

EXAMPLES OF UTILITY-IMPOSED CONSTRAINTS

SYSTEM RELIABILITY CONSTRAINTS

We just looked at a pair of constraints that have been imposed by governmental actions and which require that a utility react to the constraints. We now turn our attention to a pair of constraints that may be imposed by utilities themselves.

The first of these constraints we examine is a system reliability constraint that the utility may impose upon itself in its resource planning work. This system reliability constraint also affects specific types of resource options, but it does so indirectly. For example, while the previously discussed prohibition constraint typically affects Supply options, this self-imposed reliability constraint could affect DSM (and even renewable) options. We will focus this discussion on DSM options.

Recall from the discussion in Chapter 3 that utilities perform reliability analyses to ensure that they are able to continue to supply their customers with reliable electric service. As part of these reliability analyses, DSM options deserve special attention because DSM options have certain characteristics that differ from the characteristics of conventional fossil-fueled, nuclear-fueled, and renewable energy-based Supply options.

From a system reliability perspective, the key characteristic of DSM is the voluntary nature of most DSM options.

THE VOLUNTARY NATURE OF DSM OPTIONS

All DSM options are "voluntary" in the sense that customers must voluntarily sign up to participate in the utility's DSM program. In addition, there are at least two other voluntary aspects to specific DSM options.

---

[14] One could consider a mandate to generate 100% of a state's electricity from renewable energy sources to be the ultimate legislative/regulatory mandate and, if so, one would be correct.

First, certain DSM options require a participating customer to voluntarily sign up for either a new time-differentiated (or peak load-differentiated) electric rate or to give the utility the ability to temporarily control the operation of certain electrical equipment on the customer's premises. These types of DSM programs, referred to as load management or demand response DSM programs, were briefly discussed earlier in Chapter 6.

For purposes of this discussion, the key point is that these DSM load management programs require the *continued* voluntary participation of utility customers that have signed up for these programs. Participating customers can usually choose to drop out of these programs (particularly for programs that address residential customers) with very little notice (perhaps with only a week's notice) to the utilities. Therefore, there is always a risk that a significant number of participating customers may choose to cease participating in the utility's DSM program and leave the utility with diminished resources with which to meet electrical load, particularly during peak load hours.

As we also discussed in Chapter 6, the other basic type of DSM program is an energy conservation (or energy efficiency) program. For this type of DSM program, a participating customer is typically paid an incentive to select a more efficient appliance, a higher level of insulation, etc. The more efficient appliance, or higher level of insulation, will then be in place for a number of years. Therefore, in contrast to a load management program in which a customer may choose at any time to cease participating in the program, once a more efficient appliance, etc., is installed in a building, it will likely be there for many years without the participating customer having to make a choice about whether to participate in the same utility program.

However, even with this type of DSM program, there is another aspect of the program to consider that can, in a sense, be considered a voluntary aspect. This voluntary aspect is that customers can choose to alter their electrical usage patterns in ways that can result in thwarting the projected kW and kWh reductions that were projected for the utility DSM program.

For example, customers can decide to lower their air conditioning thermostat in summer, raise their thermostat in winter, and/or raise their water heater thermostat settings. Such actions may be the result of these customers using some/all of the savings in their monthly bills from participating in the DSM program to "purchase more comfort." Although these actions are completely understandable, the effects can become problematic if they occur on very hot or very cold days when system electrical loads are the highest because the actions increase peak-hour electrical demand.

Customers could also use the savings in their electric bill resulting from participating in the DSM program to purchase other goods/services that require electricity (such as new computers, and televisions). Both of these examples are typically discussed as "rebound" effects. While there is little question that rebound effects exist, the magnitude of rebound impacts are less well known.

The key point is that, because the projected kW and kWh reductions from conservation/energy efficiency types of DSM programs are subject to decisions and actions by customers over which the utility has no control, the projected impacts of DSM options on the utility system are inherently less certain than are the projected impacts of Supply options. The same is true for load management type DSM options because participating customers can choose to drop out of these programs with little notice given.[15]

These voluntary aspects of both types of DSM programs can lead to problems if a utility places too much reliance on DSM options to meet its system reliability criterion.

For an example of this, we again return to Florida, but this time we take a look at what happened in the late 1990s. The second largest electrical utility in the state, Florida Power Corporation (now known as Duke Energy Florida, or DEF), was enjoying a very high level of success in regard to signing up its residential customers for a load control program. This very successful program offered monthly billing credits to participants for allowing the utility to remotely control the operation of certain home equipment such as air conditioners, space heaters, and water heaters. By the late 1990s, Florida Power Corporation had signed up approximately 40% of its residential customers for this program.

At that time, the utility utilized a 15% reserve margin as one of its reliability criteria. The huge success of its residential load control program resulted in the utility deferring/avoiding the need to construct a number of new power plants. As a consequence, this DSM program accounted for a substantial portion of the utility's ability to meet its 15% reserve margin criterion.

However, during one prolonged hot spell in the summer of 1998, Florida Power Corporation needed to control the operation of participating customers' equipment much more frequently than had been the case in prior years. In particular, the customers participating in this program had their air conditioners controlled several times a week, for a number of weeks in a row.

Load control programs that involve utility control of air conditioners and/or space heating systems frequently result in participating customers facing a type of "balancing" question: *"Is the lower bill (due to monthly billing credits) worth brief periods of a slightly warmer (in summer) or cooler (in winter) home?"* Florida Power Corporation's residential load control program in the late 1990s was no exception.

Unfortunately, during that very hot 1998 summer, the tolerance level for warmer temperatures in the participating customers' homes increasingly diminished as Florida Power Corporation used its load control capability more frequently. As a result, large numbers of participating customers (approximately 70,000) choose to drop out of the program over a very short period of time. The loss of so many participating customers significantly diminished the utility's ability to lower its peak load

---

[15] In addition to the voluntary aspect of DSM programs, there is another aspect that makes their projected impacts less certain, at least in the short term, than the impacts of Supply options. That is measurement and verification (M&V) of the impact of the two types of resource options. M&V of the output of Supply options is continuous and virtually instantaneous. Conversely, M&V of most DSM options requires collecting and evaluating monitored data from relatively large numbers of customers who are participating in the DSM program. This process can take months or years to perform, particularly when monitoring impacts at the summer and winter peak hours. More than one year may be needed to gather data on seasonal peaks to ensure the data is accurate. Thus, there is a time lag in a utility knowing what it is actually getting in terms of output from DSM options compared to Supply options.

through the load-control program.[16] Consequently, the utility's actual reserve margin dropped below its 15% minimum criterion almost overnight.

The result was that the utility had to quickly add new generating units to increase its reserve margin. As for the residential load control program, the utility essentially stopped signing up new participants for a number of years to ensure that its system settled into a better balance between DSM and Supply options in regard to meeting its reserve margin criterion.

The key point of this discussion of the voluntary nature of DSM options is just that there should be a balance between DSM and Supply options. Due to the voluntary nature of DSM programs, there is inherently more uncertainty regarding the future contribution toward utility system reliability of DSM options than there is of most Supply options, particularly with fossil- and nuclear-fueled generating units.

Now that we have introduced the voluntary aspects of DSM options, what should a utility do about these aspects of DSM when performing system reliability analyses? Because the objective of reliability analyses is to ensure that the utility system can continue to be able to reliably deliver energy to its customers during all hours of the year, including peak load hours, a utility might develop its own constraint that it uses in its resource planning work to ensure that DSM is not relied upon too heavily to meet peak loads.

For example, one could apply a constraint that calculates reserve margins with and without the contribution of DSM. It could do so with the use of two reserve margin calculations (or criteria). The first reserve margin calculation, the calculation with DSM, is the utility's official reserve margin projection. Let's assume it shows a reserve margin of 20% for a given year.

The second reserve margin calculation, a calculation without DSM, will naturally show a lower reserve margin value. Let's assume the result of this second calculation is a projected "generation-only" reserve margin of 12% assuming no DSM contribution. Therefore, DSM's contribution to the utility's first reserve margin calculation is 8% of the 20% value from the first calculation (or 40% of the original 20% value).[17] This second calculation shows what the utility's reserve margin would be if all of the voluntary customer actions assumed for the projected DSM contribution to the official reserve margin projection somehow failed to occur.

Florida Power & Light (FPL) began using just such a "generation-only reserve margin" criterion, or GRM, in 2014 and, at the time the second edition of this book is written, is still using this third reliability criterion in its resource planning. FPL's use of a GRM reliability criterion is in addition to using both the conventional or "total" reserve margin criterion of a minimum of 20% and a loss-of-load-probability (LOLP) criterion of a maximum of 0.1 day per year. FPL's GRM is a minimum reserve margin of 10%. What this means in practice is that at least half of the 20% total reserve margin criterion must consist of generating capacity.

---

[16] In addition, the utility was (understandably) reluctant to continue to control the load of equipment at the remaining participants' homes due to the concern that even greater numbers of participants would drop out of the program.

[17] Note that these percentage values in this example are completely arbitrary and are used solely for illustrative purposes. The example does assume a significant DSM contribution for a utility.

To-date, FPL has not based a resource option decision solely on the 10% GRM criterion. The 20% total reserve margin criterion has always been "tripped" before either the LOLP or GRM criteria. The Florida regulatory view of this GRM criterion has been mixed. Although the FPSC has not officially sanctioned the GRM as a reliability criterion (nor has FPL requested FPSC approval of the new criterion), several of the Commissioners have made statements to the effect that the GRM offers an interesting and useful look into the "quality" of FPL's planned reserves.

I agree with these Commissioners. The use of a "generation-only reserve margin" criterion (or calculation) provides valuable insight into what types of resources a utility is depending on to remain reliable.

We now focus on another type of constraint that a utility may impose upon itself in regard to resource options.

LOAD SHAPE CONSTRAINT

The second of the utility-imposed constraints is one that we will refer to as a "load shape" constraint. There are load shape impacts on certain Supply options, most noticeably with solar options. These will be discussed in Part II of this book, but I do not consider these as a utility-imposed constraint as we are using the term in this discussion.

In the context that we are discussing utility-imposed constraints, this constraint applies only to DSM options, particularly load management (or demand response) type DSM options that we just discussed. Before we examine this constraint, we will first return to the concept of load shape that was discussed in Chapter 2.

By load shape, we are referring here to the utility's peak day load shape. For the discussion that follows, we will focus solely on the load shape for the summer peak day. You may recall that we previously presented the summer peak day load shape for our hypothetical utility system in Figure 2.1. We will now take that same load shape and focus only the hours from about noon (Hour 12) to 10 p.m. (Hour 22) using a different y-axis scale. This information is presented in Figure 8.1.

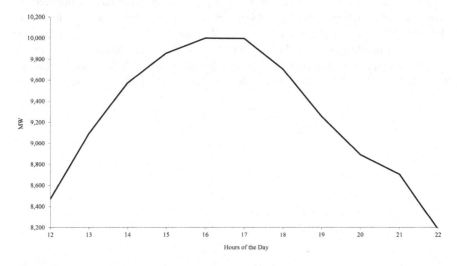

FIGURE 8.1 Summer peak day load shape (noon to 10 p.m.).

This portion of our utility's summer peak day load shape shows the portion of the summer peak day in which our utility's load is increasing from late morning to its late afternoon peak of 10,000 MW. The load then decreases as the evening hours unfold. This Hour 12 to Hour 22 portion of a summer peak day is the portion of the peak day that DSM options typically seek to address in terms of reducing system summer peak demand for our utility system, thus assisting the utility to meet its summer reserve margin criterion.

Virtually any DSM option that seeks to reduce system peak demand will affect the utility's peak day load shape during these hours. There are two basic impacts that DSM options will have and these are tied to two different types of DSM options. First, certain DSM options will tend to lower the utility's load, to some degree, over most/all of these late morning-to-evening hours on the summer peak day. Energy conservation type DSM options, such as higher efficiency air conditioning systems, are these types of DSM options that tend to lower the load curve in all/many of the hours shown in Figure 8.1. For purposes of this discussion, it is assumed that the load shape presented in Figure 8.1 has already been adjusted for the effects of existing energy conservation-type DSM programs.

The other type of DSM option, a load management option, will impact the summer peak day load shape in a more selective manner. We will discuss this by using a load control program as our example. Load control is implemented by the utility figuratively (or literally) "pushing a button" that either causes electrical power to be temporarily shut off to some electrical equipment (such as a water heater or swimming pool pump) or causes air conditioning/space heating equipment to cycle on and off in a manner different than it normally would operate when only the thermostat was controlling the equipment. (The basic effect for an individual air conditioning unit, when the utility implements load control, is that the length of time that the air conditioner would be "off" is slightly longer than would be the case if the utility were not controlling the equipment.)

The combined effect of this, when applied to tens of thousands, or hundreds of thousands, of participating load control customers, is to significantly reduce the system demand when the utility hits the load control "button." This reduction in peak load occurs very quickly on the utility system. A somewhat simplified view of what the resulting change in our utility system's peak day load shape may be if the utility "pushed the button" on 200 MW of load control is represented graphically in Figure 8.2.

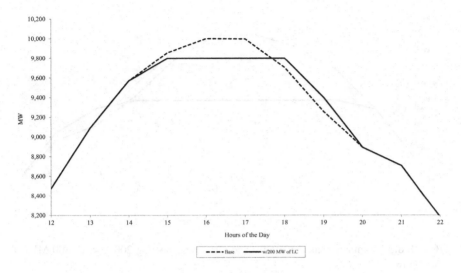

FIGURE 8.2   Summer peak day load shape after implementing 200 MW of load control.

In Figure 8.2, the dotted line now represents the original load shape on the peak day assuming no load control is implemented, *i.e.*, the same load shape that was presented in Figure 8.1. The solid line now represents the altered load shape after 200 MW of load control is implemented. We see that the former peak load of 10,000 MW has now been lowered by 200 MW to 9,800 MW, as we would expect.

However, perhaps unexpectedly, we also see that while the former peak was only 1 hour in duration (in Hour 17), the new, lower peak after load control has been implemented is a peak that now lasts for approximately 3 hours (*i.e.*, from Hour 15 to Hour 18) in order to lower the peak by 200 MW. In other words, the load management program not only lowers the peak demand, it also "flattens" the load shape.

This flattening effect occurs for two reasons. First, the objective is to lower the former peak hour's load. In order to ensure that this is accomplished, the load control button is typically pushed before the former peak hour occurs. Second, when the utility removes its finger from the button, an increase in load occurs as water heater and air conditioning thermostats switch their respective equipment back on as needed in order to bring either the water temperature in a water heater tank, or the room temperature in a home or business, back to the temperature settings on the thermostats for both types of equipment. This thermostat-driven increase in load when a utility releases load control is often referred to as a "payback" effect (which we will return to shortly).

Now suppose that our utility system is considering using 600 MW *more* of load control capability on this same day. The resulting impact when the new total of 800 MW (= 200 MW originally + 600 MW more) of load control is implemented on the summer peak day is shown graphically in Figure 8.3. For comparison purposes, this figure also shows both the original load shape (without any load control) and the projected load shape when 200 MW of load control is implemented, *i.e.*, the same information conveyed in Figure 8.2.

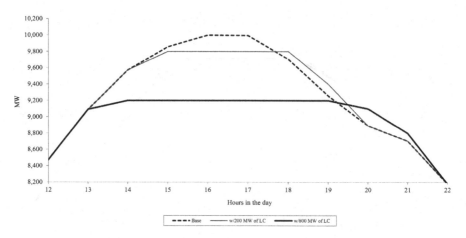

FIGURE 8.3 Summer peak day load shape after implementing 200 MW or 800 MW of load control.

We see that if the greater amount of load control is implemented, our utility now experiences a peak that has been extended so that it lasts for approximately 5 hours (*i.e.*, from Hour 14 to Hour 19). From these examples, it is clear that, the greater the amount of load control capability the utility has, the longer the duration of the utility's peak when the utility implements all of its load control capability. In other words, the more a utility tries to drive the peak load down by implementing increasing amounts of load control, the more hours that the control of the participating customers' equipment must be maintained. As a result, the utility's peak day load shape becomes even more flattened and the duration of the peak load level gets longer.[18]

This points out that there are a couple of practical issues to consider when considering the actual operation of load control. We will start with a consideration of the participating customer. Participating customers will tolerate control of their air conditioning, etc., equipment for some number of hours but will not readily tolerate longer durations of control. In other words, if the duration of load control becomes too lengthy, participating customers may find that their homes are too hot in summer (or too cold in winter) and/or their water is not warm enough. (This is exactly what happened with Florida Power Corporation's residential load control program during the summer of 1998.)

Recognizing this, utilities typically limit the duration of load control in a tariff.[19] This limit on the duration of load control is usually based on time periods that (hopefully) ensure that the tolerance levels of participating customers will not be exceeded. Absent an outright emergency on the utility system, the duration of load control for

---

[18] As we shall see in Chapter 12, a similar increase in the number of hours that must be addressed by battery storage options as more and more batteries are installed occurs.

[19] An electric utility's tariff is basically a type of "rule sheet" for the utility and its customers. In part, it states the conditions under which a specific type of electrical service (such as a load control offering) may be offered and operated.

any one participating customer cannot exceed the maximum duration of control set forth in the applicable load control tariff.

Now recall what Figures 8.1, 8.2, and 8.3 have shown us: the more load-control capability that is added, the longer the duration of the new flattened peak load is when this load control capability is implemented. What all of this leads to is the conclusion that a utility system cannot indefinitely increase the amount of load control on its system by signing up ever-increasing numbers of load control participants. As a utility signs up more load control participants, the peak day load shape becomes flatter for even more hours.

At some point, adding one more MW of potential load control capability would require that the utility's control of customers' loads would be for periods of time which are longer than the control durations allowed by the tariff. This point can be referred to as a "physical limit" to load control for the utility system. A utility *could* attempt to get around this problem by operating load control as a form of "relay race." Such an attempt would result in: (i) one participating customer being controlled for all or part of his or her allowable control duration, (ii) control of that customer then ceasing, and (iii) control of a second participating customer immediately beginning. In this way, control could be implemented for longer durations than allowed by the tariff's restrictions for any one participating customer.

However, there is a significant problem with this approach. The utility is now signing up and paying credits on the monthly electric bill to *two* customers in order to get one the same amount of demand reduction through load control that was previously achieved by signing up and paying monthly credits to just one customer. Once this point has been reached, the cost-effectiveness of adding one more participating customer in the load control program has been diminished.

One can look at this same phenomenon in a different way. Even if one were to disregard the additional cost aspect of such an approach and focus solely on the demand reduction perspective, the impact of the physical limit would show in a different way. One would reach the point at which signing up additional participating customers provided diminishing MW reductions.

In the utility example just discussed, our utility first signed up 200 MW of load management and was able to obtain 200 MW of peak load reduction. Then, our utility signed up another 600 MW of load management and was able to obtain a total of 800 MW of peak load reduction. Now let us suppose that 800 MW was the exact point at which any further flattening of the peak day load shape would result in diminishing MW reductions from any additional load management signups.

What would happen if our utility attempted to sign up another 100 MW of load management (in an attempt to have a total of 900 MW of load management) and then attempted to sign up yet another 100 of load management (in an attempt to have 1,000 MW of load management)? An example of a likely outcome is presented in Table 8.1.

TABLE 8.1

Impacts of Increasing Amounts of Load Management for Our Utility System

(1)	(2)	(3)	(4)	(5)
Incremental Amount of Load Management Signups (MW)	Cumulative Amount of Load Management Signups (MW)	Cumulative Amount of Projected Peak Load Reduction (MW)	Incremental Amount of Projected Peak Load Reduction (MW)	Percentage of Projected Peak Load Reduction Compared to Signups (%)
200	200	200	200	100
600	800	800	600	100
100	900	867	67	67
100	1,000	917	50	50

In Table 8.1, the first two rows provide in tabular format the same results presented earlier in Figures 8.2 and 8.3. Both the 200 and 600 MW increments of load management signups resulted in a projected 200 and 600 MW, respectively, of peak load reduction. However, an additional 100 MW of signups might now be projected to show diminishing returns of only 67 MW more of peak demand reduction. If our utility persisted and added yet another 100 MW of load management, the diminishing returns impact might be even more pronounced as only 50 MW more of peak demand reduction would be achieved. This is due to the flattened load becoming too long or broad to achieve 1 MW of additional demand reduction for 1 MW of additional load management signup given tariff-prescribed duration of control of individual participating customers.

Therefore, regardless of whether one thinks of this diminishing return impact from increased signups in terms of significantly reduced cost-effectiveness, or diminishing incremental peak load MW reductions, it is a characteristic of load management programs that needs to be accounted for in utility resource planning.

This characteristic of load management programs' impact on a utility system peak day load shape can be described by stating that load management programs have a "physical limit" to the programs that is driven by the utility's peak day load shape. The physical limit is reached when the amount of load management capability on the system reaches a point where one more participating customer cannot provide the same demand reduction as each previous participating customer provided. Another way to describe this is to refer to the amount of load management up to that physical limit as being the "usable" amount of load management for the utility system.

A utility that is considering a significant amount of load control as a resource option should address this physical limit aspect of load control by performing a load shape-based form of analysis in its IRP work. Such an analysis identifies the physical limit for, or the usable amount of, load management for that utility's system. Once the amount of usable load management is identified, the utility uses that as the upper limit for how much load management should be signed up by the utility. In other words, the utility imposes a "load shape constraint" on itself in regard to its load management programs.[20]

The use of a load shape constraint in a utility's resource planning eliminates a couple of potential problems. First, it prevents the utility from planning to meet its future resource needs with a load management program, the incremental signups of which, after the utility's physical limit for such programs has been reached, will be unable to provide what would otherwise be the projected demand reduction capability for a new participating customer. This eliminates a potential system reliability problem by ensuring that the utility will not overestimate the contributions from the load management programs.

Second, the use of a load shape constraint also prevents the utility from spending monies on load management programs that, once the utility's physical limit for load management has been reached, will result in greatly reduced benefits for this load management program (due to the reduced kW reduction from each incremental participating customer who is signed up after the physical limit has been reached). The use of a load shape constraint can allow a utility to increase the economic efficiency of its portfolio of various DSM programs, and other resource options, because the use of this constraint guides the utility to know when to "back off" DSM expenditures for additional load control and focus on other, potentially more cost-effective resource options.

Before we leave the discussion of the load shape constraint, a couple of additional points should be made. First, the need to use a load shape constraint may vary widely from one utility to another based on their peak day load shape.[21] Our hypothetical utility system has a quite broad summer peak day in terms of the number of hours of relatively high load. Another utility may have a peak day load shape that is more of a "spike" peak of very short duration. This is often seen with utilities that are more concerned with a relatively short-lived winter peak load than a broader summer peak load. A load shape constraint may not be needed for such a utility.

Second, let's assume that two utilities have the exact same system peak load shape. The values for the physical limits to load control for each system may vary considerably based on the type of load control program that is implemented. All else equal, a utility, whose load control programs primarily address large business customers, will typically have a higher physical limit (and can sign up more MW of incremental load control) than can a utility whose load control programs primarily address residential and small business customers. This is because the majority of the demand reduction achieved with residential and small business load control involves control of equipment that utilizes a thermostat (such as air conditioners and water heaters).

As these devices are controlled, water in the water heater tank cools and the air in the home heats up in summer or cools off in winter. When the utility ceases its control of this equipment, the equipment thermostats cause the appliance to use

---

[20] Although we have used a load control program in this explanation, the same situation occurs with other load management/demand response type programs, including special time-differentiated electric rate-based programs.

[21] A reminder of one of the many reasons we have introduced Fundamental Principle # 1 of Electric Utility Resource Planning #1.

electricity continually for an extended period of time to heat the water in the tank and to cool or heat the home. As previously mentioned, this particular amount of energy is commonly referred to as "payback" energy.

This payback energy must be accounted for in the planning and operation of residential and small business load control so that a new, artificial peak load is not created when control is released. This serves to lower that physical limit for residential and small business load control options compared to large business load control options.

In large business load control options, thermostatically driven equipment is less likely to be controlled by the utility. Instead, industrial processes, motors, etc. are more commonly controlled. Particularly for larger business customers, a participating customer may also have a backup electrical generator that allows these business functions to continue even though the utility has temporarily cut off the electrical supply to this equipment. Therefore, there is generally much less (and perhaps zero) payback energy associated with large business load control options. Therefore, there may be a higher physical limit for large business load control programs compared to residential and small business load control programs.

Finally, one may ask how one actually calculates what the load management physical limit is for a specific utility. One method that has proven successful is to use nonlinear programming (NLP) techniques involving optimization software. The basic approach is to start with a projected peak day load shape for a future year. The projected load is laid out in 15-minute interval data. Then the projected demand reduction capability impacts, and the associated payback energy impacts, are developed for each 15-minute interval if load management is implemented for that interval. (This is done separately for each type of load management program the utility offers its customers.)

The NLP optimization program then applies all possible permutations of load management operation to achieve an objective of lowering the utility's peak day load to the lowest possible peak load value.[22] In regard to the load management operation, the optimization program considers not only which load management programs may be implemented, but also what the possible starting and stopping times can be for each aspect of a load management program (*i.e.*, residential water heaters and small business air conditioners).

This type of analytical approach has proven very useful in determining the physical limits of various types of load management programs and/or time-differentiated electric rates. It can also be useful in an operational sense by providing another tool for utility system operators, *i.e.*, the folks who actually coordinate the running of all of a utility's generating units and the dispatching of load management programs, to use.

We have now completed an overview discussion of a number of potential constraints to utility resource planning. Our attention now turns to the impacts these constraints will have, or will likely have, on utility systems and their customers.

---

[22] The peak day load values are typically calculated on an hourly basis because utilities generally perform reserve margin calculations on an hourly basis. The 15-minute data is used in the optimization model's calculations in order to more accurately capture the payback impacts that can vary considerably from one 15-minute interval to the next.

WHAT ARE THE IMPACTS OF ADDRESSING THESE CONSTRAINTS?

Now that we have introduced the concept of constraints on utility resource planning analyses and have discussed six examples of constraint, the natural next question is: *"What impacts will these constraints likely have on a utility's resource planning work?"* After all, the title of this chapter includes the phrase "...can (and will) complicate resource planning analyses."

We shall begin to address this question by stating what should be obvious in regard to any situation in which one is trying to find the best solution to a complex problem (such as utility resource planning). This point is that:

All else equal, the less flexibility one is allowed to have (due the number and types of constraints imposed) when finding a solution, the less likely it is that the best possible solution will be found.

However, equally obvious is the fact that no one has complete flexibility in regard to finding solutions for certain problems. For example, when considering how best to paint a vaulted ceiling in one's home, it is not possible (except perhaps back in Michael Jordan's younger days) to leap into the air and seemingly hover for an extended period of time while you paint the ceiling. The law of gravity removes this as a possible solution.

In regard to this ceiling painting example, the law of gravity can be considered an "absolute" constraint on the problem of how the ceiling might best be painted. One then moves on to more feasible or practical potential solutions.

Similarly, a utility will typically accept the first constraint that we discussed (siting/geographic constraints) as an absolute constraint, consider what resource options remain as feasible after recognizing this constraint, and then move on with their analyses. A utility may also accept the second constraint (tightening of environmental regulations) in some degree as an absolute constraint because a utility typically has little or no direct control over such changes in regulations, especially in the short term. The utility will make its best projections of what future environmental regulations may be and then examine projected system emissions in its resource planning work to give it insight into how potential tightening of regulations may impact its resource decisions.

A key point is that not only utilities, but legislatures and regulators as well, recognize these types of constraints as "givens" when considering utility resource planning. Therefore, the resulting loss of flexibility, in regard to potential resource option decisions, that emanates from these two absolute constraints is simply not much of an issue. In other words, there is really no decision to be made regarding whether these two constraints should be used in utility resource planning (although there will likely be decisions or judgments involved in regard to how the second type of constraint is accounted for in specific resource planning work).

The same cannot be said for the remaining four constraints that we have discussed. In each case, a decision has been made by the legislature, a regulatory agency, or the utility to impose one or more of these four constraints. Let's examine these four constraints to see what their impacts are likely to be on utility resource planning and the utility's customers.

We will start by jumping to the two utility-imposed constraints that we previously discussed. The fifth and sixth constraints (that address system reliability criteria and peak day load shapes, respectively) are constraints that may be imposed by a utility on itself after consideration of the specific characteristics of its individual utility system. Keep in mind that the utility not only knows its particular system better than any other party can, but the utility also has the responsibility to serve its customers in a reliable and cost-effective manner.

Therefore, a utility would only impose the fifth and sixth constraints if it believed that it needed to do so to ensure that system reliability is maintained and that various resource options are only signed up to a level that is usable and cost-effective. In other words, the utility imposes these constraints upon itself to help it identify and select the best resource option for its customers.

Consequently, any impact that the fifth and sixth constraints will have on a utility system and the utility system's customers are likely to be beneficial for the utility's customers. This leaves us with the two legislative/regulatory-imposed constraints (the third and fourth constraints) to consider in regard to what the impacts are to utility resource planning and a utility's customers.

It is obvious that both of these two constraints (*i.e.*, the imposition of "standards"/quotas for certain resource options and the prohibition of other resource options, respectively) restrict the flexibility that a utility has in its resource planning. In effect, they limit or "squeeze" the number of resource options that can be considered in two ways. The third constraint ("standards"/quotas for specific types of resource options) dictates that a certain amount of electricity (or energy services) that a utility delivers to its customers must come from certain sources, generally renewable energy and/or DSM sources. Conversely, the fourth constraint (prohibition of specific resource options) dictates that a utility cannot add any new resources of a certain type (such as new fossil-fueled units).

Together, these two types of constraints remove a significant amount of flexibility that a resource planner would otherwise have when determining the best resource option to select for his or her utility. If certain resource options are prohibited outright, then the resource planner may be forced to select other types of resource options to, at least partially, "fill out" his or her resource plan in order to meet the imposed standards/quotas.

This reduced flexibility will, as stated earlier, make it less likely that the best possible solution for the utility's customers will be found. This is likely to lead to a non-optimum selection of resource options which, in turn, will lead to higher electric rates for the utility's customers. In addition, the selection of a non-optimum resource option may have other downsides such as lower system reliability.

Consequently, the use of the legislative/regulatory-imposed constraints ("standards"/quotas for specific resource options and prohibition of other resource options) can definitely result in negative impacts to a utility's customers.

————

This chapter concludes Part I of this book in which we have discussed the fundamentals of electric utility resource planning. In the course of this discussion, we have created a hypothetical utility system which we then used to perform a wide variety of analyses that addressed system reliability, preliminary screening of resource options, final (or system) economic analyses of resource plans, and non-economic analyses of resource plans. We closed out Part I in this chapter by looking at a number of constraints that could have altered the results presented in these analyses. One type of those constraints was a self-imposed constraint; *i.e.*, one that a utility imposes on itself.

In Part II of this book, we will examine what is perhaps the largest possible utility self-imposed constraint: a decision to achieve low- or zero-carbon operation by a certain target year. As the second edition of this book is written, many utilities have made this decision that has been spurred on by two influences. The first influence was the introduction of significant federal tax credits for zero-carbon resource options that were a part of federal legislation in 2022. The second of these influences has been the application of political, social, and investment community pressure upon electric utilities for them to each make a public announcement that the utility has a goal or ambition to achieve low- or zero-carbon operation by a certain target date.

In Part II of this book, we take a look at what it will take, both from the standpoint of resource MW needed and challenges to address, for our hypothetical utility to achieve a zero-carbon goal.

Part II

Moving toward Zero Carbon

9 An Overview of Part II

INTRODUCTION

Part I of this book discussed what I consider to be the fundamentals of integrated resource planning (IRP) for electric utilities. As part of this discussion, we first created a hypothetical electric utility ("our utility"). Then we used our utility to perform a series of analyses that focused on system reliability, economics of resource options and resource plans, and non-economic considerations of these same resource options and resource plans. We then closed Part I of this book with Chapter 8's discussion of various factors that can (and will) complicate resource planning analyses.

At the end of Chapter 8, we mentioned that, as the second edition of this book was written, there were two influences on utility resource planning that are spurring on a movement by electric utilities in the direction of zero-carbon utility operation. Those two influences are as follows:

1. The significant federal tax credits that were a part of the 2022 Inflation Reduction Act (IRA). These tax credits apply to a selected set of resource options (such as photovoltaic [PV] solar facilities) that are zero-carbon-emitting resources. The intent of these tax credits is to improve the overall economics of these favored resource options by lowering a utility's federal tax burden if the favored resource options are added to the utility system. In so doing, the end objective was to encourage U.S. electric utilities to move toward generation systems that would be low- or zero-carbon emission systems in the future.
2. The second of these is the application of political, social, and investment community pressure upon electric utilities for them to each make a public announcement that the utility has a goal or ambition to achieve low- or zero-carbon operation by a certain target date, thus ensuring utility commitment to installing zero-carbon emission resources.

These two influences, and the impacts they will have on electric utilities and these utilities' resource planning work, are a large part of the rationale for writing a second edition of this book.

I believe that both influences have been very successful in terms of meeting their objectives. The new federal tax credits for favored resource options can significantly lower a utility's overall costs if the utility chooses these options (as we shall see in Chapter 11). And the outside pressure for utilities to publicly state a carbon goal/ambition has clearly worked because numerous utilities have made such announcements. In large part due to these two influences, most, if not all, U.S. electric utilities are moving in the direction of reconfiguring their generation systems to achieve lower carbon emissions.

DOI: 10.1201/9781003301509-11

Before retiring from FPL in late 2022, my last major assignment was to lead FPL's analyses to determine what it would take to get the FPL utility to a low-or-zero-carbon position by a certain target year. These analyses contributed to FPL's parent company, NextEra Energy, making a June 14, 2022 public announcement that basically said: (i) all of NextEra Energy (including FPL) had a goal to get to zero-carbon operation by 2045, and (ii) NextEra Energy had a plan to achieve this goal.[1]

The analyses of what it would take for FPL itself to achieve zero-carbon emissions helped confirm what experienced resource planners had long suspected: that there are certain characteristics of renewable resources (such as PV) that will need to be accounted for as increasing amounts of those resources are added to a utility system. There will also be a number of issues that many, if not all, utilities will have to address in trying to reach their low- or zero-carbon goals.

Part II of this book, without revealing information not already publicly disclosed by FPL and/or NextEra Energy, discusses some of these resource characteristics and planning issues. Therefore, in these discussions, we will use the same hypothetical utility system that was used in Part I of the book to illustrate many of these topics.

A DIFFERENT APPROACH IS TAKEN IN PART II

The overall approach taken in Part II of the book is different in a number of ways than the approach taken in Part I of this book. I will briefly describe the major differences between the two approaches after first grouping those differences into two categories: focus and analyses.

1. Focus:

Part I of the book presented what is a traditional view of how a utility makes decisions regarding what type(s), and number, of resource options should be selected to best serve its customers. These decisions are typically near-term decisions. For example, in Part I of this book, our hypothetical utility was making a resource decision in the Current Year for implementation of that resource(s) by Current Year + 5.

Such an approach was taken because utilities have generally made decisions only for the near-term future year for which a decision is now needed (after accounting for the time needed to build/implement new resources). For our utility, that year was Current Year + 5. This near-term focus is based on the recognition that assumptions and forecasts used as key inputs to decision-making (such as future electrical load and costs of resources) are inherently uncertain and continually changing. However, one generally has a higher level of confidence in assumptions and forecasts for near-term values than for long-term values. Therefore, the traditional utility focus, which was used in Part I of this book, has been on the next needed decision regarding resource options in

---

[1] FYI—At the time of this announcement, FPL had slightly more than 30,000 MW of total generating capacity. Of that total, approximately 7,500 MW were from zero-carbon emission resources (~4,000 MW of PV and ~3,500 MW of nuclear.) In addition, FPL's parent company, NextEra Energy, had approximately 24,000 MW of additional zero-carbon emission resources (largely wind).

order to take advantage of the higher level of confidence on near-term values in the most current forecasts and assumptions.

In Part II of this book, the focus has significantly changed. Our utility has decided that it wants to set a low-to-zero carbon goal that will result in lower carbon emissions than would be the case if it continued its normal IRP analyses. Our utility would like to serve all of its forecasted annual GWh energy load by a certain target year (which, as we shall see, will be a couple of decades in the future) only with resources that produce zero-carbon emissions. With this decision, and because each utility will have only a relatively few new, non-nuclear zero-carbon resource options (such as solar, wind, and batteries) that are applicable in its service area, our utility has already decided the type of resources it will add through the target year (and beyond). And, with a zero-carbon goal set, our utility has also decided that it will cease operating its existing non-nuclear generating units using fossil fuels by the target year.[2]

Therefore, the focus in Part II of the book will not be on what the next resource decision will be because our utility already knows what zero-carbon resources are applicable in its service area and must be added to achieve its carbon goal. Instead, the focus in Part II of this book will be on resource planning issues our utility will have to address during what I will term the "transition period." By this term, I mean the time period starting when the utility makes its low-to-zero carbon announcement and the target year in which the carbon goal is to be achieved.

This focus will examine significant issues, and associated challenges, that our utility's resource planners, system operators, and transmission planners will have to address as the utility moves into the transition period and toward the low-to-zero carbon goal.

2. Analyses:

In Part I of this book, a number of analyses were presented and discussed. These included economic analyses of CPVRR costs and levelized electric rates for a number of competing resource plans that were based on the resource options that could be selected for Current Year + 5. Non-economic analyses were also presented for these competing resource plans that evaluated the number of years to economic "cross-over," system fuel use, and system emission levels. These analyses were all needed to provide our utility a complete picture of which competing resource option(s) would be the best selection for its customers for Current Year + 5.

However, as just mentioned above, because our utility has already announced a commitment to serve its customers with low-to-zero carbon emissions, it already knows what types of resource options it will have to select for all years up and through the target year. For this reason, and because the focus in Part II of this book is on new issues that will need to be addressed during the transition period, Part II largely presents different types of analyses than were discussed in Part I.

---

[2] Consequently, these existing generating units will either be retired at some point by the target year and/ or be converted to utilize zero-, or net-zero-, carbon-emitting fuels.

For example, there are no economic analyses in Part II that project what the costs will be for a resource plan that meets the low-to-zero carbon goal or regarding the net incremental costs of meeting this goal. There are a number of reasons the decision was made not to provide these types of cost analyses, including the following:

- Once a utility has committed to such a goal, one could argue that the costs of the resource plan required to meet that goal, and the net incremental costs of reaching that goal, are irrelevant because the utility has made a commitment (assuming, of course, that the state's regulators and/ or legislature are in agreement that this is a prudent course of action).
- In addition, even if one were to attempt such cost analyses, it is my belief that the confidence level in the accuracy of the results of such an analysis would be very low. For example, no one can forecast with accuracy what the annual costs of solar, wind, batteries, etc. will be over a time period that will (in our hypothetical utility's case) span more than a couple of decades.
- In addition, in order to determine a net incremental cost impact of a resource plan that meets our utility's low-to-zero carbon goal, that resource plan would have to be compared to another resource plan that our utility would select if it had <u>not</u> chosen to commit to a carbon goal. When developing such a "no carbon goal" resource plan, it is also true that no one can forecast, with a high level of confidence, the annual costs of new non-renewable energy resources that this second resource plan might choose over several decades in the future.[3]
- Also, no one can forecast with accuracy future changes in legislation and/or regulation that will impact costs for both resource plans.
- A major consideration in determining the net cost impact of a resource plan with a low-to-zero carbon goal is the timing of the retirement and/ or conversion (to burn zero- or net zero-carbon fuels) of each existing non-nuclear generating unit on the utility system. Each retirement/ conversion decision will be based on both economic and system reliability considerations. Those considerations will be influenced by many of the factors mentioned above.
- System reliability analyses will likely need to evolve to use perspectives different than those used in the reserve margin and LOLP criteria commonly in use today.
- Furthermore, if projected costs for solar and batteries steadily decline (as projected), it seems reasonable that the number of individual customers who will begin to acquire distributed/rooftop PV and/or batteries will

---

[3] Note that the comparison to another resource plan without a carbon goal is exactly how utilities have traditionally evaluated competing resource options: they compare competing resource plans which feature the competing resource options. However, in these traditional analyses, utilities have often assumed a common "filler unit" as needed for all years beyond the year for which the utility is making a decision. If one is trying to compare one resource plan with a carbon goal to another resource plan without the carbon goal, the common filler unit approach does not work for these plans.

steadily increase. These distributed resources will impact the electric load that will be served by the utility. Although utilities attempt to project those impacts in their load forecasting efforts, the inability to accurately forecast the annual costs of these resources a couple of decades in the future hampers the ability to forecast annual load impacts that are driven by the implementation of these resources.

- Finally, computer modeling of future transmission system needs typically address only about a decade into the future. This makes projections of future transmission costs, particularly for a utility's low-to-zero carbon resource plan which likely cannot now state with specificity where the new renewable resource additions will be sited over two decades time, highly uncertain.

For all of these reasons, any projection of costs, or net incremental costs, regarding resource plans to achieve a carbon goal that is made this early-in-the-game, are, in my opinion, of questionable value. As forecasts and assumptions change every year, the results of cost, and net incremental cost, analyses will change as well.

Therefore, instead of the types of analyses that were presented in Part I of this book, the analyses that will be presented in Part II will address key issues that utilities which set zero-carbon goals will face as they enter the transition period. A few examples of these issues (posed as questions below) include the following:

- Assuming our utility wants to go "all in" and achieve a zero-carbon goal, how many new resource MW will be needed to be added to serve 100% of our utility's forecasted GWh energy load with no-carbon emissions by the target year?
- In terms of economics, how significant are the new federal tax credits for solar (and other renewable resources) that were established in the 2022 IRA?
- As a utility adds increasing MW amounts of PV, how should PV's intermittent output be viewed in regard to the firm capacity value that these resources provide?
- How does a utility determine the needed battery duration (hours) in order for the batteries to supply firm capacity?
- How will operation of a utility's existing generating units change as increasing amounts of PV MW are added to a utility's system (and will the operation of the new PV facilities have to change)?

Analyses addressing these transition period issues are presented in the various chapters comprising Part II of this book.

HOW PART II IS STRUCTURED

In Chapter 10, we return to our hypothetical utility system and look at the amount of new resource MW it would take for our utility system to supply 100% of its forecasted annual GWh energy load by a target year (which is the last year of the analysis period used in Part I of this book: Current Year + 29). In order to keep the discussion simple, we will assume that our utility has only two resource options applicable to its service area with which to reach its zero-carbon goal: the same 120 MW PV Option that was discussed in Part I, and a battery storage option.[4] The discussion in Chapter 10 will take a look at two different estimation approaches one might use to project how many MW of new PV, and how many MW of battery storage, our utility will need to serve 100% of its target year GWh load with no-carbon emissions.

After establishing in Chapter 10 the amount of new PV MW, in conjunction with battery MW, that will be needed to serve all of our utility's forecasted GWh energy load in Current Year + 29 with zero-carbon emissions, we take a closer look at PV from a resource planning perspective in Chapter 11. Four aspects of PV options will be examined. First, we look at how the installed costs for PV have decreased in the decade leading up to the second edition of this book. Second, the economic impact of the 2022 IRA legislation's federal tax credits applicable to PV will be examined. This will be done by revisiting the economic analyses of Supply options that were presented in Part I of this book (Chapter 5) and then redoing that analysis after accounting for these federal tax credits for the PV option. Third, we take a look at the firm capacity values of PV for winter and summer peak load hours that resource planners will need to account for as they examine the addition of ever-increasing amounts of PV. This section will also include a brief discussion of system reliability analyses during the transition period. Fourth, we look at the land requirements for the amount of PV MW that our hypothetical utility projects it will need to serve all of its target year GWh energy load with zero-carbon emissions.

In Chapter 12, we take a similar look at battery storage. We will discuss the fact that battery storage, unlike virtually all other resource options, acts alternately as both a generating unit and as an electric load. We take a look at a representative projection of installed costs for battery storage at the time the second edition of this book was written. Then we will briefly discuss the 2022 IRA federal tax credit applicable to batteries and how the relative differences in those tax credits applicable to PV and to batteries may result in an unexpected outcome. We then discuss how a battery's firm capacity value is related to battery duration. A method to calculate the needed battery duration for a battery's firm capacity value to be 100% of its nameplate value is presented. Finally, a look is taken at how needed changes in battery duration will impact the cost of incremental battery additions.

We then go outside of what is typically thought of as the parameters of current IRP work in Chapter 13 to take a look at how movement toward a low-to-zero carbon goal will likely impact a utility's system operation and transmission planning functional areas during the transition period. First, we take a look at what is perhaps a surprising issue that is caused by adding ever-increasing amounts of PV MW each year in moving toward a carbon goal. That issue is the fact that PV will produce far

---

[4] The decision to limit our utility's choice of new options to just PV and batteries was made to keep the discussion as simple as possible. Although there are distinct differences between solar and wind resources, they share certain characteristics that will require resource planners to consider these resources in different ways than conventional generation is considered. Consequently, much, if not most, of the Part II discussion about PV from a resource planning perspective will also be generally applicable to wind resources.

more energy than the utility can immediately use during certain hours of the year. We will look at examples of the magnitudes of the "excess" PV MWh using our hypothetical utility system. Based on this, we next examine how our utility's system operators will have to change how they operate our utility's existing generators, as well as the new PV facilities, during a 24-hour period using our utility system. These operational changes are driven by the very real operating limitations that existing conventional generators have regarding: (i) how quickly the output of these units can be ramped up and ramped down, and (ii) the lowest MW levels at which these units can be operated. As these examples will show, deliberate curtailment of PV output may well be part of the solution.[5]

The transmission planning function of a utility will also need to deal with the need to interconnect the new PV sites with the existing transmission grid, and to then integrate the flow of electricity from those sites with an existing generation/ transmission system was designed to operate with conventional generators. Both the transmission planning and system operation functions will be significantly impacted by large additions of inverter-based resources (IBRs) of any type, such as PV. This is because current utility systems were designed to operate with "spinning rotor" generators, such as CCs, CTs, and coal units, which deliver reactive power. Conversely, IBRs do not provide reactive power to the system. We will discuss some of the challenges this presents for transmission planners and system operators.

Chapter 14 closes Part II of this book and has three distinct sections. First, a summary of what we have discussed through the first 13 chapters is presented. Second, I switch from an "instructor" mode (in which I present resource planning principles along with examples of analyses that illustrate those principles) to an "I have an opinion" mode. In this mode, I discuss approximately 20 topics using a question-and-answer (Q&A) format.[6] Third, some final thoughts about resource planning in the future are offered.

We now begin Part II of this book by estimating the amount of new resource MW it will take for our hypothetical utility to be able to serve 100% of its forecasted Current Year + 29 GWh annual energy load with zero-carbon emissions.

---

[5] These challenges will need to be addressed during the transition period until a utility begins to get reasonably close to the target year for its carbon goal and has, therefore, retired and/or converted most of its existing non-nuclear generation.

[6] A similar Q&A discussion appeared in the first edition of the book as Chapter 9. That previous discussion has now been extensively updated and includes some thoughts and opinions that have surfaced in the dozen or so years since the first edition was published. Many of the updates pertain to information discussed in Part II, *i.e.*, moving toward zero carbon.

10 Moving toward Zero Carbon

How Many MW of New Resources Will be Needed?

INTRODUCTION

We now return to our hypothetical utility only to find that it has now decided to set a goal of achieving low-to-zero carbon emissions by the target year of Current Year + 29. When we last visited our utility, it was considering a resource plan that would add two 120-MW PV facilities 5 years in the future. If that resource plan were to be selected, it would meet the projected resource need in Current Year + 5.[1] If our utility selects this resource plan, it could also be a first step in moving toward its new carbon goal.[2]

As our utility considers how to achieve the goal of serving its customers with low-to-zero carbon emission energy in Current Year + 29, it is interested in seeing what amount of new resource MW it will need to add to reach the goal. The remainder of this chapter primarily discusses how our utility might begin to answer that question by looking at a couple of estimation approaches to achieve not just low-carbon emissions, but zero-carbon emissions.

However, before we look at those estimation approaches, it will be helpful to explain the perspective that will be taken in this chapter. That perspective can be summarized by the following two points:

1. In this chapter, our utility is taking an <u>energy-only</u> (GWh) look. By that, we mean that in the target year of Current Year + 29, our utility will need to serve a forecasted number of GWh with zero-carbon-emitting resources. What our utility wants to know first is: *"How many new MW of zero-carbon emitting resources must it have operating by Current Year + 29 to produce all of that year's forecasted GWh?"* Therefore, this chapter is looking solely

---

[1] Recall that due to the PV option's 50% firm capacity value, our utility will need to add two of the 120-MW PV options, or 240 MW of PV, to meet the 120 firm MW resource need.

[2] In actual practice, a utility that announces a low-to-zero carbon goal would likely not wait 5 years before adding new zero-carbon resources. However, following our practice of keeping the discussion as simple as possible, and because it allows an easy comparison back to the discussion in Part I of this book, we will assume that our utility begins its movement toward zero-carbon in Current Year + 5. (Also, regardless of the year in which our utility begins this movement, the total new MW needed to meet a Current Year + 29 goal will not change because the forecasted GWh for Current Year + 29 does not change.)

 DOI: 10.1201/9781003301509-12

at how many new zero-carbon generation MW will be needed to supply all of the GWh to be served in Current Year + 29.

2. However, our utility will consider the resulting amount of new resource MW needed that is determined in this chapter to be the <u>minimum</u> amount of new MW it needs to add to achieve its zero-carbon goal. This is because, in addition to supplying all of its Current Year + 29 GWh carbon-free, our utility must also ensure that its system remains reliable as it adds only zero-carbon-emitting new resources, plus eventually retiring and/or converting its existing non-nuclear generating units. The topics of system reliability and unit retirement will be addressed in Chapter 11 after the concept of PV firm capacity value has been introduced. Therefore, we simply need to keep in mind that the results presented in this chapter, *i.e.*, the projected MW amount of new zero-carbon resources needed to be added to serve 100% of the annual GWh load in Current Year + 29, <u>might</u> need to be increased due to system reliability needs.

With the perspective our utility is taking now explained, our utility now begins to look at approaches with which it might quickly estimate how many MW of new resources it will need to supply 100% of the forecasted GWh to be served in Current Year + 29 with zero-carbon emissions.

THE FIRST ESTIMATION APPROACH

Our utility has heard about an approach used by others (often by non-utility organizations) to estimate the total MW amount of new zero-carbon resources that will be needed for utilities to be able to serve 100% of their energy (GWh) carbon free. This approach is simple to calculate, and our utility decides to try this estimation approach first. This estimation approach has two steps. First, it looks at how many GWh of annual energy our utility is projected to serve in a given year. In order to test this estimation approach, our utility decides to use Current Year + 5 in its calculation. Then it examines how many annual MWh are projected to be produced from 1 MW of PV.

Using those two values, our utility determines how many MW of PV are needed to serve 1% of annual energy in Current Year + 5. In addition, it takes a look at the projected output of the 2 × 120 MW PV option needed to meet the utility's resource need in Current Year + 5 if Supply Only Resource Plan 3 (PV) is selected in that year.

The second step of this estimation approach then extrapolates the amount of PV MW needed to serve 1% of the annual GWh load to the amount of PV MW needed to serve 100% of the annual GWh load. As you might guess, this is done by multiplying the MW result from the first step by 100.

Table 10.1 presents the assumptions and calculations used in the first step of this estimation approach.

TABLE 10.1

Assumptions and Calculations for the Amount of PV Needed to Supply 1% of Our Hypothetical Utility's Annual Energy in Current Year + 5

Assumptions (Shaded) & Calculation Results	Explanation
53,126	= Forecasted Annual Energy Usage in GWh for Current Year + 5
20%	= PV option average capacity factor
53,126,000	= Forecasted Annual Energy Usage in MWh for Current Year + 5 (=53,126 GWh × 1,000 MWh per GWh)
1,752	= Annual MWh from 1 MW of PV option (=1 MW × 365 days × 24 hours per day × 20%)
531,260	= 1% of Forecasted NEL in MWh for Current Year + 5 (=53,126,000 MWh × 1%)
303	= MW of PV Option needed to provide 1% of annual energy in Current Year + 5 (=531,260 MWh/1,752 MWh per MW of PV Option)
2.5	= Number of 120 MW PV Options needed to provide 1% of annual energy in Current Year + 5 (=303 MW/120 MW per PV Option)

The first and second rows of Table 10.1 are shaded to indicate that these are basic assumptions needed before the calculation can begin. These two rows show that our utility projects it will serve 53,126 GWh of energy in Current Year + 5 and that the annual capacity factor for the PV option is 20%.[3] In the third row, the 53,126 GWh of load is converted into 53,126,000 MWh of load. In the fourth row, our utility calculates that 1 MW of PV with a 20% annual capacity factor is projected to have an average annual output of 1,752 MWh (= 8,760 hours/year × 20% capacity factor).

The fifth row then shows that 1% of our utility's annual load projection in Current Year + 5 is 531,260 MWh. The sixth row divides the 531,260 MWh value by the 1,752 MWh per 1 MW of PV value to determine that our utility will need 303 MW of PV to serve 1% of its load in Current Year + 5 (= 531,260 MWh/1,752 MWh from 1 MW of PV).

Table 10.1 concludes by dividing this 303 MW of needed PV value by the nameplate rating for one PV option (120 MW). The result is that approximately 2.5 PV options of 120 MW each will be needed. Based on this calculation, it is apparent that, if selected, the 2 × 120 MW PV option will not quite serve 1% of the utility's annual energy in Current Year + 5. To do so would require 63 MW (= 303 MW – (120 MW × 2)) of additional PV. In other words, our utility has determined that if would need 303 MW of the PV option, or about 2.5 times the 120 MW nameplate rating for a single PV option, to serve 1% of its annual load in Current Year + 5 with zero-carbon emissions.

The first step of this estimation approach ended with a result that 303 MW of PV is needed to serve 1% of our utility's annual energy in Current Year + 5. The second step of this estimation approach is to extrapolate the result by multiplying this 1%-of-total-energy-amount-of-PV by 100 to derive a value of 30,300 MW (= 303 MW × 100) of the PV option. This value represents the estimated amount of PV MW that would be needed to serve 100% of its annual energy carbon-free in Current Year + 5. This projection of 30,300 MW

[3] Both these assumptions are unchanged from the analyses performed in Part I of this book for our hypothetical utility.

of new PV represents a huge amount of new resource additions for our utility, especially when we recall that our utility currently has a total generating capacity of only 12,000 MW.

However, such an extrapolation of 303 MW to 30,300 MW would apply only to the GWh load projected for Current Year + 5. Our utility recognizes that this estimation would not apply to future years in which the annual GWh load would be larger, such as in the target year of Current Year + 29. In addition, this estimation approach does not account for how our utility would serve its load in the nighttime when the sun is not shining.[4] Nor does it account for our utility's nuclear energy contribution that provides zero-carbon energy. Consequently, our utility concludes that this first approach for estimating how many MW of PV would be needed to serve all of its annual GWh loads in Current Year + 29 is an inadequate approach.

For this reason, our utility turns to a second approach for estimating how many new resources will need to be added. Our utility begins by examining the target year for its carbon goal, Current Year + 29, to see what would really be needed to serve all of our utility's GWh load in that year, including both daytime and night-time load.

THE SECOND ESTIMATION APPROACH

The second estimation approach will account separately for daytime and night-time load, plus the existing nuclear capability. Therefore, the second estimation approach should provide a more complete/meaningful view. We begin by making certain assumptions for the calculations that will then follow. These assumptions are presented in Table 10.2.

TABLE 10.2

Assumptions for the Calculation of the Amount of Resources Needed to Supply 100% of Our Hypothetical Utility's Annual Energy in Current Year + 29

Assumptions (Shaded)	Explanation
Current Year + 29	= Year assumed for example of zero-carbon calculation
PV option & battery storage	= Are the only new carbon-free resource options considered by our utility
65,750	= GWh for Current Year + 29
90%	= Assumed capacity factor for nuclear
7,884	= Annual nuclear GWh output (=(1,000 MW × 8,760 hours × 90%)/1,000 MWh per GWh)
57,866	= GWh to serve after accounting for nuclear (= 65,750 − 7,884)
20%	= PV Option average capacity factor
55%	= % of NEL during daylight or daytime hours
90%	= "Round trip efficiency" for battery storage
4	= Duration (hours) of battery storage effective full discharge during night-time before recharging the next day from PV

---

[4] There will be other instances when available solar energy will be reduced (heavy cloud cover, rain/snow, etc.). These instances are primarily addressed when projecting a PV facility's annual energy output, *i.e.*, capacity factor. Therefore, the main concern in this estimation exercise is accounting for how to serve the utility's load in hours when it is certain that there will be no available solar energy to utilize, *i.e.*, night-time hours.

The first assumption is the target year for our utility's carbon goal: Current Year + 29. Then, in a continued effort to keep the discussion as simple as possible, yet still demonstrate the key concepts and principles, we will restrict our utility's choices of resource options with which it can pursue a zero-carbon goal to just two different resources: the previously used 120 MW PV option and a battery storage option with a 4-hour duration.

Then, using the same load forecast that was used in all of the Part I analyses of this book, we find that the forecasted annual energy to be served in Current Year + 29 is 65,750 GWh.

Recall from Part I of this book that our utility currently has 1,000 MW of nuclear generating capacity on its system. It makes the assumption that the nuclear generation will still be operating in Current Year + 29 and assigns a 90% annual capacity factor of this nuclear capacity. As a result, nuclear generation is projected to serve 7,884 GWh of the annual load (= (1,000 MW × 8,760 hours/year × 90% capacity factor)/1,000 MWh/GWh) and will do so with zero-carbon emissions.

Consequently, the utility will only have to serve the "remaining" load (after accounting for nuclear's 7,884 GWh) with new zero-carbon emission resources. In other words, our utility can ignore these 7,884 GWh when determining how many additional new resource MW it will need to achieve zero-carbon emissions. The remaining energy load that must be served by these new resources is 57,866 GWh (= 65,750 − 7,884).[5]

Next, the table shows our utility retains its earlier assumption that the annual capacity factor assumption for the PV option is 20%.

Because our utility will have to serve load both during daylight hours and at night, an assumption is needed for what percentage of the annual hourly load will be in daytime hours. We assume that 55% of the annual GWh load will occur in daylight hours. This means that the remaining hours, or 45% of the annual GWh, will occur at night. Our utility will need resources to serve load during these night-time hours when the sun isn't shining.[6]

Our utility will use battery storage to address the night-time load. However, batteries are not 100% efficient. What that means is that if one charges a battery with 100 units of electricity, the battery will not be able to discharge all 100 units of electricity when called upon to provide electricity to the system. This "round trip" efficiency for batteries is roughly in the 85% to 90% range at the time the second edition of this book is written. Therefore, we will assume that our utility's battery storage option has a 90% round trip efficiency. In other words, our utility will need to charge the batteries with 100 MWh of energy in order to be able to discharge 90 MWh from the batteries.[7]

---

[5] Later in this chapter, after we have determined the amount of new resource MW our utility will need to serve all of its target year GWh with zero-carbon emissions after accounting for nuclear's contribution, we will see how many additional new resource MW would be needed if the nuclear capacity was not present in Current Year + 29.

[6] This assumption is based both on review of years of FPL actual loads and from discussions with resource planners at other utilities. However, this 55%/45% assumption for daytime/night-time load will vary from one utility to another (which once again reminds us of the Fundamental Principle 1 of Resource Planning).

[7] In Part I of the book, we discussed that the output of PV facilities degrade over time. Battery conversion efficiency also degrades over time although ongoing expenditures can be made to maintain (replenish) a battery's efficiency and capacity. In our ongoing effort to keep the discussion simple, we will assume that such expenditures are made and that our utility's batteries continue to have a 90% round trip efficiency throughout the analysis period.

The last of our assumptions is how long each battery can be counted on to dis-charge electricity to the utility system. Again, to keep the discussion simple, we will assume that all of our utility's batteries can discharge their maximum output (which we shall call the battery's capacity) for a maximum of 4 hours. I will refer to this as the battery having a 4-hour "duration." If the battery is discharged at its maximum capacity for these 4 hours, then the battery will need to be recharged from the elec-trical grid before it can again provide electricity.

At that point in time, the battery becomes an additional electric load while it is being recharged. And, assuming that we are discussing the target year, if a battery is used solely during night-time hours, it cannot be recharged from PV until the next morning when the sun is again shining (keeping in mind that our utility wishes to serve its customers at all hours with zero-carbon emissions which precludes recharg-ing the batteries with conventional non-nuclear generating units and the nuclear units are already directly serving the load).

With these assumptions in place, we will estimate the amount of PV and battery storage resources needed to enable our utility to fully meet its Current Year + 29 annual energy needs with zero-carbon emissions. This second estimation approach uses a four-step calculation:

- *Step 1*: Determine how many MW of PV are needed to serve the remaining GWh load (after accounting for nuclear) during the daytime hours;
- *Step 2*: Determine how many MW of PV are needed to serve the remaining GWh load during the night-time hours by charging batteries;
- *Step 3*: Combine the results of Steps 1 and 2 to determine how many total MW of PV are needed to serve the total GWh load remaining after nuclear in all hours; and
- *Step 4*: Determine how many MW of 4-hour batteries are also needed to serve the GWh load remaining after nuclear in the night-time hours.

The Step 1 calculation is presented in Table 10.3.

TABLE 10.3
Calculation Step 1: How Many MW of PV Are Needed to Serve All Daytime Hours in Current Year + 29?

Calculation Results	Explanation
57,866,000	= Annual MWh to be served after accounting for nuclear (=57,866 GWh × 1,000 MWh per GWh)
31,826,300	= Remaining MWh to be served in daytime hours after accounting for nuclear (=57,866,000 MWh × 55%)
1,752	= Annual MWh from 1 MW of PV option (=365 × 24 × PV Option average capacity factor of 20%)
18,166	= MW of solar needed to serve daytime hour MWh (=31,826,300 MWh/ 1,752 MWh per 1 MW of PV Option)
151	= Number of 120 MW PV Options needed to serve daytime hour NEL (=18,166 MW of PV needed for daytime hours/120 MW of PV Option)

The first value shown in Table 10.3 is the conversion of the remaining load (after accounting for nuclear's contribution) in Current Year + 29 of 57,866 GWh into 57,866,000 MWh. Then that MWh value is multiplied by the percentage of annual hours that are assumed to be the electric load served during daytime hours (55%) to determine the number of daytime MWh of load. That value is 31,826,300 MWh (= 57,866,000 MWh × 55%) which is the amount of daytime MWh load that PV will need to supply.

Next, we repeat a previous calculation from Table 10.1 to again determine that 1 MW of PV option can be expected to produce 1,752 MWh each year. Then we divide the 31,826,300 MWh that PV must supply in daytime hours each year by the 1,752 MWh that each 1 MW of the PV option is projected to supply.

The result is shown on the next-to-last row of the table: 18,166 MW (= 31,826,300 MWh/1,752 MWh per PV MW) of PV will be needed to serve our utility's annual energy load during daytime hours in Current Year + 29.[8] In terms of how many 120 MW PV options this equates to, the answer is 151 as shown in the last row in Table 10.3 (= 18,166 MW of PV/120 MW of our utility's PV option).

Now we turn our attention to Step 2 of the calculation in which we determine the amount of PV needed to serve load during night-time hours. We present that calculation in Table 10.4.

TABLE 10.4

Calculation Step 2: How Many MW of PV Are Needed to Serve All Night-Time Hours in Current Year + 29?

Calculation Results	Explanation
26,039,700	= Remaining MWh to be served in night-time hours after accounting for nuclear (= 57,866,000 MWh × 45%)
14,863	= MW of PV needed to meet remaining load during night-time hours before accounting for battery efficiency (=26,039,700 MWh/1,752 MWh per 1 MW of PV Option)
16,514	= MW of solar needed to meet remaining load during night-time hours after accounting for battery efficiency (= 14,863 MW/0.9)
138	= Number of 120 MW PV options needed to serve remaining load during night-time hours (= 16,514 MW/120 MW of PV Option)

The Step 2 calculations are similar to the calculations in Step 1. Table 10.4 first calculates how many MWh of remaining night-time load will need to be served in Current Year + 29. This result is 26,039,700 MWh (= 57,866,000 MWh × 45%). Then this value is divided by the same 1,752 MWh of PV annual output per 1 MW of PV option that was used previously in Step 1. The result is 14,863 MW (= 26,039,700 MWh/1,752 MWh per 1 MW of PV) as shown in the second row of Table 10.4.

---

[8] Note that in this estimation, we are assuming that our utility has decided it will not be converting (then operating) any of its fossil-fueled generating units to burn zero- or net zero-carbon fuels in Current Year + 29. If that assumption were to change, this estimation approach would need to account for that after assuming when, daytime or night-time, the converted units would likely be running.

However, this calculation has not yet accounted for the fact that batteries are not 100% efficient. More energy must be used to charge the battery than the desired discharged energy from the battery. Using our assumption that the round trip efficiency of our batteries is 90%, the 14,866 MW of PV option must be increased to account for this characteristic of battery storage. This is done by dividing the 14,866 MW value by the round trip efficiency value of 90%, resulting in 16,514 MW (= 14,866 MW/0.9) of PV needed to serve the night-time load through charging the batteries. This value equates to approximately 138 additional 120 MW PV options being needed as shown on the last row of the table (= 16,514 MW/120 MW of PV option).

Step 3 of the calculation then sums the amounts of MW of PV needed to serve the daytime hours and the PV MW needed to serve the night-time hours. This is shown in Table 10.5.

TABLE 10.5

Calculation Step 3: How Many Total MW of PV are Needed to Serve All Hours in Current Year + 29?

Calculation Results	Explanation
34,680	= Sum of MW of PV needed to serve remaining MWh load after accounting for nuclear (= 18,166 MW + 16,514 MW)
289	= Number of 120 MW PV Options needed to serve remaining MWh load (= 34,680 MW of PV needed/120 MW of PV Option)

Table 10.5 presents the simple derivation of how many total PV MW will be needed in Current Year + 29 to achieve zero-carbon emissions. This is done by adding together the PV MW needed to serve the annual MWh load during daytime hours (18,166 MW) and the PV MW needed to serve the annual night-time MWh load (16,514 MW). The result is that our utility will need 34,680 MW of PV option by that year. This equates to 289 of the 120 MW PV options.

Next, we present the fourth and final step of our calculation in which our utility estimates how many MW of battery storage will also be needed for our utility to meet its zero-carbon goal. This calculation is presented in Table 10.6.

TABLE 10.6

Calculation Step 4: How Many MW of Battery Storage Are Needed to Serve All Night-time Hours in Current Year + 29?

Calculation Results	Explanation
26,039,700	= Remaining MWh to be served in night-time hours after accounting for nuclear (= 57,866,000 MWh × 45%)
1,460	= Annual hours of discharge for 1 MW of battery storage (= 365 nights × 4 hours duration of batteries)
17,835	= Total MW of 4-hour batteries needed to meet remaining night-time load (= 26,039,700 MWh/1,460 hours of discharge per night per battery)

The starting point for the Step 4 calculation presented in Table 10.6 begins with the number of remaining (after accounting for nuclear) night-time MWh that must be served in Current Year +29 with zero-carbon emissions: 26,039,700 MWh that was shown previously in Table 10.4. Then, assuming that a 4-hour battery storage device will only operate once per night for each of the 365 nights in a year before it can be recharged the next morning by PV-produced energy[9], an annual energy discharge of 1,460 MWh per 1 MW of battery storage is derived (= 365 nights × 4 hours per night).

The 26,039,700 MWh annual night-time load in Current Year + 29 is then divided by the 1,460 MWh annual discharge from 1 MW of battery storage. The result is a projection that our utility will need 17,835 MW of 4-hour battery storage (= 26,039,700 MWh/1,460 MWh per 1 MW of battery).

These MW of battery resources are in addition to the 34,680 MW of PV that will also be needed, resulting in a total of 52,515 MW (= 34,680 MW of PV + 17,835 MW of batteries) of new resources needed for our utility to serve 100% of the GWh in Current Year + 29 with zero-carbon emissions.

This total MW amount of needed new resources, 52,515 MW, is certainly a lot of new resources for a utility system that currently has 12,000 MW of resources. In fact, it shows that our utility will have to install new resource MW that are approximately 4.4 times as much as its current generation MW (= 52,515/12,000). And, assuming our utility starts the process of adding these new resource MW in Current Year + 5, it will have only 25 years to make all these additions by Current Year + 29.

A logical question that one might ask at this point is as follows: *"How does this 52,515 MW of new resources needed to supply 100% of the energy load in Current Year + 29 compare to the firm MW amount of new resources that would be needed to solely meet the 20% total reserve margin requirement through Current Year + 29?"* This is an interesting question, but it is not one that is easy to answer.

To see why this is the case, let's restate the question as follows: *"If our utility did not set a zero-carbon goal for itself, how many firm MW of new resources would it need to build to maintain its 20% reserve margin?"* In trying to answer this restated question, the difficulty comes in knowing how many MW of PV and/or batteries it might build based solely on economics, especially considering the 2022 IRA's federal tax credits. As we shall discuss in the next chapter, the federal tax credits that resulted from the 2022 IRA have made PV (and other zero-carbon-emitting resources) much more economically attractive to utility systems. Therefore, utilities will almost certainly build more of such resources based on economics than they would absent these federal tax credits.

However, for purposes of trying to answer the question, we will temporarily ignore the economic aspect of resource planning. Instead, we will look at our utility's projected growth in summer peak load over the analysis period. Let's assume a scenario in which our utility does not set a carbon goal and will not retire any of its

---

[9] This once-per-night assumption for batteries is necessary given that our utility would want to recharge the batteries with zero-carbon emission energy and its only sources for doing so are nuclear (which has already been accounted for in the calculations of remaining load) and solar that does not operate at night. Consequently, the recharging of batteries by solar will need to wait until the next morning when the sun comes up and PV is again producing energy.

existing conventional generation by Current Year + 29. We will then see how many MW of new firm capacity would be needed to just meet the projected load growth. We do so in Table 10.7.

TABLE 10.7

Comparison of the New Resource MW Needed to Reach Our Utility's Zero Carbon Goal vs Firm MW Needed to Maintain a 20% Summer Reserve Margin

Calculation Results	Explanation
34,680	= Total PV nameplate MW needed to supply 100% of Current Year + 29 GWh with zero carbon emissions
17,835	= Total battery storage MW needed to supply 100% of Current Year + 29 GWh with zero carbon emissions
52,515	= Total new resource MW needed to supply 100% of Current Year + 29 GWh with zero carbon emissions (= 34,680 MW of PV + 17,835 MW of battery storage)
9,600	= Summer peak load in Current Year
12,500	= Summer peak load in Current Year + 29
2,900	= Growth in summer peak load during the analysis period (= 12,500 – 9,600)
3,480	= Firm capacity MW needed to meet 20% reserve margin (= 2,900 × 1.20)
15.1	= Ratio of total new resource MW needed to supply 100% of Current Year + 29 GWh vs total firm capacity MW needed to meet 20% reserve margin (= 52,515/3,480)

The first three rows of Table 10.7 simply repeat the number of new resource MW that our utility projects that it will need to reach its zero-carbon goal in Current Year + 29. As previously discussed, our utility will need 52,515 MW of new resources to supply 100% of Current Year + 29's GWh with zero-carbon emissions.

In the fourth and fifth rows, we see that the summer peak loads for our utility are projected to be 9,600 MW for the Current Year and 12,500 MW for Current Year + 29.[10] Taking the difference between these two values gives us the projected growth in summer peak load over that time period of 2,900 MW (= 12,500 − 9,600) as shown on the sixth row. In the seventh row, we multiply the projected summer peak load growth by our utility's 20% total reserve margin criterion to derive the amount of new firm capacity MW that will be needed solely to maintain system reliability. The result is 3,480 MW (= 2,900 × 1.20) of new firm capacity that would be needed to simply maintain a 20% summer reserve margin.

On the last row of Table 10.7, we divide the amount of new resource MW needed to supply 100% of Current Year + 29 GWh with zero-carbon emissions by the amount of new firm capacity MW that would be needed solely to meet the reserve margin criterion. The result is 15.1 (= 52,515 MW/3,480 MW). Therefore, our utility will need to build more than 15 times the MW amount of new resources to supply

---

[10] These projected summer peak load values were previously shown in Table 5.4, Column (2), in Part I of this book.

100% zero-carbon energy in Current Year + 29 than it would need to simply maintain a 20% reserve margin through that year.[11]

Figure 10.1 provides a graphic comparison of these values, plus our utility's current generation MW.

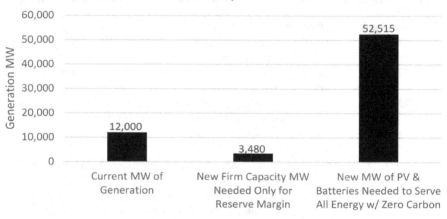

FIGURE 10.1 Generation MW: current MW versus two perspectives of needed new MW by Current Year + 29.

Based on the values shown in Figure 10.1, one might be tempted to immediately conclude that the 52,515 MW of new resources would easily meet the 3,480 MW of firm capacity needed to maintain our utility's reserve margin criterion. However, recall that the reserve margin calculation looks at only the peak hour in summer for our utility (which is a summer-peaking utility). Because the 17,835 MW of batteries are assumed to be operated only at night-time when the sun isn't shining, these MW cannot contribute to meeting our utility's reserve margin criterion. This leaves only the 34,680 MW of new PV "in play" regarding reserve margin calculations.

Furthermore, we have already discussed the fact that the PV option for Current Year + 5 has only a 50% firm capacity value. If this 50% firm capacity value applies to all new PV, then our utility would have 17,340 MW (= 34,680 × 0.50) of firm capacity supplied by the new PV. But our utility wonders if the firm capacity values for PV might change in the future.

---

[11] This very large MW amount of new resources being added by a utility, particularly if there is no federal or state mandate to get to zero carbon, might become an issue in regard to obtaining regulatory approval for these new resource MW. This issue is discussed in Chapter 14.

We will address the issue of system reliability in regard to PV in Chapter 11 after we first examine certain characteristics of PV. For now, we return our focus to the magnitude of the new resource MW needed to supply 100% of the Current Year + 29 GWh with zero-carbon emissions.

Faced with adding 15 times more resource MW than would be needed solely to maintain system reliability, our utility certainly faces a challenge from a resource acquisition perspective. Another way our utility might view this from a resource acquisition perspective is to look at having to add 52,515 MW of new resources over a 25-year time period (starting in Current Year + 5 through Current Year + 29). This equates to slightly more than 2,100 MW (= 52,515 MW/25 years) of new resource additions per year, each year, for 25 years. Each annual resource addition of 2,100 MW equates to more than one-sixth of our utility's 12,000 MW current total of existing generation capacity. Stated another way, every 6 years our utility will be adding as many new generating resources as it currently has today.

Sticking with this resource acquisition perspective, and focusing only on the PV additions, one might wonder what the land requirements would be for siting this amount of PV. Table 10.8 presents an estimate of this land requirement.

TABLE 10.8

Assumptions and Calculations for the Amount of Land Needed for Enough PV to Supply All of Our Hypothetical Utility's Annual Energy in Current Year + 29

Assumptions (Shaded) & Calculation Results	Explanation
8	= acres per MW of PV Option
640	= acres in a square mile
277,440	= acres for needed PV Options (= 34,680 MW of PV needed × 8 acres per 1 MW of PV)
434	= square miles needed for PV (= 277,440 acres needed for PV/640 acres per square mile)
21	= (assuming a square land area) miles on each side of the land area needed for the PV (= square root of 434)
27,000	= Walt Disney World (WDW) resort acreage (approximate)
10.3	= Number of WDW-equivalent areas required for needed PV MW (= 277,440 acres needed for PV/27,000 acres for WDW resort area)

The first assumption we make is to estimate the amount of land in acres it takes to site 1 MW of PV. The assumption our utility has chosen is 8 acres/MW as shown on the first row of Table 10.8.[12] We note that this value is intended to represent the

---

[12] This acreage value assumption is based on information publicly available in annual filings by Florida utilities regarding more than 4,000 MW of PV that has been installed PV through 2022, plus conversations with resource planners at a number of utilities. However, as has often been stated, different utilities will have different assumptions.

total amount of land that might be purchased when siting a PV facility. Typically, the actual footprint of the PV facility itself will require less land, perhaps 5 to 6 acres/MW. However, land is usually sold in parcels that the owner is willing to sell, and those parcels frequently are larger than what the utility will actually use for the PV. In addition, the land parcel may contain areas that are unsuitable for siting PV panels and other PV-related infrastructure due to the existence of streams/wetlands, elevation changes, timber, etc. Consequently, a utility often has to purchase more land area than would be needed if all land parcels were perfect for PV (flat, dry, and without trees), and if landowners were willing to sell only the exact amount of land that is needed for PV.

The second row simply reminds us that there are 640 acres in a square mile. The third row calculates the number of acres of land that is projected to be purchased for our utility's 34,680 MW of PV needed to reach the zero-carbon goal. That acreage is 277,440 acres (= 34,680 MW × 8 acres/MW). Because most people do not typically think of land in terms of acres, we translate the needed land area in terms of miles and square miles.

On the fourth row we divide the 277,440 acres by 640 acres per square mile to show that the required land area for our utility's needed amount of PV equates to approximately 434 square miles (= 277,440 acres/640 acres per square mile). Then, for illustrative purposes, let's assume that this total land area is in the shape of a square. With that assumption, the fifth row shows that the square would be about 21 miles per side. Thinking of it another way, if one were to drive around all four sides of this square land area, one would have to drive 21 miles per side × 4 sides, or 84 miles of total driving.

Yet another way to put the size of this land area into perspective is to compare it to another piece of land with which one might be familiar. To do this, we have chosen the Walt Disney World (WDW) resort in Florida.[13] At the time the second edition of this book is written, the WDW resort (theme parks, hotels, roads, etc.) encompasses approximately 27,000 acres as shown on the sixth row of Table 10.8. On the last row of the table, we return to the 277,440 acres of land our utility will need for the amount of PV necessary to supply all of our utility's energy with zero-carbon emissions in Current Year + 29 and then divide that acreage by the 27,000-acre size of WDW. The result is that our utility will need slightly more than ten times the size of Disney World's entire land area for the 34,680 MW of PV needed to supply 100% of our utility's Current Year + 29's GWh with zero-carbon emissions.

Therefore, looking at what it will take for our utility to serve 100% of its Current Year + 29 GWh with zero-carbon emissions from both a MW acquisition and a land acquisition perspective, our utility has a challenge on its hands in regard to the MW amount of PV resources it projects it will need.

Faced with such a large new resource MW projection, one might be tempted to ask: *"But how accurate is this projection that used the 2nd estimation approach?"*

We succumb to the temptation and decide to take a look.

---

[13] As one who has gotten lost more than once on the Walt Disney World property, I can attest that it is a very large tract of land.

HOW ACCURATE DO WE BELIEVE THE SECOND ESTIMATION APPROACH IS?

The second estimation approach is certainly an improvement from the original estimation approach that was based on determining how many MW of PV would it take to serve 1% of the annual load in 1 year, then multiplying that number of MW by 100. The second estimation approach is an improved one because it accounts for growth in annual load out to the target year, the need to serve night-time load that PV alone cannot do, and it incorporates the contribution of our utility's nuclear capability.

However, the second estimation approach is based solely on a projection of a single value: annual GWh sales. This method does not account for ongoing changes in a couple of key parameters (for example, load and amount of sunlight) that vary from hour-to-hour, day-to-day, or month-to-month. To do so would require the use of a sophisticated resource planning computer model that addresses monthly and hourly changes in both electric load and solar output. In addition, this type of model would determine when batteries should be charged and discharged on an hour-by-hour basis. In effect, the use of a resource planning model allows a more complete view of important discrete time interactions between resources and loads than an estimation approach based only on annual GWh sales is capable of providing.

Therefore, the question becomes: *"Would the use of such a resource planning model result in a significantly different projection of new resources needed compared to the results from the second estimation approach?"*

There are two ways we could go about answering this question. One way would require running a resource planning model in an analysis of our hypothetical utility system. The second way is to: (i) first examine a published result for an actual utility that used a resource planning model to determine how many MW of new resources would be needed to reach its zero-carbon goal, (ii) for that utility, apply the second estimation approach we have been using to develop an estimate of needed new MW of PV and batteries, and (iii) then compare the published results that used the utility's resource planning model with the results from our second estimation approach. I have decided to use the second way and to use information regarding FPL that was made public by NextEra Energy in its June 2022 announcement of a zero-carbon goal for the year 2045.[14]

As part of its announcement, NextEra Energy publicly issued a roughly 40-page document titled: "Zero Carbon Blueprint™" ("Blueprint"). This document addressed FPL as well as NextEra Energy's other companies. In the portions that discussed FPL, the Blueprint provided the following information regarding how many new MW of PV and battery storage were projected to be needed by FPL to achieve zero carbon by a 2045 target year. The relevant statements in the Blueprint are as follows:

- For PV: *"By no later than 2045, FPL would significantly expand its solar capacity ... to more than 90,000 MW ...."*

---

[14] This decision was made easier by the fact that, upon retirement, I no longer have ready access to a resource planning computer model (and, more importantly, my wife does want one in the house).

- For battery storage: *"Achieving Real Zero would add more than 50,000 MW of battery storage to FPL's grid ...."*

Other relevant statements in the Blueprint included the following:

- *"FPL presently has more than 3,500 MW of nuclear capacity on its system, and the plan intends for that capacity remain stable."*
- *"FPL would convert 16 GW of existing natural gas units to run on green hydrogen."*
- *"To ensure additional generation is available for reliability purposes, FPL would be able to generate up to 6,000 MW of CO_2-neutral power from renewable natural gas (RNG)."*
- *"FPL uses (a) ... commercial electric utility model ... in all of its resource planning work, including all of the low-carbon analyses. The model contains many data points relating to how the utility system operates (load forecast, unit data, new unit costs, etc.)."*

In regard to these bullet points, NextEra Energy's Blueprint provided only two of the input assumptions (*"data points"*) needed for our second estimation approach: a 2045 target year, and 3,500 MW of nuclear capacity. Unfortunately for our purposes, the Blueprint did not provide numeric values for several of the assumptions FPL used in its modeling, which would have been helpful in applying the second estimation approach. These values include (i) the forecasted GWh load in 2045, (ii) solar capacity factor, (iii) battery duration, and (iv) battery round trip efficiency.

However, we can estimate what a forecasted annual GWh load for FPL in 2045 might be using another document that is publicly available: FPL's Ten Year Power Plan Site Plan 2022–2031 (TYSP) that shows a number of forecasts and assumptions for the years 2022–2031. One of those forecasts is FPL's GWh forecast that showed projected values of 136,705 GWh for 2022 and 149,499 GWh for 2031. Using the 136,705 GWh value as a starting point, we calculate that a constant average annual growth rate of slightly less than 1% would result in the stated 2031 value of 149,499 GWh. We then assume a continuation of that same growth rate to extrapolate out to 2045. That projection is 171,821 GWh for 2045 that we shall use in applying the second estimation approach.

Because the Blueprint document does not provide the other three assumptions listed above that FPL used in its zero-carbon modeling, we will simply use the same values for these factors that were used in the previous estimation for our hypothetical utility: 20% annual capacity factor for PV, a 4-hour battery duration, and a 90% battery efficiency.[15]

---

[15] Using these prior assumptions will almost certainly result in the comparison we are making, between the results from a resource planning model and the second estimation approach, being less accurate than it would be if those inputs were known. However, this is of minor importance because what we are really after in making this comparison is a ballpark feel for how close the results of the two approaches might be.

After inserting these values for FPL into the second estimation approach tables[16], the projected amounts of PV and battery MW needed for FPL to achieve a zero-carbon goal in 2045 are 86,438 MW of PV, 44,454 MW of batteries, and 130,892 MW of new resources in total. Table 10.9 summarizes the projections of new resource MW needed for FPL to serve all of its 2045 energy with zero-carbon emissions using our second estimation approach and those stated by NextEra Energy using FPL's resource planning model.

TABLE 10.9

Comparison of Needed New Resources for FPL to Reach Zero Carbon by 2045: Using FPL's Resource Planning Model versus Using 2nd Estimation Approach

New Resource MW Needed	Using FPL's Resource Planning Model*	Using the 2nd Estimation Approach
New PV Needed (MW)	90,000	86,438
New Battery Storage Needed (MW)	50,000	44,454
Total New Resources Needed (MW)	140,000	130,892

* NextEra Energy's 2022 "Zero Carbon Blueprint™" document listed these values as "... *more than 90,000 MW* ..." for PV and "... *more than 50,000 MW* ...." for batteries.

We can draw two conclusions from this comparison: (i) the second estimation approach appears to have gotten us "in the ballpark" compared to the results from FPL's detailed computer modeling, and (ii) however, the second estimation approach underestimated the MW amount of new resources needed for both PV and batteries. This estimation approach underestimated the new MW needed by approximately 4% for PV and 11% for batteries. Overall, the second estimation approach resulted in significantly underestimating the total MW of new resources needed by more than 9,000 MW compared to FPL's resource planning model.

A logical follow-up question at this point would be: "*How might FPL's other projections – the ability to generate 6,000 MW using renewable natural gas and planning to have 16,000 MW (16 GW') of natural gas-fueled generation converted to run on green hydrogen – change the comparison results shown in Table 10.9?*"

This is a very good question and one we cannot answer with certainty given only the information presented publicly in the Blueprint. However, we will take a common sense approach in an attempt to answer this follow-up question. In doing so, we will address the two other projections separately.

The previously quoted passage from the Blueprint states that these 6,000 MW are to "... *ensure additional generation is available for reliability purposes* ...." This language makes it appear that this resource will be used primarily in times of very high loads and/or reduced output from other generating resources. If so, then

---

[16] The calculation steps are not shown in order to save time. However, the reader is invited to perform these calculations on his/her own.

this resource would appear to operate similarly to peaking capacity such as can be provided by a CT, or in a zero-carbon scenario, as similar to additional battery storage. Consequently, it seems a reasonable assumption that, if this resource was not an option (as is the case with our hypothetical utility), FPL's computer modeling might have selected even more battery storage MW.

We now take a look at how a utility might use the 16,000 MW of existing natural gas-fueled generating units that are planned to be converted to burn hydrogen as a fuel source. The Blueprint states that these units will operate on "*green hydrogen*." This term basically refers to hydrogen gas that is created using renewable (or other zero-carbon) energy to power the creation of hydrogen.[17] This newly created hydrogen would then be used as fuel to be burned in the converted generating units.

In order to simplify the discussion, we will ignore the process that produces the hydrogen (except to note that this process is almost certain to be less than 100% efficient). For our purposes, it will be sufficient to focus only on the efficiency with which the hydrogen fuel will be burned in the converted generating unit. Most existing natural gas-fueled generating units have a fuel-to-energy conversion efficiency of less than 60%.[18] Thus, at least 40% of the energy value in the hydrogen fuel will be unused by burning hydrogen in the generating units.

Therefore, it would seem logical to assume that FPL would try to use its two zero-emission energy sources, PV and nuclear, to supply energy from these sources directly to customers whenever possible, rather than primarily converting this energy into hydrogen to be burned in converted generating units. Because nuclear can supply zero-emission energy around the clock, it will be operated around the clock. But PV can only operate during daylight hours. Therefore, a logical approach would be to build enough PV to serve all of the non-nuclear load during daylight hours, then build even more PV to "feed" the hydrogen conversion process, and then store the hydrogen for hours when PV cannot directly supply the necessary energy.

Thus, it appears reasonable to assume that the primary use of green hydrogen would be during night-time hours or at other times when solar insolation is less than expected. If this assumption is correct, then the use of green hydrogen would also be used primarily as additional energy storage. Consequently, it also seems reasonable to assume that, if this green hydrogen resource was not an option (as is the case with our hypothetical utility), FPL's computer modeling might have selected even more battery storage MW.

Based solely on the information supplied in NextEra Energy's Blueprint document, we cannot know if our assumptions and conclusions are correct. However, if they are correct, and if these other two options had been assumed not to be available in FPL's modeling, it is likely that FPL's projection of "... *more than 50,000 MW* ..." of needed new batteries would have been significantly higher.

---

[17] This topic is addressed in Chapter 14.

[18] For example, a highly efficient CC unit with a heat rate of 6,200 BTU/kWh has a conversion efficiency of approximately 55% (= 3,413 BTU/kWh/6,200 BTU/kWh). Most existing generating units have a worse (higher) heat rate.

More importantly for the discussion of our hypothetical utility's projected need for new resources, this could mean that—using this same second estimation approach—our hypothetical utility <u>may</u> have underestimated its needed new resource MW and, perhaps, the associated land that would be needed.

Although our utility takes note that the second estimation approach may underestimate the amount of new resource MW needed, it believes that the second estimation approach appears to result in a projection that is at least in the ballpark. Therefore, our utility now pushes forward to take a look at how its projected new resource MW need using the second estimation approach would change if it did not have its nuclear capacity in Current Year + 29.

A "WHAT IF" ANALYSIS: HOW MANY ADDITIONAL NEW RESOURCE MW WOULD BE NEEDED IF OUR UTILITY'S NUCLEAR CAPABILITY WAS NOT AVAILABLE?

Earlier in this chapter we mentioned that, once we had completed our analyses of what new resources our utility would have to add to serve all of its Current Year + 29 energy with zero-carbon emissions, we would revisit those results assuming that our utility would <u>not</u> have its 1,000 MW of current nuclear capability available in Current Year + 29. Based on the results previously discussed, and using the second estimation approach for projecting the MW amount of new resources needed, we compare the estimates both with and without this nuclear capability in Table 10.10.

TABLE 10.10

Comparison of Needed New Resources for Our Utility to Achieve its Zero-Carbon Goal in Current Year + 29: With and Without its 1,000 MW of Current Nuclear Capability

Parameter	With Nuclear Capability	Without Nuclear Capability
Total load to be served in Current Year +29 with new zero-carbon emitting resources (GWh)	57,866	65,750
New PV Needed (MW)	34,680	39,405
New Battery Storage Needed (MW)	17,835	20,265
Total New Resources Needed (MW)	52,515	59,670
Square Miles Needed for New PV	434	493

Without our utility's nuclear capability (which provides energy with zero-carbon emissions during both daytime and night-time hours), our utility will have to serve its entire GWh load in Current Year + 29 with new resources. Therefore, as shown in the first row of Table 10.10, instead of having to serve only the remaining load (after accounting for the nuclear energy contribution) of 57,866 GWh, our utility would now have to serve all of the forecasted load of 65,750 GWh. This represents an increase in GWh of almost 14%.

After rerunning the same calculations previously discussed with a 14% higher load value, we find that the MW amount of new resources needed, and the corresponding land area needed for PV, have also increased by roughly the same 14% percentage. In regard to new PV MW, an additional 4,725 MW of new PV would be needed (= 39,405 − 34,680). An additional 2,430 MW (= 20,265 − 17,835) of new batteries would also be needed. Therefore, the total new resource MW that would be needed absent the nuclear contribution to our utility's carbon goal is 7,155 MW (= 59,670 − 52,515). Finally, in terms of land area needed for new solar, an additional 59 square miles (= 493 − 434) would be needed.

Therefore, from a resource acquisition perspective, our hypothetical utility's current nuclear capability is definitely an advantage in its quest to achieve a zero-carbon goal.

SUMMARY: KEY TAKEAWAYS REGARDING THE MAGNITUDE OF NEW RESOURCE MW NEEDED TO REACH A ZERO-CARBON GOAL

From a resource planning perspective, the key takeaways from this discussion of the magnitude of how many MW of new resources will need to be added by our hypothetical utility to enable it to supply 100% of the target year's GWh with zero-carbon emissions are as follows:

1. By looking at the amount of new resource MW that are projected to be needed to meet a zero-carbon goal for both our hypothetical utility and from the portion of NextEra Energy's 2022 Blueprint that pertained to FPL, it is clear that the needed MW amount will vary, and can vary significantly, from utility-to-utility.[19]
2. However, it is also clear that meeting the goal of supplying 100% of a utility's GWh with zero-carbon emissions with new resources, such as PV and batteries, will require an <u>enormous</u> amount of new resource MW. A utility can expect to have to acquire a MW amount of new resources that is some multiple of its total existing generation MW. Using our hypothetical utility system both with and without its nuclear capability, the MW amount of needed new resources was about 4.4 times our utility's current 12,000 MW of generating capacity assuming continued availability of our utility's existing nuclear capability, or about five times (= 59,670/12,000) assuming that nuclear capability was no longer available in Current Year + 29.[20]
3. When viewed from the perspective of comparing the new resource MW needed to supply all of our utility's energy with zero-carbon emissions by Current Year + 29, with the new resource firm capacity MW needed solely to continue to meet a 20% total reserve margin criterion, the zero-carbon

---

[19] Once again, Fundamental Principle 1 of resource planning, "All Electric Utilities are Different," comes into play.

[20] NextEra Energy's 2022 Zero Carbon announcement indicated that, for FPL, the MW amount of new resources needed to reach zero carbon was roughly five times its existing generation capacity of roughly 30,000 MW (=("*more than*" 50,000 MW of PV + "*more than*" 90,000 MW of battery + 16,000 MW of existing unit conversion plus hydrogen production facilities)/30,000 MW existing generation).

resource MW needed will be an order of a magnitude or more higher, *i.e.*, at least ten times more MW. Using our hypothetical utility system, the zero-carbon resource MW needed was approximately 15 times the new firm capacity MW needed solely to meet its reliability criterion.

4. However, a more meaningful way to project the magnitude of the needed new resources is through the use of a resource planning computer model. Other, simpler estimation approaches to developing projections of the needed new resource MW that have been used (primarily by non-utility entities) use incomplete information such as not examining the hour-to-hour changes in solar output, load, and when batteries should be charged/discharged. Although the second estimation approach appears, based on a single comparison effort, to provide results that are "in the ballpark" with those from a resource planning model, it also appears that the second estimation approach *underestimates* the needed new resource MW.

5. In addition to the needed amount of new resource MW, the land area that would be required for the needed PV MW will also be significant. For example, our hypothetical utility would need more than 430 square miles for its needed PV MW.

In this chapter, we have taken a look at two of the challenges facing a utility that sets a zero-carbon goal: the amount of new resource MW, and associated land, that will be needed to supply 100% of our utility's GWh in Current Year + 29 with zero-carbon emissions. In the next chapter, we will focus in more detail on PV resources. In so doing, we will examine certain characteristics of solar that our utility's resource planners will need to take into account, including the firm capacity value of PV. We will also discuss system reliability and unit retirement during the transition period.

11 Moving toward Zero Carbon

Resource Planning Considerations for Solar (PV) Resources

INTRODUCTION

In the previous chapter, we saw that, in order to supply 100% of its Current Year + 29 GWh load with zero-carbon emissions, our hypothetical utility was projected to have to add an enormous amount of new PV (34,680 MW) even using the second estimation approach that, as we discussed, may underestimate the amount of needed new resource MW. Because our utility will have to add so much new PV, this chapter focuses solely on PV resources.

This chapter begins with an examination of trends in solar installed costs in the years leading up to 2022 when the second edition of this book was written. Then we will look at what impact the 2022 Inflation Reduction Act's (IRA's) tax credits have on the economics of the PV option we previously analyzed in Chapter 5. Most of the remainder of this chapter then focuses on PV's firm capacity values for winter and summer peak days. These firm capacity value characteristics of PV resources are ones that resource planners will have to account for as our utility installs very large amounts of solar each year. In addition, the topics of system reliability and existing unit retirement during the transition period will be discussed.

SOLAR INSTALLED COSTS ($/KW): 2009–2022

In Chapter 5, one of the Supply options our hypothetical utility analyzed to meet its resource needs in Current Year + 5 was a 120 MW PV option. Based on the assumptions used in those analyses, the Supply Only Resource Plan 3 (PV) emerged as the second best Supply option resource plan (behind the Supply Only Resource Plan 1 (CC)) in the economic evaluation. Once the analyses were expanded to include two With DSM resource plans, the Supply Only Resource Plan 3 (PV) slipped to the third best resource plan in the economic evaluation (behind With DSM Resource Plan 1, followed by the Supply Only Resource Plan 1 (CC)).

In those analyses, the installed cost of the PV option had been updated from what was assumed in the first edition of this book. When the first edition of the book was written in 2011 (and published in 2012), the installed cost for the PV option was then

DOI: 10.1201/9781003301509-13

assumed to be $5,000/kW. That assumption was based on PV installations that had occurred up to that time by a number of utilities (including FPL). Now, using FPL information that is in the public domain regarding its PV installed costs, let's take a look at how PV costs have changed from 2010 to the present.

In 2010, FPL installed two PV facilities in response to Florida legislation that encouraged Florida utilities to gain experience with utility-scale (*i.e.*, multi-MW) solar facilities. This legislation also ensured that the costs for these PV facilities, which were clearly not cost-effective, would be fully recoverable from a utility's customers. The installed cost of these two PV facilities ranged from approximately $6,900/kW to $7,900/kW.

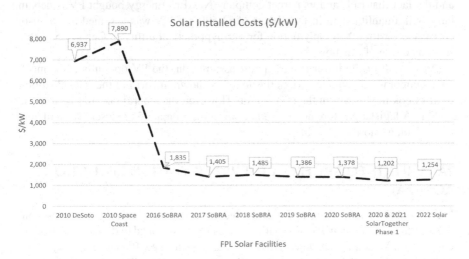

FIGURE 11.1 Solar installed costs ($/kW): FPL PV installations from 2010 through 2022. Costs are in nominal dollars in the installation year.

Since that time, the $/kW installed costs of PV have decreased significantly. Again, using FPL PV cost information that is in the public domain, Figure 11.1 presents the installed costs of a number of PV installations FPL made through 2022.[1]

By 2016, the installed costs of PV had declined significantly to the point where FPL's IRP analyses showed that PV installations were becoming cost-effective on FPL's system. As a result of these analyses, FPL built several approximately 75 MW PV facilities at different sites at an average installed cost of roughly $1,800 per kW. FPL also began including additional future PV facilities in its official resource plans at that time as shown in its annual Ten Year Power Plant Site Plan filings.

---

[1] These cost values include all PV-related equipment (panels, inverters, etc.) as well as costs for inter-connecting the PV facility to the existing transmission system. For PV installations in the earlier years only, there were zero costs for land (because the land had been purchased previously and those costs were already included in FPL's rate base) and essentially zero costs for integrating the output from the PV facility into the entire transmission system (due to the size of the PV installations and the location of the sites).

FPL added more PV in subsequent years as PV's installed costs continued to decline. Those costs declined to approximately $1,200 per kW through 2021. However, in 2022, the installed costs for PV edged upward to approximately $1,250 per kW largely due to supply chain issues aggravated by the COVID-19 pandemic and to somewhat higher land costs as increasing amounts of land were acquired for what was clearly a long-term plan by FPL to continue to build PV.

The key point regarding the information presented in this figure is that actual PV costs have significantly declined in the last decade or so since the book's first edition used an installed cost value of $5,000 per kW. With that in mind, and after accounting for transmission integration costs, increasing costs for land suitable for PV siting, and the fact that FPL and its parent company NextEra Energy bought PV panels in huge bulk quantities, an installed cost of $1,500 per kW was assumed as an all-in cost value for the PV Supply option for our hypothetical utility in Chapter 5 of this second edition of this book.[2]

As a result of the lower installed cost assumption, the PV option became much more economically competitive in the analyses shown in Part I of the second edition of this book than it was in the first edition. However, the federal tax credits from the 2022 IRA legislation were not accounted for in the Chapter 5 analyses. We will now take a look at those tax credits.

TAX CREDITS FOR PV FROM THE 2022 INFLATION REDUCTION ACT (IRA)

For a number of years prior to 2022, the federal government allowed an Investment Tax Credit (ITC) for new PV installations. The ITC was applied to the capital/installed cost of the PV facility. The tax credit itself was 26% for new PV facilities that would be installed through the year 2023. After that year, the tax credit level began dropping in a step-like manner to 22% in 2024, and 10% for 2025-on. The intent of the federal tax credits was to encourage electric utilities to build new solar facilities both earlier, and in greater quantity, than would be the case if the ITC did not exist.

However, as previously mentioned, the federal government passed a massive piece of legislation in 2022, the IRA. Included in this legislation were significant new federal tax credits for resource options that are zero-carbon-emitting resources. Among those resource options that are eligible for these tax credits are solar resources, including PV.

As previously mentioned, the new 2022 IRA tax credits were not accounted for in the updated economic analyses that appear in Chapter 5 of this second edition of the book. There were a couple of reasons for this. The first reason is, as explained at the beginning of the book, the intended audience of this book includes folks who are not energy professionals. Regarding the Supply options that were used in

---

[2] It is probably useful to again remind readers that costs for all resource options, as well as for fuel and environmental compliance, are constantly changing. In order to write this book, assumptions for these cost values had to be selected. So don't get hung up on whether one agrees with a particular cost assumption. The cost assumptions are best thought of as placeholders, which allow the narrative about how resource planning analyses are performed to proceed.

Chapter 5 for our hypothetical utility—CC, CT, and PV options—the new federal tax credits apply only to the PV Supply option and do not apply to either the CC or CT options.

Therefore, part of the reason the decision was made to not account for the new tax credits in Chapter 5 was to keep the Chapter 5 discussion of resource planning analyses as simple as possible by keeping all of the Supply options on a level playing field (*i.e.*, no new tax credits). Then, using the information and analyses presented in Chapter 5 as a starting point, the new 2022 IRA tax credits that apply only to the PV option would be discussed in Part II of the book.

The second reason that the IRA's new tax credits for PV were not included in the analyses presented in Chapter 5 is that any tax credits (and/or penalties) are legislatively driven and, therefore, subject to potential change in the future. Tax instruments such as the IRA's new credits could be modified, or even eliminated, by future administrations due to changes in the U.S. economy and/or simply due to changing political winds. In addition, some utilities may now, or in the future, have such low federal tax obligations that they cannot take full advantage of these federal tax credits.

By waiting to discuss the impact of the tax changes separately in Part II of this book, the reader is shown the economic impacts of the PV option without the new tax credits in Chapter 5, then with the tax credits in this chapter. This better allows a reader to see what the impacts of the new tax credits are.

Now let's address the IRA's tax credits for new PV facilities in very simple terms. These new tax credits come in two basic types, and a utility can choose to utilize one, but not both, of the two types of credits. One type is basically a modification of the previously available ITC. The IRA's new ITC for solar increased to 30% starting in 2022 and remains at that level through at least the year 2041.

The IRA's other type of tax credit is a new one: a Production Tax Credit or PTC. Rather than being based on the capital cost of a new PV facility, the PTC is based on the annual MWh produced annually by the new PV facility. The greater the number of MWh produced by a PV facility, the greater the amount of the tax credit. As the second edition of this book is written, it appears that the PTC will be the type of tax credit that many, if not most, utilities will use because the new PTC provides more economic benefit to utilities than does the revised ITC (particularly if PV capital costs continue to decline).

Our discussion will begin with a look at how the PTC would economically impact 1 year of our utility's projected costs assuming the Supply Only Resource Plan 3 (PV) was selected. Then we will take a look to see how the CPVRR cost and level-ized electric rate for this resource plan would change after accounting for the PTC for all 25 years of the prior analyses, *i.e.*, for Current Year + 5 (when the PV option would be added) through Current Year + 29 (the end year of the analysis period). The explanation and examples will be simplified ones.

The IRA's PTC basically starts by providing a tax credit of approximately \$28 for each MWh produced from eligible new PV in the 1ˢᵗ year the PV facility operates. However, evaluations of how the tax credit could actually be applied that took place as the second edition of this book was written indicated that at least some electric utilities may be able to "gross up" the PTC to approximately $37 per MWh.[3] With that in mind, we will use both sets of values, $28/MWh and $37/MWh, in the examples to follow. With either scenario, after the PV facility's 1st year of operation, the PTC $/MWh value escalates with inflation each year thereafter.

Going back to our PV option in Chapter 5, we are reminded that this PV option was assumed to have a nameplate rating of 120 MW and an annual average capacity factor of 20%. Therefore, the annual MWh that is projected to be produced by this PV facility for a typical year would be:

$$120 \text{ MW} \times 8,760 \text{ hours/year} \times 20\% \text{ capacity factor} = 210,240 \text{ MWh per year}$$

If we assume that our utility can utilize the full tax credit, and that the PTC's starting value is $28 per MWh, the PTC for the first full year of the PV option's operation would amount to:

$$210,240 \text{ MWh per year} \times \$28/\text{MWh} = \$5,886,720 \left(\text{or approximately } \$5.89 \text{ million} \right)$$

If we assume that the PTC's starting value is $37 per MWh, the PTC for the first full year of the PV option's operation increases as follows:

$$\$5.89 \text{ million} \times \left(\$37/\$28 \right) = \$7.78 \text{ million}$$

The astute (and engaged) reader might well ask: *How do these PTC annual tax credit amounts compare to the annual fuel savings that the PV option is projected to have on our hypothetical utility system?* That is a good question because it helps to put the purely tax-driven, "artificial" value of the PTC in context with "real" fuel cost savings that the PV option will actually provide.

The answer would depend in part on what year in the 25-year analysis time period we are using to answer the question because the projected cost of our utility's marginal fuel (natural gas) increases each year. For the purpose of answering this question, we will assume we are looking at the PV facility's 1st year of operation in Current Year + 5. The projected cost of natural gas, which started at $6.00/mmBTU in the Current Year, is projected to have escalated to $6.62/mmBTU in Current Year + 5. Let's further assume that: (i) all of the MWh that the PV facility would produce that year displaces MWh that would have come from the existing CC units on the utility system, and (ii) the average heat rate of these existing CC units on the system is 7,300 BTU/kWh.

---

[3] This approach was explained to me as follows: the utility grosses up the listed ($28/MWh) PTC by dividing that value by (1 minus the utility's overall tax rate). For example, if the utility's overall tax rate is 25%, then $28 / (1 − 0.25) = approximately $37/MWh. When the explanation then addressed why it is appropriate to gross up a tax credit using the tax rate, I began to get a headache. Therefore, for the sake of my own health, and possibly yours, no additional explanation will be attempted. (You are welcome.) In addition, the overall tax rate will vary from one utility to another (as Fundamental Principle # 1). For the sake of simplicity, we will (conservatively) use a 25% tax rate for our hypothetical utility that results, if applicable, in a grossed up PTC of $37 per MWh.

With those assumptions, we calculate the 120 MW PV option's system fuel savings for Current Year + 5 as follows:

210,240 MWh of PV output × 1,000 kWh/MWh × 7,300 BTU/kWh × 0.000001 mmBTU/BTU = 1,534,752 mmBTU of existing CC unit output displaced by PV in Current Year + 5;

then,

1,534,752 mmBTU × $6.62/mmBTU = $10,160,058 (or approximately $10.16 million) in projected fuel savings from the 120 MW PV option

Therefore, for Current Year + 5, and assuming a $28 per MWh PTC starting value, the projected PTC of $5.89 million represents an additional, solely tax-driven benefit to the utility that is approximately 58% (= $5.89/$10.16) of the projected annual fuel cost savings of $10.16 million for PV. If we assume a $37 per MWh PTC starting value, the projected PTC of $7.78 million represents an additional benefit to the utility that is approximately 77% (= $7.78/$10.16) of the projected annual fuel cost savings of $10.16 million.

Regardless of the PTC value assumption, the additional benefit to the PV option granted by the IRA's federal tax credits is a significant one.

IMPACTS OF PTC OVER THE 25-YEAR ANALYSIS PERIOD

We will now extend the calculation of the PTC for the PV option in the Supply Only Resource Plan 3 (PV) for all of the entire 25-year period used in our previous analyses in Chapter 5. But let's first refresh our memory of how this resource plan, and the PV option upon which it is based, fared in the analyses discussed in Chapter 5. Table 11.1 is a condensed, summary version of previously presented Table 5.15 that showed the CPVRR cost economic evaluation results of the three Supply Only Resource Plans. (As a reminder, the new PTC for the PV option was not accounted for in the previous analyses whose results are shown here.)

TABLE 11.1

Summary of Economic Evaluation Results of Supply Only Resource Plans Without PTC for Solar: CPVRR Costs

Resource Plan	Total Costs (Millions, CPVRR)	Difference from Lowest Cost Supply Only Plan (Millions, CPVRR)
Supply Only Resource Plan 1 (CC)	41,810	0
Supply Only Resource Plan 2 (CT)	41,883	73
Supply Only Resource Plan 3 (PV)	41,829	19

As shown in Table 11.1, the Supply Only Resource Plan 3 (PV) was projected to have higher CPVRR costs than Supply Only Resource Plan 1 (CC). The difference in CPVRR costs was $19 million.

The calculation of PTC for a 120 MW PV facility for a single year (Current Year + 5) that was just discussed is now extended by assuming that the escalation for the PTC starting value is 2.5% per year. Using this escalation assumption, the same 1-year calculation is now performed for future years, then all of those annual PTC nominal values are discounted back to the Current Year using the same 8% discount rate we used in all of the Part I analyses. The calculation also accounts for the fact that our utility has to add two 120 MW PV facilities to meet its 20% reserve margin criterion in Current Year + 5. The resulting net present value sums of the PTC over the 25-year analysis period are:

Assuming a $28 per MWh starting PTC value, the PTC total benefit = **$115 million**;

and,

Assuming a $37 per MWh starting PTC value, the PTC total benefit = **$152 million**.

We now take those projected PTC total net present value sum of tax benefits for the PV option and apply them as CPVRR cost savings for the Supply Only Resource Plan 3 (PV). The results are presented in Table 11.2, which is an extension of Table 11.1.

TABLE 11.2

Summary of Economic Evaluation Results of Supply Only Resource Plans With and Without PTC for Solar: CPVRR Costs

Resource Plan	Total Costs (Millions, CPVRR)	Difference from Previous Lowest Cost Supply Only Plan (Millions, CPVRR)
Supply Only Resource Plan 1 (CC)	41,810	0
Supply Only Resource Plan 2 (CT)	41,883	73
Supply Only Resource Plan 3 (PV)—w/o PTC	41,829	19
Supply Only Resource Plan 3 (PV)—w/ $28 per MWh PTC	41,714	(96)
Supply Only Resource Plan 3 (PV)—w/ $37 per MWh PTC	41,677	(133)

In Table 11.2, the first three rows are identical to the three rows presented in Table 11.1, but a fourth and a fifth row have been added. The fourth row shows the projected CPVRR results assuming a PTC with a starting value of $28 per MWh is applied to the Supply Only Resource Plan 3 (PV). The fifth row shows the projected CPVRR results assuming a PTC with a starting value of $37 per MWh is applied to the same resource plan.

As indicated in Table 11.2, application of the PTC using either of the $ per MWh starting values now results in a projection of the Supply Only Resource Plan 3 (PV) having the lowest CPVRR value of any of the three Supply Only resource plans.

Assuming a $28 per MWh PTC starting value, Supply Only Resource Plan 3 (PV) is now projected to be $96 million CPVRR less expensive than the previous lowest CPVRR cost Supply Only Plan (Supply Only Resource Plan 1 (CC)). If a $37 per MWh PTC starting value is assumed, then Supply Only Resource Plan 3 (PV) is projected to be $133 million CPVRR less expensive than the previous lowest CPVRR cost Supply Only resourced plan.

However, recognizing that all of the Supply Only resource plans, including Supply Only Resource Plan 3 (PV), will still need to compete with the two With DSM resource plans on a levelized system average electric rate basis, Table 11.3 now extends Table 11.2 by adding another column, which presents the respective electric rate values for the Supply Only plans.

TABLE 11.3

Economic Evaluation Results of Supply Only Resource Plans With and Without PTC: CPVRR Costs and Levelized System Average Electric Rates

Resource Plan	Total Costs (Millions, CPVRR)	Difference from Previous Lowest Cost Supply Only Plan (Millions, CPVRR)	Levelized System Average Electric Rate (cents/kWh)
Supply Only Resource Plan 1 (CC)	41,810	0	12.053
Supply Only Resource Plan 2 (CT)	41,883	73	12.064
Supply Only Resource Plan 3 (PV)—w/o PTC	41,829	19	12.056
Supply Only Resource Plan 3 (PV)—w/ $28 per MWh PTC	41,714	(96)	12.040
Supply Only Resource Plan 3 (PV)—w/ $37 per MWh PTC	41,677	(133)	12.035

As expected with the lower CPVRR costs (recalling that a utility's costs comprise the numerator in an electric rate calculation), and no change in the number of GWh of sales (the denominator in the electric rate calculation), the two "with PTC" versions of Supply Only Resource Plan 3 (PV) now have lower projected levelized system average electric rates than they did without the PTC.

As one would expect, these two "with PTC" resource plans have the lowest projected electric rates of any of the Supply Only resource plans. Those projected electric rates are, respectively, 12.040 cents/kWh (assuming a $28 per MWh starting PTC value) and 12.035 cents/kWh (assuming a $37 per MWh starting PTC value). Both of these projected electric rates are lower than the electric rate for the Supply Only Resource Plan 1 (CC), which previously had the lowest electric rate of 12.053 cents/kWh.

All that is left to do is to see if either, or both, of these two "with PTC" versions of Supply Only Resource Plan 3 (PV) have a lower projected levelized system average electric rate than the previous winner from the electric rate perspective: With DSM Resource Plan 1. That information is provided in Table 11.4.

In Table 11.4, new fourth and fifth rows have been added. These new rows present the economic information for the two With DSM resource plans.

TABLE 11.4

Economic Evaluation Results for All Five Resource Plans With and Without PTC: CPVRR Costs and Levelized System Average Electric Rates

Resource Plan	Total Costs (Millions, CPVRR)	Difference from Previous Lowest Cost Supply Only Plan (Millions, CPVRR)	Levelized System Average Electric Rate (cents/kWh)
Supply Only Resource Plan 1 (CC)	41,810	0	12.053
Supply Only Resource Plan 2 (CT)	41,883	73	12.064
Supply Only Resource Plan 3 (PV)—w/o PTC	41,829	19	12.056
With DSM Resource Plan 1	41,597	(213)	12.048
With DSM Resource Plan 2	41,376	(434)	12.093
Supply Only Resource Plan 3 (PV)—w/ $28 per MWh PTC	41,714	(96)	12.040
Supply Only Resource Plan 3 (PV)—w/ $37 per MWh PTC	41,677	(133)	12.035

As shown in Table 11.4, assuming either a $28 or $37 per MWh PTC starting value, the projected electric rate for Supply Only Resource Plan 3 (PV) is lower (12.040 or 12.035, respectively) than the 12.048 cents/kWh electric rate of the With DSM Resource Plan 1.

This outcome would make Supply Only Resource Plan 3 (PV) the overall economic winner among all of the resource plans that our hypothetical utility analyzed because it would have the lowest levelized system average electric rate. The 2022 IRA's new tax credits moved Supply Only Resource Plan 3 (PV) from third place to first place in the economic comparison.

WHAT DO WE CONCLUDE FROM THIS DISCUSSION OF THE 2022 IRA'S PTC IMPACT ON SOLAR RESOURCE OPTIONS?

The clear intent of the portion of the 2022 IRA that pertains to new federal tax credits for solar and other zero-carbon-emitting resource options was to make these resource options more attractive to electric utilities by lowering the overall cost of the options to the utility. By doing this, utilities can more easily justify moving toward a low- or zero-carbon-emitting utility system by incorporating increasing amounts of these resources.

As seen in the preceding discussion of how the economics of the PV option would change for our hypothetical utility by applying the PTC, the projected economics of PV is significantly enhanced with the PTC. Therefore, one can conclude that this one objective of the 2022 IRA legislation—to make solar and other renewable energy options more attractive economically—has been met. One would then assume that, assuming all else is equal, many more PV MW would be added by electric utilities across the United States based on PV's improved economics for the utility systems. In addition, one would expect that the PV additions would occur earlier than would be the case without the new federal tax credits.

Speaking of many more MW of PV, we saw in Chapter 10 that, because our hypothetical utility wants to serve 100% of its GWh load in Current Year + 29 with zero-carbon emissions, it will need to add a very large amount (at least 34,680 MW) of new PV. As stated in Chapter 10, this PV MW amount was based solely on how to supply all of the energy needed in Current Year + 29. This estimated amount of new PV MW needed to supply this amount of GWH did not address whether this amount of PV would also allow our utility to continue to be a reliable system.

However, before we address system reliability during the transition period, we need to look at certain characteristics of PV resources in regard to firm capacity values for PV that resource planners will need to take into account, especially as ever-increasing amounts of PV are added to a utility system. For that reason, we will next take a look at the firm capacity values of PV.

A MORE COMPLETE LOOK AT THE FIRM CAPACITY VALUE ASPECT OF SOLAR

In Chapter 5, we assumed that the firm capacity value of our utility's PV option at the utility's summer peak hour was 50% of the 120 MW nameplate rating. In other words, only 60 MW (= 50% × 120 MW) of the PV option could be counted on to be supplied at our utility's summer peak load hour. As a consequence, in order to meet our utility's 120 MW summer resource need in Current Year + 5, our utility had to install two of the 120 MW PV options, and this was accounted for in our economic and non-economic analyses.

In so doing, we accounted for the firm capacity value contribution of the PV option for (i) the utility's summer peak hour only (but not for its winter peak hour), and (ii) for only the 240 MW addition of PV in this 1 year (Current Year + 5). Therefore, we have not yet addressed what the PV option's firm capacity value contribution may be as our utility analyzes its winter reserve margin. Nor have we examined whether the PV option's summer firm capacity value will remain at 50% for all additional solar increments that the utility will have to add in future years in order to meet its zero-carbon goal. We will take a look at both of those issues now, starting with the winter firm capacity value for PV.

FIRM CAPACITY VALUE OF PV AT OUR UTILITY's WINTER PEAK HOUR

We have not yet introduced our hypothetical utility's winter peak load or the shape of the utility's hourly load on the winter peak day as we previously did for the summer peak day in Figure 2.1.[4] By applying the same approach that was used to create the utility's summer peak day, and assuming an annual increase in winter peak load of 90 MW, our utility is assumed to have a winter peak day load in Current Year + 5 of

---

[4] In an effort to avoid potential confusion, recall that Figure 2.1 shows a summer peak day with a peak hour load of 10,000 MW. This actually represents the projected summer peak load for Current Year + 4, which occurs just before a projected 100 MW increase in summer peak load results in our utility needing to add resources in Current Year + 5.

9,750 MW (compared to a summer peak load of 10,100 MW in the same year as was shown previously in Table 5.4 in column (2)). The hourly load shape on the winter peak day for Current Year + 5 is presented in Figure 11.2.

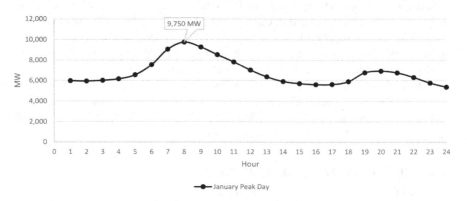

FIGURE 11.2 Representative winter peak day load shape for a hypothetical utility with a winter peak load of 9,750 MW.

When comparing the basic shapes of the load for peak days in summer versus winter, two things are apparent.[5]

First, the utility's summer peak day highest hourly load of 10,100 MW for Current Year + 5 is definitely greater than its winter peak day highest hourly load of 9,750 MW for the same year. Thus, we can surmise that our utility is a summer-peaking utility (although it may occasionally have an extreme winter load that results, for that year, in the winter peak load exceeding the summer peak load). This means that our utility will typically have its future resource needs driven by summer reserve margin, and the utility will typically perform its resource planning work based on meeting summer reserve margin. This was the assumption used in the analyses of our utility system previously discussed in Chapters 5 and 6.

However, the utility will always need to also keep the winter reserve margin criterion in mind as it looks into the future in its reliability analyses. To do so, a utility will perform winter reserve margin analyses that examine both its winter peak load, plus the winter firm capacity values of its existing resources and potential future resource options that may be added.

This brings us to the second thing we notice in Figure 11.2: the basic shapes of the hourly load on the two seasonal peak days are very different. Our utility's winter peak load hour occurs in the early morning in hour 8. This is significantly different from the summer peak load hour that occurs in hour 17. This leads to the logical question: "*What is the PV output at the much earlier winter peak load hour?*" To answer that question, we turn to Figure 11.3.

---

[5] Note that we have assumed that the basic ratios of load from any one hour to its adjacent hours remain the same over the years as the peak load grows. One set of ratios is used for summer peak days, and another set of ratios is used for winter peak days. Therefore, when comparing the basic summer shape of the hourly load to the winter shape of the hourly load, the differences in the load shapes apply regardless of the year in question.

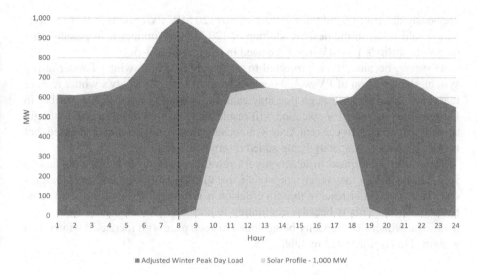

FIGURE 11.3   Adjusted winter peak load and profile for 1,000 MW solar.

In order to make it easier to view PV's output over all hours of our utility's winter peak day, our utility's hourly load has been adjusted so that the peak hour's load has been reduced from 9,750 MW to 1,000 MW. The peak load still occurs in hour 8. Then all of the other hours on the utility's winter peak day are adjusted accordingly using the same "rationing" approach. This adjusted hourly load is shown in the dark shaded area in the figure. Finally, the projected hourly output of 1,000 MW (nameplate) of PV has also been shown in this figure by the lighter shaded area. The objective of Figure 11.3 is to make it easier to see how our utility's winter peak day hourly load shape matches up with PV's hourly output on our utility's winter peak day.

The result shown in Figure 11.3 is that there is virtually no expected PV output at the utility's winter peak hour 8. Only a very small sliver of PV output can be seen at this hour. Therefore, our utility would assign a winter firm capacity value for the PV option of 0 MW, or close to 0 MW, in its winter reliability analyses.

This would not be the case with the other Supply options (CC and CT) that the utility also considered in Part I. These other options typically have fairly consistent firm capacity values for both summer and winter, *i.e.*, the CC and CT options will have winter firm capacity values fairly close to their respective 500 MW and 160 MW summer firm capacity values. CC and CT resources typically have relatively slight differences in their seasonal firm capacity values occurring due to differences in ambient air temperatures in winter versus summer. (Cooler, denser air in winter typically results in slight increases in projected output—and slightly higher winter firm capacity values—for most conventional types of generators when comparing winter versus summer outputs.)

Even ignoring for a moment our utility's commitment to be carbon-free by Current Year + 29, from an economic-only perspective, and assuming all else equal, the solar PTC would tend to lead our utility to select Supply Option Resource Plan 3 (PV) for meeting its summer reserve margin criterion in Current Year + 5 as previously discussed. And the selection of two 120 MW PV options in that resource plan would result in the desired addition of 120 MW of summer firm capacity that meets our utility's 120 MW resource need in Current Year + 5.

However, because PV is projected to contribute 0 MW in winter for our utility system, the selection of PV would contribute no MW to our utility's winter reserve margin in that year. Although that may cause no reliability issues for our utility in Current Year + 5, the PV selection will result in the utility's winter reserve margin actually decreasing in Current Year + 5 because the winter load has increased, but no new winter firm capacity being added if only PV is selected.

If our utility continues to select only PV resources in future years as it pursues its zero-carbon goal, it may reach a point that, due to decreasing winter reserve margins each year, the winter reserve margin criterion is now driving the utility's projected resource needs. This is because continuing to add only PV resources results in no additional winter firm capacity being added to keep pace with increased winter load growth. This is illustrated in Table 11.5.

TABLE 11.5

Winter Peak Reserves and Total Reserve Margins for Our Utility If No Additional Firm Winter Generating Capacity Is Added from Current Year + 5—On

	(1)	(2)	(3) = (1) + (2)	(4)	(5) = (3) − (4)	(6) = (5) / (4)	(7) =((4* 1.2) − (3)
Year	Existing Winter Genera-tion Firm Capacity (MW)	Additional Winter Genera-tion Firm Capacity (MW)	Projected Winter Genera-tion Firm Capacity (MW)	Forecasted Winter Peak Load (MW)	Winter Reserves (MW)	Winter Reserve Margin (%)	Additional MW of Firm Winter Peak Generation Needed (MW)
Current Year + 5	12,400	0	12,400	9,750	2,650	27.2%	(700)
Current Year + 6	12,400	0	12,400	9,840	2,560	26.0%	(592)
Current Year + 7	12,400	0	12,400	9,930	2,470	24.9%	(484)
Current Year + 8	12,400	0	12,400	10,020	2,380	23.8%	(376)
Current Year + 9	12,400	0	12,400	10,110	2,290	22.7%	(268)
Current Year + 10	12,400	0	12,400	10,200	2,200	21.6%	(160)
Current Year + 11	12,400	0	12,400	10,290	2,110	20.5%	(52)
Current Year + 12	12,400	0	12,400	10,380	2,020	19.5%	56
Current Year + 13	12,400	0	12,400	10,470	1,930	18.4%	164
Current Year + 14	12,400	0	12,400	10,560	1,840	17.4%	272

Column (1) of this table shows the projected winter firm generating capacity value from our utility's existing generation system. The value shown for Current Year + 5 is projected to be 12,400 MW, which is slightly greater than the summer firm generating capacity value of 12,000 MW that we have discussed previously. (This is due to the just discussed fact that utility systems typically have a somewhat higher firm capacity value for their generation systems in winter compared to summer. Consequently, we have assumed that our utility's firm capacity value is 12,400 MW, or 400 MW higher than the 12,000 MW summer firm capacity value for the same year as was shown earlier in Table 5.3 in column (1).) Because this winter firm capacity value is for existing generating units only, this value will not change in subsequent years.

Assuming that our utility system selects the PV option for Current Year + 5, then adds only PV in subsequent years as it pursues its zero-carbon goal, column (2) shows that no additional winter firm generation will be added. Column (3) then shows that resulting total of existing and new winter firm generation will remain constant at 12,400 MW.

Column (4) presents the forecasted winter peak load. As previously mentioned, the winter peak load in Current Year + 5 is 9,750 MW. The winter peak load is projected to grow by 90 MW per year as shown in this column.

Columns (5) and (6) then show the calculation of winter reserves (MW) and winter reserve margin (%), respectively. Finally, column (7) shows the results of a calculation that shows how many MW of winter generation firm capacity would have to change in order to get to a 20.0% winter reserve margin.

The key result in Table 11.5 is that our utility, which began to add only PV in Current Year + 5 and continues to do so in subsequent years, is projected to fail to meet its minimum 20% winter reserve margin criterion 7 years later in Current Year + 12 as indicated by the shaded values in the table. It will be 56 MW of winter firm capacity short of meeting the 20% criterion in Current Year + 12. It will then fail to meet this reliability criterion in each subsequent year by an increasing amount.

Clearly our utility will need to add resources that have non-zero winter firm capacity values by Current Year + 12 (or before) in addition to the PV additions needed in moving toward its zero-carbon goal. In Chapter 10, we assumed that our utility had both PV and battery storage resource options with which to pursue its goal. With this assumption, our utility will need to add batteries not only to address its future night-time load but also to ensure that the peak load on winter peak days can be served.[6]

We also note that our utility experiences its winter peak load in the early morning hours. If a battery has discharged its energy during the previous evening to serve night-time load, it would be unable to be recharged by solar prior to being called upon to serve the next morning's peak load on a winter peak day. This suggests that, in the future, our utility may need to install one "set" of batteries whose purpose is to serve night-time loads, plus another "set" whose purpose would be to help address early morning peak loads in winter.[7]

---

[6] One might ask: *"Could additional DSM be used to address the growing winter peak?"* The answer is *"Perhaps, but only if the additional DSM was projected to be cost-effective."* However, we are focusing now solely on the winter impacts of continuing to only add PV. A few thoughts regarding DSM are offered in Chapter 14.

[7] Note that this is one reason why a resource planning computer model might project a significantly larger MW amount of batteries needed to reach a carbon goal than is projected by the second estimation approach as we saw in Chapter 10. Note also the second "set" of batteries would be available to serve load if the utility experienced higher-than-forecasted summer loads and if/when the utility had unexpected outages of generating units.

SUMMER FIRM CAPACITY VALUES AS INCREASING AMOUNTS OF PV ARE ADDED IN THE FUTURE

Having looked at the addition of ever-increasing amounts of PV from a winter reliability perspective, we now turn to the question of whether our utility can expect future PV additions after Current Year + 5 to continue to provide 50% of their nameplate ratings as summer firm capacity. We start with Figure 11.4 which is the summer equivalent of Figure 11.3, which presented an adjusted view of our utility's winter peak day load shape and the projected winter day output of 1,000 MW of PV.

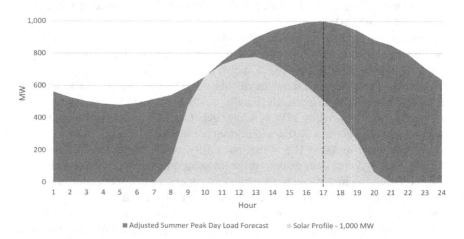

FIGURE 11.4   Adjusted summer peak load and profile for 1,000 MW solar.

The same approach used to adjust the winter hourly load for Figure 11.3 was used again for Figure 11.4, but this time the summer peak day hourly loads have been adjusted down so that the peak hour load is 1,000 MW. All of the other hours on the utility's summer peak day are adjusted accordingly using the same ratioing approach. The adjusted hourly loads are depicted by the darker shaded area. Then the projected hourly outputs of 1,000 MW (nameplate) of PV for a summer day are graphed and appear as the lighter shaded area. (Note again that the solar profile for a summer day is different than the profile for a winter day. PV output profiles typically vary from one month or season to the next.)

There are two important pieces of information that are shown in Figure 11.4. First, at our utility's summer peak hour 17 (which is denoted by the vertical dotted line), the PV expected MW contribution is approximately 500 MW or 50% of the 1,000 MW nameplate rating. This is consistent with the firm capacity value assumed for summer for the PV option of 50% in the analyses presented in Chapter 5.

The other important piece of information is presented to the right of the peak hour's dotted vertical line where we see the downward movement of the PV output profile line. Here we see that PV's projected output (and, therefore, its firm capacity values) decline steadily as time moves past hour 17. At hour 18, PV's firm capacity value drops to approximately 40% (as shown by the solar profile line being at about the 400 MW level). At hour 19, PV's firm capacity value has dropped further to approximately 25%.

Extending our look one more hour to hour 20, PV's firm capacity value drops to approximately 6%. This steady decline is driven by the fact that the sun is getting lower in the sky as the afternoon hours roll by, thus delivering less solar energy to the PV panels.

At this point, the logical question is: *"OK, but why is this important?"* Before we answer that question, we need to continue our discussion of firm capacity values for PV as ever-increasing amounts of PV are added to our utility's system.

Recall that other Supply resource options, such as CCs or CTs, can produce the same amount of energy in virtually every hour they operate. Consequently, these other Supply resource options have an essentially constant firm capacity value, or MW output, for every hour in which they operate. Therefore, these other Supply options have the same impact in every hour when our utility performs its reliability analyses. As shown in Figure 11.4, this is clearly not the case with the PV option.

We begin to analyze what this means in regard to PV summer firm capacity values in Table 11.6, which assumes our utility has added 240 MW of PV.

TABLE 11.6
Summer Peak Day Hourly Loads and Hourly Loads Remaining after PV: 240 MW of PV

(1)	(2)	(3)	(4) = (2) − (3)
Hour-Ending	Utility Summer Peak Day Hourly Load	Hourly Output of 240 MW of PV	Remaining Load After PV Contribution
1	5,681	0	5,681
2	5,327	0	5,327
3	5,081	0	5,081
4	4,921	0	4,921
5	4,854	0	4,854
6	4,961	0	4,961
7	5,213	0	5,213
8	5,460	30	5,430
9	5,981	113	5,868
10	6,624	160	6,464
11	7,583	172	7,410
12	8,408	182	8,227
13	9,076	183	8,893
14	9,532	175	9,357
15	9,815	159	9,656
16	10,039	142	9,897
17	10,100	120	9,980
18	9,909	97	9,813
19	9,517	62	9,455
20	8,916	15	8,900
21	8,596	0	8,596
22	8,013	0	8,013
23	7,177	0	7,177
24	6,440	0	6,440

Column (1) in Table 11.6 lists the 24 hours of our utility's summer peak day. (Note that all of the values in all of the columns that correspond to our utility's highest loads that are about to be discussed in this table, and in a few tables that follow, have been shaded.) Column (2) presents our utility's projected hourly load for that day. As expected, we see that in hour 17, our utility's peak load in Current Year + 5 is 10,100 MW as was assumed in all of the analyses in Chapters 5 and 6 of this book.

Column (3) shows the projected MWh output for each hour from the original 240 MW of PV that the utility assumed would be added if Supply Only Resource Plan 3 (PV) had been selected.

We consider this projected PV output as the firm capacity value for the PV option for each hour. As previously discussed, the output of 240 MW of PV at the peak hour 17 is 120 MW, which fully meets our utility's 120 MW resource need in Current Year + 5.

Finally, column (4) shows the "remaining" load to be served after the PV contribution is accounted for. The values in column (4) are important when examining system reliability in regard to PV. We will now explain why.

The "remaining" load to be served after PV's output is accounted for is an important concept because PV's output differs from hour-to-hour as shown in column (3) of Table 11.6 and in Figure 11.4. For a utility such as ours whose reliability analyses are driven by its summer reserve margin criterion, the utility needs to keep track of the "remaining" load to be served as increasing amounts of solar are added. In Table 11.6, with only 240 MW of PV on our system, the original forecasted peak load, and the highest load remaining after accounting for PV's output, both occur in hour 17. However, as we shall see, not only can the peak hour of the remaining load differ from the original peak hour 17, but the firm capacity value of PV can change, as increasing amounts of PV are added to a utility system.

We will first examine how the utility's peak hour remains after accounting for changes in PV's output due to increasing amounts of solar being added by taking a look at Table 11.7.

TABLE 11.7

Summer Peak Day Hourly Loads and Hourly Loads Remaining After PV: With the Original 240 MW and an Additional 1,920 MW of PV

(1)	(2)	(3)	(4) = (2) − (3)	(5)	(6) = (4) − (5)
Hour-Ending	Utility Summer Peak Day Hourly Load	Hourly Output of 240 MW of PV	Remaining Load After PV Contribution	Hourly Output of Additional PV	Remaining Load After Addlitional PV
8	5,460	30	5,430	238	5,192
9	5,981	113	5,868	902	4,966
10	6,624	160	6,464	1,284	5,180
11	7,583	172	7,410	1,378	6,032
12	8,408	182	8,227	1,453	6,773
13	9,076	183	8,893	1,468	7,425
14	9,532	175	9,357	1,399	7,958
15	9,815	159	9,656	1,274	8,382
16	10,039	142	9,897	1,134	8,763
17	10,100	120	9,980	963	9,016
18	9,909	97	9,813	775	9,038
19	9,517	62	9,455	494	8,961
20	8,916	15	8,900	123	8,777

Table 11.7 is similar in design to Table 11.6, but with two changes. First, the hours 1 through 7, and hours 21 through 24, have been removed because there is no solar output projected for those hours. Second, Table 11.6 presented values that assumed only the "original" 240 MW of PV had been added to our utility system. Table 11.7 now assumes that 1,920 MW (or 8 × 240 MW) of PV have been installed, in addition to the original 240 MW of PV. Therefore, the total PV on our utility system is now assumed to be 2,160 MW (= 240 + 1,920). To address this, columns (5) and (6) have been added in this table. (Note that the values for the hours shown in columns (1) through (4) are unchanged from Table 11.6.)[8]

What this table shows is that, although our utility's actual peak load hour remains in hour 17 (as shown in Column 2), the utility's peak load that remains after accounting for PV's output from 2,160 MW of total PV has now moved from hour 17 to hour 18 (as shown by the shaded values in Column 1 and 6). The logical question at this point is whether this is a one-time-only occurrence or whether an even greater amount of additional solar will cause the utility's summer peak hour that remains after accounting for PV's output to move again.

---

[8] To keep the discussion as simple as possible, we will keep the hourly loads shown in Column (2) the same regardless of the amount of solar that is projected, *i.e.*, we are assuming all of this takes place in the same year. In reality, as the utility adds vast amounts of PV, it would be in a future year and the loads would be different. I have chosen to "freeze" the load to avoid having the Column (2) values changing as we move from one table to the next. The principle we are discussing is not affected by using this approach.

To answer that question, we turn to Table 11.8, which assumes that even more PV MW are added to our utility system.

TABLE 11.8

Summer Peak Day Hourly Loads and Hourly Loads Remaining After PV: With the Original 240 MW and an Additional 2,640 MW of PV

(1)	(2)	(3)	(4) = (2) − (3)	(5)	(6) = (4) − (5)
Hour-Ending	Utility Summer Peak Day Hourly Load	Hourly Output of 240 MW of PV	Remaining Load After PV Contribution	Hourly Output of Additional PV	Remaining Load After Additional PV
8	5,460	30	5,430	327	5,103
9	5,981	113	5,868	1,241	4,628
10	6,624	160	6,464	1,765	4,699
11	7,583	172	7,410	1,895	5,515
12	8,408	182	8,227	1,998	6,229
13	9,076	183	8,893	2,018	6,875
14	9,532	175	9,357	1,924	7,433
15	9,815	159	9,656	1,752	7,904
16	10,039	142	9,897	1,559	8,338
17	10,100	120	9,980	1,324	8,655
18	9,909	97	9,813	1,065	8,747
19	9,517	62	9,455	679	8,776
20	8,916	15	8,900	170	8,731

In Table 11.8, we now assume that 2,640 MW (or 11 × 240 MW) of PV have been installed, in addition to the original 240 MW. Therefore, the total PV on our utility system is now assumed to be 2,880 MW (= 240 + 2,640). The table shows that, with these additional MW of PV, our utility's peak load hour that remains after accounting for PV's output has again moved to later in the day: to hour 19. From these results, we can safely conclude that the movement of our utility's peak hour that remains after accounting for PV's output to later in the day from additional PV MW (as shown in Table 11.7) was not a one-time occurrence. Further movement of the "remaining" peak hour to even later in the day is shown in Table 11.8, as even more PV is added.

But we are still left with the question of *"What does this really mean?"* in terms of the firm capacity value of future increments of PV if more and more PV installations are made. The answer to that is presented in Table 11.9.

TABLE 11.9

Firm Capacity Values for 240 MW Increments of PV

	(1)	(2)	(3)	(4)	−(5) = (4) Previous − (4) Current	(6) = (5) / 240
Row Number	No. of Incremental PV Additions (Multiples of 240 MW)	Total MW of PV on System Incld. Original PV (MW)	Peak Hour for Remaining Load	Remaining Peak Load to be Served (MW)	Remaining Peak Hour Load Served (MW)	Firm Capacity Values for 240 MW Increments of PV (%)
1	6	1,680	17	9,257	–	–
2	7	1,920	17	9,137	120	50%
3	8	2,160	18	9,038	99	41%
4	9	2,400	18	8,941	97	40%
5	10	2,640	18	8,844	97	40%
6	11	2,880	19	8,776	68	28%
7	12	3,120	20	8,715	61	25%
8	13	3,360	20	8,700	15	6%
9	14	3,600	20	8,685	15	6%

This table builds upon the previously presented tables, and it is designed to summarize the impacts that occur as our utility adds ever-increasing amounts of PV to its system in 240 MW increments.

Column (1) shows a range of incremental PV additions that will be on top of the original 240 MW of PV. The values shown represent multiples of 240 MW. The multiple values range from 6 to 14. The first row's value of "6" indicates 6 incremental additions of 240 MW of PV, or 6 × 240 MW = 1,440 MW, are assumed to be added to the original 240 MW of PV.

Column (2) then shows our utility's total PV MW for each year. For the first row, the total PV MW is 1,680 MW (= 1,440 + 240). Information for the remaining rows in columns (1) and (2) is developed in a similar fashion. As columns (1) and (2) indicate, we are examining scenarios for our utility in which 240 MW increments of PV are added sequentially. As a result, the total amount of PV shown in these scenarios ranges from 1,680 MW to 3,600 MW.

Column (3) shows the hour in which the load that remains after accounting for PV's contribution is at its peak. This "remaining load" peak hour is the same as the original forecasted peak hour 17 until our utility adds at least 8 new PV increments of 240 MW each. At that point, the remaining load peak hour load moves to hour 18 if our utility adds 8, 9, or 10 increments of 240 MW of additional PV. (We previously saw this result of the remaining load peak hour moving to hour 18 in Table 11.7 when 8 increments of 240 MW each were added to the original 240 MW of PV.)

With 11 increments of 240 MW PV, the remaining load peak hour moves again, this time to hour 19 (a movement which we also saw in Table 11.8). And, if 12, 13, or 14 increments of 240 MW PV are added, our utility's remaining load peak hour moves yet again to hour 20.

Column (4) presents the remaining peak load to be served after accounting for the total firm MW of PV on our utility system. Then, by comparing this remaining peak load for one row with the remaining peak load for the row above it, column (5) shows how many MW of remaining peak hour load were actually served by a next increment of 240 MW of PV.

For example, looking at the values in columns (4) and (5) for the first two rows, it is apparent that the 240 MW increment of PV added in the second row served 120 MW of the remaining peak load. Finally, column (6) calculates the summer firm capacity value for each 240 MW increment of PV.

Note that the 240 MW increment added in the second row served 120 MW of our utility's summer peak load, which indicates a 50% firm capacity value for this next 240 MW increment of PV. What we conclude from examining columns (5) and (6), through the second row, is that each increment of 240 MW of PV, starting at 0 MW and continuing up to a total PV MW value of 1,920 MW, is projected to serve 120 MW of remaining peak load. In addition, each of the 240 MW increments of PV that have been added to reach 1,920 MW of total PV will have had a 50% firm capacity value for summer.

However, as columns (5) and (6) show for the third, fourth, and fifth rows, the utility has reached a point where at least 2,160 MW of PV has been added and the peak hour of the remaining load has moved to hour 18 (as shown previously in Table 11.7). At that point, each subsequent 240 MW increment of PV serves only 99 MW of remaining peak load (in the third row) and then drops to 97 MW (in the fourth and fifth rows. These values equate to summer firm capacity value of only 41% (= 99/240), or 40% (= 97/240), respectively, for these next three 240 MW increments of PV.

The information presented in columns (5) and (6) for the sixth through the eighth rows show a continuing decline in both MW of remaining peak load served and in the firm capacity value for each subsequent 240 MW increment of PV. As indicated in the eighth row, by the time our utility has 3,360 MW of total solar on its system, the last 240 MW increment of PV only serves 15 MW of the remaining peak load and has a firm capacity value for summer of approximately 6% (= 15/240).[9]

A graphic view of this same information is presented in Figure 11.5.

---

[9] Once enough PV has been added to a utility system to shift the peak hour of the utility's remaining load out to approximately hour 20, the firm capacity value for additional PV increments is likely to either not decline further or to decline by only a small amount. For purposes of our discussion, we are assuming that the firm capacity value of even more PV will remain at 6%.

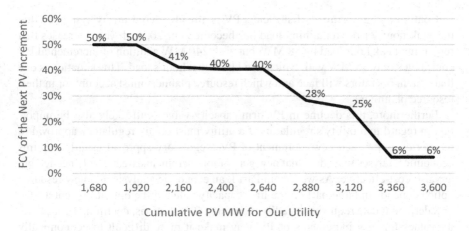

FIGURE 11.5 Firm capacity value (FCV) of the next PV increment based on our utility's cumulative PV MW.

The key takeaway from Table 11.9, and Figure 11.5, is that, once our utility has established a summer firm capacity value for its first increment of PV, it cannot expect that firm capacity value to remain constant for all subsequent PV increments. Because PV's output declines hour-by-hour as the afternoon hours go by, eventually the peak hour for the load that remains after accounting for PV output will shift to later hours. As this remaining peak load hour shifts later, the firm capacity value contribution for the next PV increment diminishes. If our utility were to continue to try to meet its future summer reserve margin criterion solely with PV, it would have to add much larger MW amounts of PV to provide the same amount of incremental firm capacity that the first increment of PV provided.

Now let's answer the question of why it is important to consider the firm capacity values for incremental PV in regard to the utility's peak hour for load to be served <u>after</u> accounting for all prior PV installed on the utility system. We do so first from the perspective of our utility's system operators who must dispatch the utility's generating units to match the load at all hours.

On the summer peak day, it is reasonable to assume that the system operators want as much generation available for their dispatch as possible. Therefore, they will welcome PV's full output in each hour and use their remaining generating units during the transition period to serve the remaining load in each hour.

As we have seen, as our utility steadily adds increasing amounts of PV, the peak load that the system operators must match after accounting for PV's output will move from hour 17 to later hours (for example, to hour 20). In addition, our utility's load in <u>all</u> hours increases each year due to load growth. Beginning in Part I of this book, we have assumed that our utility's load in hour 17 on the summer peak day is growing at 100 MW per year. From Table 11.6, we see that when hour 17's load was projected to be 10,100 MW, the projected load for hour 20 is projected to be 8,916 MW. From those values, we assume that if hour 17 load grows by 100 MW per year, the annual growth in hour 20's load on the summer peak day will be approximately 88 MW (= 8,916/10,100).

Continuing our example, if sufficient PV is already on our utility system so that the peak hour for the remaining load has become hour 20, each year increases that remaining peak hour load by 88 MW, but each 240 MW addition of incremental PV contributes only 15 MW with which to address that hour's load. This situation is one that system operators will face and which resource planners must account for in their resource planning.

Furthermore, this decline in PV firm capacity values will likely also be important in regard to a utility's regulators if a utility must obtain regulatory approval to recover costs for each new increment of PV. Regulatory approval usually requires economic analyses that show that new generation increments are not only needed but are also cost-effective. As we saw in Part I, an economic analysis of a new resource option depends in large part on the firm capacity value of the option. Absent a state or federal mandate requiring steadily increasing PV additions, declining firm capacity values for new increments of PV may make it more difficult to economically justify these increments.

HOW SERIOUS ARE THE CHALLENGES THAT UTILITY RESOURCE PLANNERS WILL FACE FROM THESE TWO PV FIRM CAPACITY VALUE CHARACTERISTICS?

In regard to the first PV characteristic, the low-to-zero winter firm capacity value of PV for utilities whose winter peak hour occurs in early morning, this does pose a problem for these utilities. However, resource planners regularly perform system reliability analyses that examine both the summer and winter peak hour loads. Therefore, resource planners will certainly recognize that additional, non-PV resources will need to be added to ensure their system remains reliable during winter peak loads.

Their challenge then becomes <u>what</u> other resources to add, particularly if these winter capacity resources are needed well before the target year in which the utility has stated its low-or-zero-carbon goal will be met. Does the utility add batteries that will likely be useful in later years to serve night-time load from zero-carbon resources, or does it choose other resources (such as CTs) which will operate only a small amount of the time, but which will have carbon emissions in the hours they do operate?[10]

Due to a utility having publicly made a low-to-zero-carbon goal announcement, this decision will be based not only on economics but also on whether the new resource(s) selected to address winter peak load is consistent with the utility's carbon announcement. Although this may be a difficult decision, the challenge posed by PV's low-to-zero winter firm capacity value is one that must be resolved, but it is definitely solvable by adding some additional resource(s) that supply firm capacity at the utility's winter peak hour.

---

[10] Other options also exist such as performing winter-only capacity upgrades to existing CC units. But such a solution also has potential public relations problems because a utility that has announced a zero-carbon goal may find it difficult to explain why it is increasing the winter capability for fossil-fueled generators (unless it has announced plans to eventually convert these existing units to run on zero-carbon, or carbon-neutral, fuels).

Regarding PV's other characteristic, a decreasing summer firm capacity value for future increments of PV as more and more PV is added to the system; this is a different system reliability issue. This is particularly true for a utility (such as our hypothetical utility) whose system reliability is driven by its summer reserve margin criterion.

Therefore, before our utility attempts to determine if the 52,515 MW of new resources needed to supply 100% of its Current Year + 29 GWh is also sufficient to maintain a reliable electric system, let's take a broader look at the topic of system reliability during the transition period (including consideration of retiring and/or converting existing non-nuclear generating units).

SYSTEM RELIABILITY ANALYSIS DURING THE TRANSITION PERIOD

Once a utility announces a carbon goal, thus entering its transition period, it seems reasonable to assume that it will continue to use its existing system reliability criteria for some period of time. This is because, in the early years of the transition period, much (if not most) of the utility's generating capability, upon which its system reliability depends, will be its existing conventional generating units. The traditional reserve margin and LOLP reliability criteria have worked well for maintaining system reliability for more than a half-century for utility systems' conventional types of generation.

Although seldom articulated this way, these two reliability criteria are based on the fact that the output of conventional generators, whether nuclear, coal, or natural gas, were capable of being essentially constant from hour-to-hour, for as long as a generator's fuel sources (primary and backup, as applicable) were available. The only exceptions to this were in regard to planned and unplanned generation unit outages. And, because planned outages were scheduled in advance, the planned outages were not a major concern in system reliability analyses.

Therefore, the key concerns, when analyzing the reliability of a utility system using either of these two reliability criteria, have been: (i) uncertainty of weather-driven load, (ii) potential unplanned outages of generating units, and (iii) unanticipated actions by large groups of customers that affect electric load (such as we discussed in Chapter 8 regarding the former Florida Power Corporation and residential load control).

These three key concerns will still be present not only at the beginning of, but throughout, a utility's transition period. However, as more and more renewable energy resources are added to the utility system, a fourth concern emerges that will also have to be accounted for in system reliability analyses: the uncertainty of resource output from moment-to-moment, and hour-to-hour, due to the inherent intermittency solar and wind.[11]

––––––––

[11] The intermittency issues with solar and wind are different. With solar, a utility knows its output "pattern" will include (i) no output during nighttime, and (ii) during daytime, the output will usually resemble a rough bell curve shape from morning to evening. Wind does not have this same pattern and, therefore, could be viewed as having more uncertainty than solar in regard to hour-to-hour output.

(Author's Note: At this point, an opinion of the author is about to enter the discussion.)

I do not believe that reserve margin and LOLP criteria, as they are currently structured, will do an adequate job of analyzing system reliability once a utility gets far enough into its transition period to the point when renewable resources such as intermittent solar and wind provide something approaching 50% of the utility's annual energy. This is because neither of these reliability criteria, as currently structured, examine all hours of the year. Instead, they examine only one peak hour per season (reserve margin), or one peak hour per day (LOLP). Even with LOLP's use of the peak hour for each day, this criterion is only focused on approximately 4% (= 365/8,760) of the hours in a year.

I believe that these existing criteria will need to be restructured and/or augmented or replaced by a new criterion, which accounts for uncertainty in both load and intermittent generation for more/all hours of the year.

This statement is based on an assumption that all of a utility's non-nuclear generating units will eventually be retired by the utility's carbon goal target year and that all non-nuclear zero carbon energy will come from intermittent renewable resources. However, as indicated in NextEra Energy's 2022 Blueprint, FPL is planning to convert 16,000 of existing CC generation to run on green hydrogen. As we discussed in Chapter 10, the Blueprint does not provide a clear description of what hours of the day the converted generators are planned to be operated, only that these conventional generators will continue to be operated but with green hydrogen as the fuel.

It is conceivable that such an approach could somewhat minimize the concern regarding reliance on uncertain intermittent renewable resources to serve load continually from moment-to-moment. However, the hydrogen that is to be used as a fuel for these converted generating units will need to be produced using those intermittent renewable resources (in order to be considered as "green" hydrogen). For this reason, although the use of green hydrogen as fuel in converted generating units may diminish the hour-to-hour concerns about intermittency of renewable resources, this concern is not eliminated.

As the second edition of this book is written, the electric industry has begun work on how to evaluate system reliability during a utility's transition period (and beyond), and a few potential approaches have been put forward. However, I have not yet seen a new approach that I believed combined the desired attributes of: (i) fully addressing all the uncertainty in both load and generation, (ii) being easy to understand and utilize, and (iii) being easy to explain to regulators and other interested parties. I look forward to seeing the development of such a new approach and/or the needed modification of the current two reliability criteria.[12]

(Author's Note: End of Discussing this Opinion)

What this means is that our utility cannot now answer with confidence the question of whether the estimated 52,515 MW of new resources needed to supply 100% of

----

[12] Assuming I am not busy golfing at the time. Did I mention I am now retired?

its GWh in Current Year + 29 will be sufficient to also maintain system reliability. Therefore, for the remainder to this book, our utility will (warily) assume that these 52,515 MW, developed using the second estimation approach, <u>might</u> be sufficient both for supplying all the needed GWh in Current Year + 29 with zero carbon emissions and to maintain system reliability.

SUMMARY: KEY TAKEAWAYS FOR RESOURCE PLANNING REGARDING PV RESOURCES

There are at least five key takeaways from this discussion of PV resource options:

1. The 2022 IRA legislation's inclusion of a PTC for new PV additions has been successful in making PV options very attractive for a utility system in regard to total utility economics due to the PTC's ability to lower the utility's federal tax obligations. Because of this, and assuming all else equal, electric utilities will be inclined to build more PV, and to do so earlier, than would have been the case without the PTC. This will also make it more likely that electric utility executives will make public announcements of goals/ambitions that their utility will get to low- or zero-carbon operations, which will further drive a push for more/earlier solar.

2. However, the utilities' resource planners will have to deal with PV firm capacity value characteristics as a utility adds ever-increasing MW amounts of PV. The first of these characteristics is that PV facilities typically provide more firm capacity value in summer than they do in winter. For certain summer peaking utilities whose winter peak load hour occurs in the early morning (as is the case with our hypothetical utility), the winter firm capacity contribution is likely to be 0 MW or close to 0 MW. These utilities will face declining winter reserve margins each year if solar is the only type of new resource added (due to PTC economics and/or chasing a carbon emission goal). At some point, the utility may well become a utility whose resource needs are now being driven by its winter reserve margin with the realization that further solar additions alone cannot address those resource needs. If that is the case, then the utility will need to add even more, and different, resources that supply sufficient winter firm capacity MW to maintain system reliability during high winter loads.

3. The second characteristic of PV firm capacity value is that, as ever-increasing increments of solar are added, at some point the summer firm capacity value of the next increment of solar, in relation to the load remaining to be served after accounting for PV, will be less than the firm capacity value of the previous solar increments. The summer firm capacity values for subsequent increments of solar will then continue to decline in a stepwise manner as discussed previously. The result will be that new increments of solar will contribute less to meeting the utility's summer reserve margin regarding this remaining load than the previous increment contributed.

At some point, the summer firm capacity value of solar will level out to an essentially constant value. However, that constant value will be at a very low percentage of the solar nameplate value (as shown by the 6% value for our hypothetical utility).

4. The diminishing firm capacity values for the next increments of PV may result in a utility having difficulty showing that the next PV increment is more economic than other resource options. This is especially true if the utility operates in a state that does not have a carbon goal and/or renewable mandate. This could pose a challenge in future regulatory proceedings.

5. It is the author's opinion that the two most currently used reliability criteria, reserve margin and loss-of-load-probability (LOLP), as they are currently structured, will not be sufficient to adequately predict system reliability as utilities move further into their transition periods. A new criterion, and/or modifications to one or both of the two current criteria, will likely be needed.[13]

In the next chapter, we switch our focus to battery storage resources. We will take a look at these options from several perspectives, including cost, the relative impact of the 2022 IRA's federal tax credits for batteries compared to PV, and how to determine the duration that is needed for the firm capacity of batteries to be 100% of the battery's nameplate rating.

---

[13] The topic of system reliability is discussed again in Chapter 14.

12 Moving toward Zero Carbon

Resource Planning Considerations for Battery Storage Resources

INTRODUCTION

In Chapter 10, we assumed that our hypothetical utility had two new resource options with which to meet its zero-carbon goal: PV and battery storage. The analyses performed in Chapter 10 showed that, based on the second estimation approach, our utility would need to install 17,835 MW of 4-hour duration batteries to serve the night-time load in order to achieve its goal. Then, in Chapter 11, we discussed the fact that ever-increasing amounts of PV alone would not be able to allow our utility to continue to meet its winter total reserve margin criterion because our utility's winter peak load occurs in the early morning hours, and PV MW output was essentially zero in those hours. Therefore, our utility would likely have to add even more batteries in the future (or some other option that provides winter peak hour firm capacity) to ensure system reliability on winter peak days.

In this chapter, we will take a look at several aspects of battery storage. We start with a projection for the installed costs of batteries. Next, we take a brief look at the 2022 Inflation Reduction Act's (IRA's) new federal tax credits for batteries. Then the rest of this chapter will focus on one other aspect of batteries: battery duration and how battery duration relates to firm capacity values of batteries. We will explain how a utility can determine what the minimum battery duration should be if the battery is intended to provide firm capacity that equates to 100% of the battery's nameplate rating. And, as we shall see, the needed duration for a battery has cost implications as well.

A REPRESENTATIVE PROJECTION OF BATTERY STORAGE INSTALLED COSTS ($ PER KW): 2023–2030

In Chapter 11, we discussed the installed costs ($ per kW) of PV installations that had occurred from 2010 through 2022 using PV additions made by Florida Power & Light Company (FPL) as an example. What this discussion showed is that PV installed costs had significantly declined through the year 2021. This cost decline was driven by competition, plus improvements in manufacturing and fabrication of solar components. However, FPL's PV installed costs leveled out in 2022 due to supply chain problems and, to a lesser extent, increased costs for land.

Over this same time period, there have been far fewer utility-scale battery MW manufactured and installed, especially in the United States, compared to the amount of PV MW installed. In addition, these relatively few battery installations ranged in size from less than 1 MW to just over 400 MW. For these reasons, a look at the historical cost of utility-scale batteries over the last decade is not as meaningful or informative as it is for PV. Therefore, our discussion will focus instead on the projected installed costs of batteries. We will first do so without accounting for the new federal tax credits from the 2022 IRA.

The downside of discussing the projected installed costs of batteries is that one can find a very wide range of projected cost values, and these projections change regularly. However, the one feature that virtually all of these projections have in common is that future costs will decline. Figure 12.1 shows a representative projection of installed costs for battery storage for the years 2022 through 2030 issued in 2022 by the National Renewable Energy Laboratory.

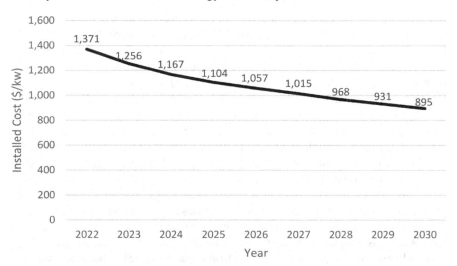

FIGURE 12.1   Projected installed costs ($/kW) for 60 MW$_{dc}$, 4-hour duration batteries.

(Source: National Renewable Energy Laboratory 2022 Annual Technology Baseline, 7/21/2022, Moderate Scenario, 30-Year Cost Recovery Period, **Lithium-ion batteries, costs in 2020 U.S. dollars)**

Two things can be seen from Figure 12.1. First, the figure projects a significant, ongoing cost reduction from 2022 through the year 2030. Second, as indicated by the figure's title, these cost projections are based on batteries that have a certain set of characteristics. Two key battery characteristics listed for the values in this figure are as follows: 60 MW$_{dc}$ of capacity (maximum battery output) and 4-hour duration (the amount of time the battery's full MW capacity can be delivered before recharging is needed). A change in any of these assumptions would change the projected installed costs. (We note that another key characteristic not listed for these values is how often the battery is projected to be used, for example, 365 days per year, 50 days per year, 10 days per year, etc. This characteristic also has implications for battery cost.)

This points out one aspect of battery storage options: the need to consider the battery's capacity, duration, and how it will be used before assuming the future cost of batteries. For example, assuming all else equal, the shorter the duration of the battery, the lower the cost.[1] Similarly, the less frequently the battery will be used during the year, the lower the total cost.

We will return to the duration of batteries later in this chapter to discuss how battery duration is related to the firm capacity value of batteries and to battery cost, but first we will take a brief look at how the 2022 IRA legislation's federal tax credits affect the cost of batteries.

TAX CREDITS FROM THE 2022 INFLATION REDUCTION ACT (IRA)

The 2022 IRA also included a new level of federal tax credits for batteries. Prior to the passage of the IRA, batteries were eligible for an investment tax credit (ITC) of 10%. The 2022 IRA did not offer a PTC for batteries as it did for solar, but it did increase the ITC for batteries to 30%. Therefore, from a total utility cost perspective, the cost of adding new batteries decreased from what it was prior to the enactment of the 2022 IRA because of increased federal tax savings for a utility that installs new batteries.

When comparing the impact of this new ITC for batteries with the PTC for solar, a utility will likely find that the solar PTC will lower the overall cost of adding solar more than the battery ITC will lower the overall cost of adding batteries. This is due in large part to the fact that, for a new resource addition, the ITC applies only once to the installed cost of the resource in the year the resource is installed. On the other hand, the PTC applies to the annual PV output for at least 10 years.

In addition, the PTC $ per MWh starting point value for solar escalates each year, thus further increasing the PTC's benefits. The ITC percentage value for batteries does not increase over time. Furthermore, the PTC is tied only to the annual MWh output of solar, not to solar installed cost. As a result, the PTC value is unaffected if PV installed costs decrease over time as expected. Conversely, the ITC for batteries is tied directly to the installed cost for batteries so the impact of the battery ITC will diminish over time as battery costs decrease as projected in Figure 12.1.

For these reasons, and assuming all else equal, a utility will likely be tempted to install PV instead of installing batteries, or at least to install more PV MW than battery MW, due to the greater tax credit benefits for PV compared to batteries. This could contribute to the situation we discussed in Chapter 11 in which a utility that solely adds PV will likely face declining winter reserve margins if the utility's winter peak hour occurs in the early morning. In the next chapter (Chapter 13), we will discuss other system challenges, including challenges to how the utility's existing fleet of generating units is operated, that will likely arise if a utility begins adding many more PV MW than it adds battery MW.

---

[1] As battery duration changes (either increases or decreases), the cost changes for several types of components in a battery facility. The primary cost driver is the battery "cabinet" itself (simplistically, think of a number of individual small batteries that are housed together) required to provide energy for the desired duration. Another cost driver is the cost of the electric balance of system components that are tied to the battery cabinet.

As the second edition of this book is written, it is too early to see if a number of utilities' resource plans are showing the building of far more PV MW than battery MW. If such a trend emerges, it will be due, at least in part, to the difference in overall system economics driven by the 2022 IRA's differences in federal tax credits between PV and batteries. If this becomes apparent in the next few years, it will be interesting to see if the 2022 IRA legislation is "tweaked" to address this.

We now turn our attention to how batteries need to be considered in regard to system reliability. In a continuing effort to simplify the complexities of utility resource planning in this book, we will be looking only at "stand-alone" batteries (*i.e.*, not a battery-and-PV combination) in regard to the system reliability aspect of battery resources. Such battery-and-PV combinations can add complications, and these can vary from state-to-state due to state laws/regulations. However, the key messages that will be presented for stand-alone batteries regarding system reliability still apply with such combinations. We will return to briefly discuss battery-and-PV combinations, from the perspective of system reliability, near the end of this chapter.

A GRAPHICAL LOOK AT THE FIRM CAPACITY VALUE ASPECT OF BATTERY STORAGE

Assuming a battery is fully charged, it can serve as a generator that can provide the designed MW output for the designed duration of the battery. Therefore, from a system reliability perspective, a 50-MW, 3-hour duration battery can function as a 50-MW generator for 3 hours. It can also serve as a 25-MW generator for 6 hours or any other combination of MW and time that is restricted only by the battery's designed full MWh output of 150 MWh (= 50 MW × 3 hours). Once that battery's charge of 150 MWh has been used, it can no longer serve as a generator. Once this amount of energy has been discharged to the utility system, the battery must now be recharged by the utility system.[2]

In regard to system reliability, the most important characteristic of batteries is battery duration. And the key question is as follows: *"What duration (hours) must a battery have in order to receive 100% firm capacity value for its MW capacity?"* Another way to state the question, using an example, is as follows: *"What is the needed duration for a 50 MW battery to ensure that the battery is credited with the full 50 MW in utility reserve margin analyses?"* This question is one that many utilities have wrestled with, and we will discuss how it can be answered.

We begin to do this by looking at our utility's summer peak day load shape, which has the now familiar 10,000 MW summer peak hour load in Hour 17 for the Current Year. Then we assume that our utility is considering adding a 200-MW battery that it will dispatch as the last generation resource on its summer peak days. Our utility's objective is to serve all of its load above 9,800 MW solely with the 200 MW battery. We look at a graphical approach to answering this question in Figure 12.2.

---

[2] Therefore, a battery storage facility is first an electric load during the time the battery is being charged by the utility system. Then the battery can serve as a generator as it provides energy and capacity back to the utility system. Then the cycle of changing "roles," electric-load-followed-by-generator, for the battery is repeated.

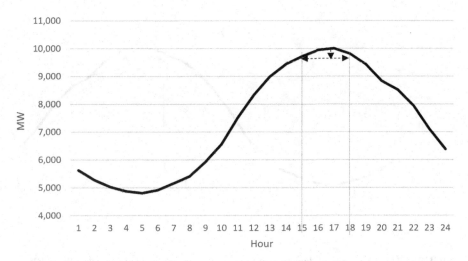

FIGURE 12.2 A graphical depiction of the battery duration needed to serve 200 MW of load on our utility's summer peak load day.

Because the battery's capacity is 200 MW, we first draw a vertical arrow down from the 10,000-MW peak load in Hour 17 to a point that represents a remaining load in that hour of approximately 9,800 MW.[3] Then, from that 9,800-MW load point, we draw horizontal arrows in both directions. The horizontal lines intersect the load shape curve at approximately Hours 15 and 18. These two intersecting points on the load shape curve also indicate 9,800 MW of load. From this sketch, we see that the 200-MW battery appears to need to operate in 3 consecutive hours (from Hour 15 to Hour 18) in order to fully serve all of our utility's load on the summer peak day that is above 9,800 MW. In other words, this 200-MW battery appears to need a 3-hour duration.

We next explore how the projected duration might change for another increment of battery storage that would be added and operated after the initial 200-MW battery increment is in place. To examine this, we assume another battery increment of 800 MW is added (resulting in 1,000 MW of total battery capacity). Such a scenario is presented in Figure 12.3 that appears on the next page.

Figure 12.3 begins with the same information shown in Figure 12.2. Then another battery increment of 800 MW is assumed to be added. Our utility still views batteries as a generation resource of last resort, *i.e.*, it will be dispatched only after all other generation has been used. Therefore, our utility's objective now is to serve all load above 9,000 MW with batteries.

What Figure 12.3 shows is that the second increment of battery resources, with a capacity of 800 MW, will need to operate for what appears to be at least 6 (and maybe 7) consecutive hours in order to fully serve all remaining load above

---

[3] The term "approximately" is used because we are sketching the vertical and horizontal lines on an already developed load shape curve. Consequently, the exact points of the vertical and horizontal lines are approximate values. The use of this graphical approach is to develop a conceptual idea of how to think about battery duration.

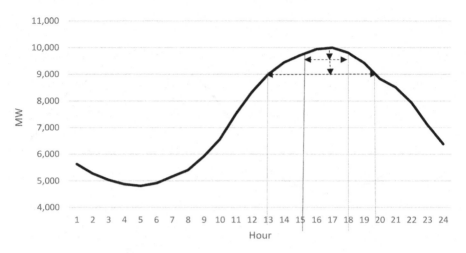

FIGURE 12.3 A graphical depiction of the battery duration needed to serve 1,000 MW of load on our utility's summer utility's peak load day.

9,000 MW on our utility's summer peak day. Note that the vertical lines down to the x-axis do not intersect this axis precisely at "on the hour" points, thus making it a bit difficult to judge accurately the number of hours in which the battery will need to operate. It appears that this second battery increment of 800 MW will need to operate from about Hour 13 to just before Hour 20.

Recognizing the lack of precision inherent in using this graphical approach, and the fact that a graphical approach is a somewhat awkward method to use in resource planning work, we seek a more precise, and easier-to-use, method with which to calculate the needed duration for batteries. This method is presented in the next section.

A BETTER METHOD TO DETERMINE THE NEEDED DURATION OF BATTERY STORAGE OPTIONS

We begin by going back to take another look at Figure 12.2. The area between the load shape curve and the horizontal line represents the MWh of energy that the initial 200-MW battery will have to serve if this battery is to serve all load above 9,800 MW. In Figure 12.3, the area between the two horizontal lines represents the remaining MWh of energy that the 800-MW battery will then have to serve if all of the utility's load above 9,000 MW is to be served solely by batteries.

The key point is that, in order to determine the needed duration of batteries, our focus should include the <u>MWh</u> a battery can serve before it must be recharged, *i.e.*, a focus on energy (MWh) as well as on capacity (MW). (Note that this perspective does not apply to fuel-based resource options, such as CC, CT, or nuclear units, that can continue supplying energy and capacity as long as their fuel supply is available.) Now, after establishing the importance of considering the maximum MWh output of a battery before it must be recharged, we turn to two new tables to continue the discussion. We start with Table 12.1 that examines the addition of 200 MW of battery capacity.

TABLE 12.1

Calculation of Needed Battery Duration for Our Utility's Summer Peak Load Day: For Initial 200-MW Battery Increment

Row Number	1st Battery Increment: 200 MW	Forecasted Hourly Load on Summer Peak Day (MW)							
		Hour 13	Hour 14	Hour 15	Hour 16	Hour 17	Hour 18	Hour 19	Hour 20
1	Hourly load (MW):	8,986	9,438	9,718	9,939	**10,000**	9,811	9,423	8,828
2	1st battery increment applied to peak hour (MW):	0	0	0	0	200	0	0	0
3	Remaining hourly load to serve (MW):	8,986	9,438	9,718	**9,939**	9,800	**9,811**	9,423	8,828
4	Needed battery contribution in other hours (MW):	0	0	0	139	0	11	0	0
5	Remaining hourly load to serve (MW):	8,986	9,438	9,718	**9,800**	**9,800**	**9,800**	9,423	8,828
6	Hours in which battery operates (1 = Operates):	0	0	0	1	1	1	0	0
7	Energy served by battery by hour (MWh):	0	0	0	139	200	11	0	0
8	Total energy served by battery (MWh):	**350**							
9	Minimum battery duration needed (hours):	**1.75**	(= 350 MWh/200 MW)						

In row (1) of Table 12.1, the hourly loads from Hour 13 to Hour 20 for our utility's summer peak day are presented. The peak hour load of 10,000 MW is presented in boldface under "Hour 17." In row (2), we assume that the 200 MW battery is operated at its full hourly output of 200 MW. However, in row (2), we assume that the battery is discharged only in Hour 17 (the peak load hour).

In row (3), we calculate what the remaining load is that still needs to be served for each hour. In Hour 17, the remaining load to be served has been reduced by the 200 MW of battery discharge, but the remaining load to be served for all other hours remains unchanged. What we now see on row (3) is that we have 2 hours, the load of which is now higher than the remaining load to be served of 9,800 MW in Hour 17. These are Hour 16 (with a load of 9,939 MW) and Hour 18 (with a load of 9,811 MW).

Because our utility desires to serve all load above 9,800 MW with the 200 MW battery, it needs to operate the battery in these 2 hours as well. As shown in rows (4) and (5), the battery will need to supply 139 MW in Hour 16, and 11 MW in Hour 18, in order to ensure that all of the utility's load on the summer peak day above 9,800 MW is served solely by the battery. As we see by the boldface values in row (5), we now have 3 hours with an identical remaining load to be served of 9,800 MW. The remaining load to be served in all other hours is less than 9,800 MW. Therefore, the battery does not have to be operated in these hours.

In order to summarize how the battery must be operated to meet our utility's objective, rows (6) and (7) show that the batteries will need to operate in these 3 consecutive hours, and that the needed discharge from the battery in those hours is 139 MWh, 200 MWh, and 11 MWh, respectively.

Row (8) sums up the needed amount of total energy (MWh) to be served to ensure all load above 9,800 MW is served solely by the battery. That total is 350 MWh (= 139 + 200 + 11). Then, row (9) divides the needed 350 MWh of battery-supplied energy by the capacity of the battery, 200 MW, to derive the needed duration of the battery, from an energy perspective, of 1.75 hours (= 350 MWh/200 MW).

Two things are worth noting at this point. The first is that the number of hours in which the battery must operate (in 3 different hours in this case) can be a different value than the needed duration of the battery (1.75 hours). The second point is that our calculation assumes perfect advance knowledge of what the load will be in each hour, and perfect execution by the system operators to know when to begin discharging the battery and for how long to discharge it. In other words, this calculation assumes a "perfect knowledge" scenario.

In real life, system operators are always facing an unknowable load in the next minute, much less in the next hour. Faced with loads that will soon be upon them which they can only estimate at the current moment, there is no way that system operators can know exactly when they should begin dispatching the battery, or for how long they will need to use it.

This means that, in this example, the calculated 1.75 hour needed duration of the battery is likely too short. This is particularly true if a utility plans to use the battery as the last generation resource to be dispatched as we have assumed for our utility. Therefore, our utility would likely conclude from the calculations presented in Table 12.1 that its initial battery increment should have a minimum duration greater than 1.75 hours to account for the uncertainty that the system operators will face. Our utility may decide that a minimum duration of at least 2.00 hours (or more) is needed to provide some "cushion" needed for system operation.

We now use the same calculation approach in Table 12.2 (that appears on the next page) to see what it tells us about the needed duration for the second battery increment of 800 MW.

This table uses the identical approach that was used in Table 12.1. However, the results for another 800 MW of battery storage are considerably different than they were for the first 200 MW battery increment.

TABLE 12.2

Calculation of Needed Battery Duration for Our Utility's Summer Peak Load Day: For a Second Battery Increment of 800 MW

Row Number	2nd Battery Increment: 800 MW	Forecasted Hourly Load on Summer Peak Day (MW)							
		Hour 13	Hour 14	Hour 15	Hour 16	Hour 17	Hour 18	Hour 19	Hour 20
1	Remaining hourly load to serve (MW):	8,986	9,438	9,718	**9,800**	**9,800**	**9,800**	9,423	8,828
2	2nd battery increment applied to peak hours (MW):	0	0	0	800	800	800	0	0
3	Remaining hourly load to serve (MW):	8,986	**9,438**	**9,718**	9,000	9,000	9,000	**9,423**	8,828
4	Needed battery contribution in other hours (MW):	0	438	718	0	0	0	423	0
5	Remaining hourly load to serve (MW):	8,986	**9,000**	**9,000**	**9,000**	**9,000**	**9,000**	**9,000**	8,828
6	Hours in which battery operates (1 = Operates):	0	1	1	1	1	1	1	0
7	Energy served by battery by hour (MWh):	0	438	718	0	0	0	423	0
8	Total energy served by battery (MWh):	**3,979**							
9	Minimum battery duration needed (hours):	**4.97**	(= 3,979 MWh/800 MW)						

In row (1) of Table 12.2, the hourly loads that we start the calculation with are the remaining loads to be served from row (5) of Table 12.1. Therefore, the starting point for examining the second battery increment of 800 MW is one in which Hours 16 through 18 have 9,800 MW of remaining load. Our utility wants to utilize the 800 MW battery to serve all remaining load above 9,000 MW. To do so, we assume in row (2) that the 800 MW battery is operated at its full hourly output of 800 MW. However, in row (2), we assume that the battery would be discharged only in the 3 hours, the remaining load of which is 9,800 MW, *i.e.*, in Hours 16, 17, and 18.

In row (3), we calculate what the remaining load is that still needs to be served for each hour. The remaining load to be served for these 3 hours is now reduced to 9,000 MW as desired. However, we now have 3 other hours, the remaining load of which to be served is greater than 9,000 MW. These are Hour 14 (with a load of 9,438 MW), Hour 15 (with a load of 9,718 MW), and Hour 19 (with a load of 9,423 MW).

Because our utility desires to serve all load above 9,000 MW (and below 9,800 MW) with the 800 MW battery, it needs to operate the battery in these other 3 hours as well. As shown in rows (4) and (5), the battery will need to supply 438 MW in Hour 14, 718 MW in Hour 15, and 423 MW in Hour 19, in order to ensure that all of the utility's load on the summer peak day above 9,000 MW is served solely by the battery.

To summarize the needed operation of this second battery, rows (6) and (7) show that the second battery increment (800 MW) will need to operate in 6 consecutive hours, and that the needed discharge from the battery in those hours is 438 MWh, 718 MWh, 800 MWh, 800 MWh, 800 MWh, and 423 MWh, respectively.

Row (8) sums up the needed amount of energy to be served to ensure all load above 9,000 MW is served solely by the battery. That total is 3,979 MWh (= 438 + 718 + 800 + 800 + 800 + 423). Then, on row (9), the needed 3,979 MWh of battery-supplied energy is divided by the capacity of the second battery increment, 800 MW, to derive a needed duration of the second battery increment of 4.97 hours (= 3,979 MWh/800 MW).

As mentioned before, our utility recognizes the load uncertainty that the system operators will be dealing with on the summer peak day from hour-to-hour and will consider the 4.97-hour duration to be a "perfect knowledge" scenario value (as was the case in Table 12.1 which provided a 1.75-hour duration result). Therefore, the second increment of battery storage of 800 MW capacity would be more likely to be deemed to have a minimum needed duration of 5.5 or 6.0 hours.

From this discussion, it is clear to see that as a utility continues to add battery MW, especially stand-alone battery MW, with the primary objective of maintaining system reliability, the needed durations of battery increments will increase as more and more battery MW are added. (Alternatively, more batteries of lesser duration could be added.) This fact will have another impact on our utility's resource planning because it impacts the cost of future battery additions. We examine that in the next section.

THE RELATIONSHIP BETWEEN THE NEEDED DURATION OF BATTERY STORAGE OPTIONS AND THE COST OF BATTERIES

In the "perfect knowledge" calculations we just discussed, our utility was projected to need a minimum duration of 1.75 hours for its initial battery increment that was assumed to offer 200 MW of firm capacity. Then its second battery increment, this time with a capacity of 800 MW, would require a minimum battery duration of 4.97 hours. For ease of discussion, let's round these two minimum duration values to 2 hours and 5 hours. (For this discussion, we will also ignore that our utility would likely increase those durations due to uncertainty in the hour-to-hour load that system operators will have to deal with, especially if the battery MW are used as the last generation resource on system peak load days.)[4]

---

[4] In striving to keep the explanation as simple as possible, we also ignore the fact that if the second battery increment is added several years after the first battery increment is added, the battery duration calculation will start with higher hourly MW values. Consequently, the calculation results for the second battery increment might be higher (or lower) than what was calculated in Table 12.3. This can be ignored for the discussion we are now having about cost impacts because it does not change the basic point of the discussion.

What our utility now sees is that it can add 200 MW of battery capacity as a firm capacity resource as long as the needed battery duration is 2 hours. Then its next battery increment of 800 MW will need a duration of 5 hours that is 3 hours longer than the first increment. What impact, if any, does that have on the installed $ per kW cost of the second battery increment?

In Figure 12.1, we previously presented a 2022 NREL projection of installed costs for batteries. This cost projection was for a 60-MW_{dc} battery with a 4-hour duration. For purposes of this discussion, let's use the battery costs from Figure 12.1 to see what happens to battery costs as a utility adds an increasing amount of battery MW. If we take the values for two of the years shown, this figure shows that the projected cost is $1,104 per kW in 2025 and $895 per kW in 2030.

Now suppose that the results of our utility's calculations just discussed had been different. We now assume that our utility plans to add 60 MW_{dc} of batteries in 2025 and the needed battery duration for this initial battery increment is 4 hours.[5]

We further assume that our utility plans to add another 60 MW_{dc} of batteries in 2030, but the needed duration for this second battery increment is now 7 hours. (These assumptions give us another situation in which the second increment of batteries has a 3-hour longer duration than the first increment of batteries.)

Figure 12.1 has given us NREL's 2022 projection of 4-hour battery costs, but not the costs for a 7-hour battery. Fortunately, the same NREL document that Figure 12.1 was based on also provides a graph showing the comparative costs for a 4-hour, 6-hour, and 8-hour battery for various years. A visual examination of that graph shows the following approximate installed cost values in 2030 for a 60-MW_{dc} battery of either 6-hour or 8-hour duration, compared to the installed cost for a similar battery of 4-hour duration[6]:

- For the 6-hour duration battery, the installed costs are approximately 25% more than an equivalent size battery of 4-hour duration.
- For the 8-hour duration battery, the installed costs are approximately 50% more than an equivalent size battery of 4-hour duration.

Because no cost projection was given for a 7-hour duration battery, we extrapolate between these 2 percentage values and assume that a 7-hour battery will cost approximately 37.5% (= (25% + 50%)/2) more than an equivalent size battery of 4-hour duration.

Continuing to use the cost values for Figure 12.1 in this discussion, our utility sees the projected installed cost for a 4-hour duration battery drop from $1,104 per kW in 2025 to $895 per kW in 2030. However, because our utility would need a 7-hour battery in 2030, the projected cost for the 7-hour battery in 2030 is now projected to be approximately $1,230 per kW (= $895 × 1.375).

---

[5] Which (very conveniently) we have a projected cost for, from Figure 12.1.
[6] The NREL document did not provide the specific projected numeric $/kW cost values from which the line graph was constructed. Therefore, one has to visually estimate what those annual cost values were. We did so and then approximated the percentage differences between the values. As a result, the percentage values are approximate.

Therefore, the second battery increment's installed cost in 2030 is actually higher ($1,230 per kW) than the first battery increment's installed cost in 2025 ($1,104 per kW) due to the needed for the second battery increment to have a longer duration.

This results in two implications for this utility's resource planning work. First, the projected installed cost of the second battery increment did <u>not</u> decrease over time to $895 per kW as one would project from Figure 12.1 when looking only at 4-hour duration batteries. Instead, due to the need for an additional 3 hours of battery duration, the cost of a second MW battery increment to be installed in 2030 increased to $1,230 per kW. In other words, due to the need to increase battery duration as more batteries are added to a utility system, the effective cost of batteries needed to address our utility's summer peak hour load will likely not decline as indicated in Figure 12.1.

The second implication is that the battery ITC will be a larger factor with the installed cost of the second battery increment at $1,230 per kW than it would be if this battery's cost was $895 per kW. However, the ITC impact is only 30% and, therefore, will not fully offset the $335 per kW (= $1,230 − $895) additional cost driven by the need for a 3-hour longer duration. The ITC will only reduce approximately $100 per kW (= $335 × 30%) of the $335 per kW higher cost.

A BRIEF LOOK AT BATTERY-AND-PV COMBINATIONS IN REGARD TO SYSTEM RELIABILITY

We now return to briefly discuss the combining of batteries and PV into a hybrid battery-and-PV facility from the perspective of system reliability. As discussed in Part I of this book, system reliability for a utility system is currently most often measured in regard to either a reserve margin criterion and/or a loss-of-load-probability (LOLP) criterion. Both reliability criteria use the projected MW output values of generating units at the utility's peak hour. The summer peak hour(s) and the winter peak hour are used in calculations based on the reserve margin criterion. The daily peak hour is used in calculations based on an LOLP criterion.

Up to this point in our discussion, our utility has viewed PV and batteries as separate, stand-alone resource options. For example, a 120 MW PV facility that had a 50% firm capacity value would be projected to provide 60 MW of firm capacity in Current Year + 5. Likewise, a separately located 50 MW battery facility, with sufficient duration to have a 100% firm capacity value, would be projected to provide 50 MW of firm capacity. Together, these two separate resources would be projected to provide 110 MW (= 60 + 50) of firm capacity.

Now suppose these two resources were located at the same site and that the battery was charged directly from the 120 MW PV installation. From a system reliability perspective, would the total firm capacity of this hybrid facility be any different than the total firm capacity of the two, separately located facilities?

The answer is "no." Assuming that the battery has already been adequately charged by the PV installation, then the battery can provide its 50 MW designed capacity at the peak hour, but no more. And, assuming that the PV installation is not charging the battery at a peak hour, the PV installation would still be projected to provide its projected 60 MW at a peak hour, but no more. Therefore, one would not expect a hybrid battery-and-PV facility to be able to provide more firm capacity at a peak hour than a combination of two, separately located PV and battery facilities would provide.[7]

SUMMARY: KEY TAKEAWAYS FOR RESOURCE PLANNING REGARDING BATTERY STORAGE RESOURCES

From a resource planning perspective, the key takeaways from this discussion of battery storage resource options are as follows:

1. Batteries must be viewed as having two "roles" on a utility system. First, they act as an additional electric load when they are being charged from the utility system. Second, they act as a generator (but a "limited" generator with design limits as to how many MW and how many MWh it can produce before it must be recharged) when their stored energy is being discharged into the utility system. These two roles are then repeated many, many times over the life of the battery.
2. The projected installed $ per kW installed cost of batteries is expected to decline. (However, just as with the installed cost of PV, there is uncertainty regarding how far, and how fast, these costs will decline.)
3. The 2022 IRA established a 30% ITC for batteries. This tax credit further decreases the overall cost of battery additions to a utility which installs new batteries by lowering the utility's federal tax burden. However, the battery ITC does not provide as much economic incentive for adding new batteries as the new PTC does for adding new solar. As a result, and assuming all else equal, a utility will be tempted to pursue more new solar MW than new battery MW.
4. The firm capacity value of batteries is based on battery duration, *i.e.*, the length of time it can discharge its full capacity back to the utility system before the battery must be recharged by the utility system.
5. There is no one-size-fits-all battery duration value that will ensure that a battery can receive a 100% firm capacity value in resource planning. The duration value will vary from utility-to-utility depending upon many factors, including a particular utility's peak day hourly load shape.[8]
6. In addition, the more stand-alone battery MW a utility plans to add to its system to maintain summer reserve margins, the longer the needed durations of the incremental battery additions.

---

[7] However, there can be economic advantages to a hybrid battery-and-PV facility versus separate facilities depending on the design of the hybrid facility. An example is the ability to use a single, shared bidirectional inverter if the hybrid facility utilizes a DC coupling of the PV and battery components. Another example could be potential reduced costs for a single, larger transmission interconnection to the existing transmission grid instead of two separate transmission interconnections. Both of these could result in some reduction in total installed costs.

[8] Once again, Fundamental Principle # 1 of resource planning, "All Electric Utilities are Different," keeps coming into play.

7. The need for increasing battery durations as more and more batteries are added to a utility system to act as a generation source for meeting peak loads will result in higher $ per kW installed costs for these battery increments than would be the case if there were no changes in the needed battery duration. (These effective $ per kW cost increases will be offset, but only in part, by the battery ITC.)

In the next chapter, we step outside of what has typically been considered the resource planning function of electric utilities to also take a look at other challenges that are likely to surface for utilities that have set a low-to-zero carbon goal. These challenges will primarily be faced by a utility's functional areas of system operations and transmission planning.

13 Moving to Zero Carbon
Ramifications for System Operations and Transmission Planning

INTRODUCTION

In order to successfully plan and operate their electric systems, utilities have always needed effective coordination among a number of functional areas of the utility. For example, regular ongoing coordination is particularly needed among three functional areas: resource planning, transmission planning, and system operations. This coordination takes many forms and addresses a variety of issues. A few of these issues are listed below (in no particular order):

- Ensuring that transmission interconnection and integration costs associated with different generation options are accounted for in IRP analyses;
- Analyzing competing generation and transmission options, particularly when faced with load growth and/or retirement of existing generation in a specific load pocket or region of the utility system;
- Performing scenario analyses to help determine how the generation fleet and the transmission system would be operated under potential future situations or scenarios; and
- Checking the appropriateness of reliability criteria currently being used to plan the utility system and, if necessary, modifying one or more of those criteria and/or developing a new reliability criterion.

For many years, the coordination activities between these three functional areas have been well understood and only occasionally resulted in novel challenges. However, that comfortable situation is now changing as increasing amounts of PV and batteries (plus other zero-carbon resource options) are analyzed and installed on utility systems. New challenges are emerging that these three functional areas will have to address.

In this chapter, we first take a look at two of these challenges: solar curtailment and how changes will have to be made in how existing conventional generators are operated. As we shall see from the discussion, these two challenges are related. Later in this chapter, we will address system stability issues with which the transmission planning and the system operation functional areas, in particular, will have to deal with. As a result of these challenges and issues, coordination among a utility's resource planning, system operations, and transmission planning functions will need to be enhanced as a utility moves toward a zero-carbon goal.

DOI: 10.1201/9781003301509-15

SOLAR CURTAILMENT[1]

Conventional generating units, such as a CC or CT unit, are correctly thought of as dispatchable resources. With either of these conventional types of generators, the system operator has to start up the unit before that unit can begin to deliver electricity to the utility system. The CC or CT unit is then operated at different output levels according to the system needs at a particular moment in time. Eventually, that generating unit is turned off by the system operator when its contribution is no longer projected to be needed for the immediate future.

Conversely, PV panels begin producing electricity shortly after the sun comes up. The PV facility then produces electricity in different amounts each hour as the sun travels across the sky (all else equal, producing more electricity as the sun climbs higher in the sky, followed by less electricity output as the sun gets lower in the sky). Solar then ceases to produce electricity when night-time comes. In this regard, the PV is sometimes thought of as being on "automatic pilot," and therefore, PV is often referred to as a non-dispatchable generation resource. The system operator does not need to "operate" a PV resource for it to produce electricity in the same sense he/she does with conventional generating units.

At least that has been the commonly held view of PV, as relatively small amounts of PV have been added to utility systems. However, as ever-increasing amounts of PV are analyzed in resource planning models, and as system operators manage their systems with these growing amounts of PV resources, this perception of PV as a resource that is permanently on automatic pilot has faded. This is especially so when considering months that have relatively low electrical load, but substantial solar insolation.[2] The Spring and Fall months are typically good examples of this combination of relatively low electric load and significant solar output.

Let's return to our hypothetical utility system and take a look at the hourly electrical load for an average day in April in Current Year + 5. We note the electrical load during April is considerably lower than the loads on peak days in summer and winter. Our utility's average April load is shown by the solid line in Figure 13.1.

---

[1] "Solar curtailment" refers to actions taken during time periods (minutes/hours) in which there is more solar-generated energy (MWh) being produced than the utility's instantaneous load (or remaining load) that must be served. The "excess" solar-generated energy must then either be blocked from entering the utility system, stored, or sold to other parties. The use of the term "curtailment" is used as shorthand to denote that the utility must take one of these actions to deal with the excess solar MWh.

[2] The discussion of solar curtailment that follows is applicable, to a degree, in all months of the year. However, given the relatively low MW levels of PV that have been installed to-date for many utility systems, solar curtailment is generally most pronounced in low load months. As the total amount of PV MW increases, this curtailment issue will be seen in other months as well.

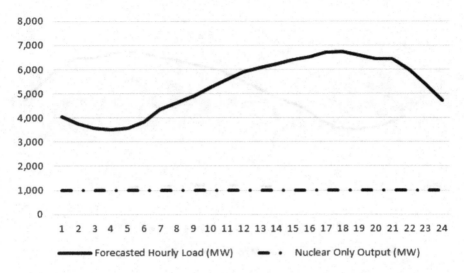

FIGURE 13.1   April day in Current Year + 5: Hourly load versus hourly output nuclear only (assuming no PV) (MW).

Also shown in this figure is a dash-dot line that represents the projected output of our utility's 1,000 MW nuclear capacity on this April day, *i.e.*, from our utility's existing zero-carbon-emission-generating capability. As previously mentioned in Part I of this book, nuclear units typically have the lowest $ per MWh operating cost of any type of conventional generation. Partly due to their low operating cost, nuclear units are not typically ramped up or down during a day. For these reasons, our utility is fully operating its 1,000 MW nuclear capacity at all hours of this April day as shown in Figure 13.1.

We note two other items regarding this figure. First, the area between the hourly load (the solid line) and the nuclear output (the dash-dot line) represents energy that the other, non-nuclear generating units on our utility system will have to supply. Second, Current Year + 5 is the year in which our utility needs to add new generation capacity to maintain a 20% total reserve margin criterion due to growing load. Because Figure 13.1 does not show any solar contribution, the figure basically represents a scenario in which our utility has decided not to select the PV Option we discussed in Part I of this book, but it has decided to select one of the other resource options instead.

But how would the picture change if our utility had selected Supply Only Resource Plan 3 (PV), which features 240 MW of PV, to address the capacity need in Current Year + 5? This new picture is presented in Figure 13.2.

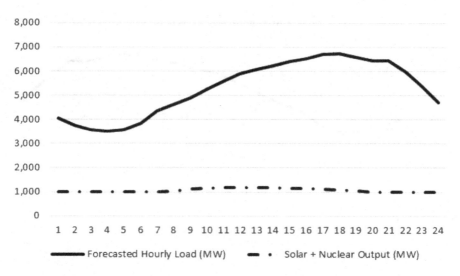

FIGURE 13.2 April day in Current Year + 5: Hourly load versus hourly output nuclear and solar only (assuming 240 MW PV) (MW).

In Figure 13.2, the hourly load line (the solid line) has not changed. The dash-dot line now represents the combined output of the 1,000 MW of nuclear capacity and the 240 MW of PV. However, by comparing the dash-dot lines in Figures 13.1 and 13.2, we see that the change in the dash-dot line is very small. Only a slight upward bump in Figure 13.2's dash-dot line during the hours from about Hour 9 to Hour 18 is evident.

From this, we conclude that adding 240 MW of PV does not result in a significant change in the amount of energy between the solid and dash-dot lines that remains to be served by our utility's conventional non-nuclear generating units. Consequently, the way in which the utility operates its existing fleet of generating units will probably not be affected much by adding 240 MW of PV to its system.

Now suppose that our utility has set a goal of being carbon-free by Current Year + 29, *i.e.*, 25 years after Current Year + 5. Let's also assume that our utility has decided to install PV in equal annual amounts starting in Current Year + 5 and continuing on through Current Year +29. Using the projection from our utility's second estimation approach to determine how many MW of new resources will need to be added to reach its carbon goal, our utility-projected 34,680 MW of PV would be needed by Current Year + 29 as discussed in Chapter 10.

Therefore, assuming PV is added in equal annual amounts over this 25-year period starting in Current Year + 5, our utility will need to add approximately 1,387 MW (= 34,680 total MW of PV/25 years) per year.

By starting the 1,387 MW/year additions in Current Year + 5, our utility will have added 8,322 MW of total PV by Current Year + 10 (= 1,387 MW/year × 6 years).[3]

---

[3] Yes, the 8,322 MW of new PV in just 6 years is a lot of PV. However, recall that our utility needs 34,680 MW of new PV in 25 years. Therefore, the 8,322 MW of new PV represents only about 24% of what our utility will ultimately need to add by Current Year + 29.

Based on these new assumptions, we now move ahead in time to see how the information presented in Figures 13.1 and 13.2 for Current Year + 5 are projected to change in Current Year + 10 when more than 8,300 MW of PV have been installed on our utility system. We do so in Figure 13.3.

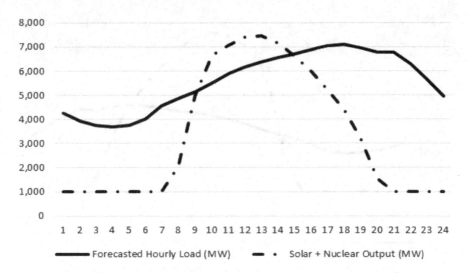

FIGURE 13.3 April day in Current Year + 10: Hourly load versus hourly output nuclear only (assuming 8,322 MW PV) (MW).

By moving out in time to Current Year + 10, both lines on the graph have changed. The hourly load for an average April day shown by the solid line has increased due to load growth during the additional 5 years. However, the load only increases by roughly 1% per year so the load in Current Year + 10 has not increased by a significant amount compared to the load in Current Year + 5 (as shown in Figures 13.1 and 13.2).

The same cannot be said for the dash-dot line that represents the projected hourly MW output of our utility's 1,000 MW of nuclear and its 8,322 MW of PV. The combined nuclear-and-PV MWh output now exceeds the projected hourly load for about five continuous hours: from approximately Hour 10 through Hour 14.

At this point, a reader's reaction might be as follows: *"Whoa! I did not see this coming. What does it mean for our utility?"* The main takeaway from Figure 13.3 is that the amount of energy (MWh) supplied by PV[4] that is shown by the area above the hourly load solid line and below the dash-dot PV output line cannot be used by our utility to serve its own load during those hours. This is because there is more electricity being generated than the load our utility must instantaneously serve in those hours.

---

[4] The output shown is actually from both nuclear and PV. However, the nuclear output has remained at 1,000 MW/hour. It is the output from the rapidly increasing amount of PV that has resulted in the projected hourly output being larger than the projected hourly load. For this reason, and to keep the narrative simple, we will refer only to the PV output from this point going forward.

One might then ask if this situation is exacerbated in later years as more and more PV is installed on our utility system. The answer is "yes" as shown by Figure 13.4 which provides a projection for Current Year + 15.

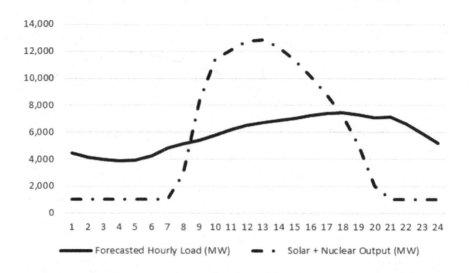

FIGURE 13.4 April day in Current Year + 15: Hourly load versus hourly output nuclear only (assuming 15,257 MW PV) (MW).

In Figure 13.4, we continue to assume that our utility is pursuing its zero-carbon goal with a target year of Current Year + 29. Therefore, our utility has continued to add 1,387 MW of additional PV for five more years, thus ending up with 15,257 MW of installed PV in Current Year + 15 (= 8,322 MW + [1,387 MW/year × 5 years]).

Our utility's projected hourly load to serve (the solid line) has again increased slightly due to five more years of load growth. However, as one might expect from Figure 13.3, the increase in generation output from adding 1,387 MW of new PV per year for five more years significantly enlarges the area between the hourly generation (the dash-dot line) and hourly load line. (Note the increase in the scale of the y-axis.) The increase in PV output also results in increasing the number of hours in which PV output exceeds load. This number has increased from 5 hours in Current Year + 10 shown in Figure 13.3 to 9 hours in Current Year + 15 shown in Figure 13.4. These hours now stretch from approximately Hour 9 through Hour 17.

These last two figures portray the gap between "excess" PV output and hourly load for a number of hours on an average April day. However, the figures do not provide a detailed view of the magnitude of the excess PV outputs by hour. In order to have a clearer picture of the hourly excess PV output, we return to Current Year + 10 in Figure 13.5.

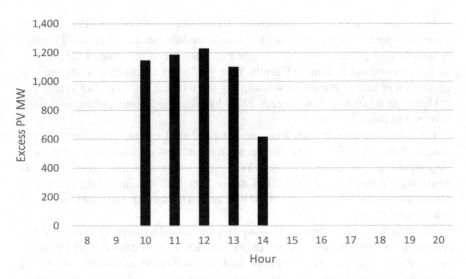

FIGURE 13.5 Current Year + 10: Excess generation on an average April day (assuming 8,322 MW PV) (MW).

This figure shows that excess PV output for an average April day in Current Year + 10, driven by the ever-increasing amounts of PV being added to our utility system, varies between roughly 1,000 MW and slightly more than 1,200 MW/hour for Hours 10 through 13, then the excess PV output decreases to about 600 MW in Hour 14.

We next take a similar look at projected excess generation in Current Year + 15 in Figure 13.6.

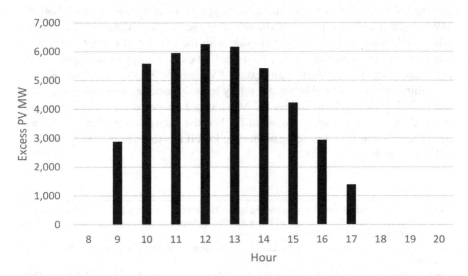

FIGURE 13.6 Current Year + 15: Excess generation on an average April day (assuming 15,257 MW PV) (MW).

In this figure, we see that not only has the number of hours in which excess PV output is experienced grown from 5 to 9 hours, the MW amount of excess PV output by hour has greatly increased. (Note again the increase in the scale of the y-axis as we have moved out in time to Current Year + 15.) In Current Year + 10, the largest amount of excess generation was slightly more than 1,200 MW. In Current Year + 15, excess generation now exceeds 5,000 MW during 5 hours and exceeds 6,000 MW in 2 of those 5 hours.

If we look at these 2 years from the perspective of excess total energy (MWh) instead of excess MW by hour, we get another indication of how much the excess PV output has increased. In Current Year + 10, the sum of the excess generation during the 5 hours is 5,261 MWh. In Current Year + 15, the sum of the excess generation during the 9 hours is 40,781 MWh, approximately an eight-fold increase in just 5 years' time.[5]

Faced with a daily situation of having ever-increasing amounts of excess PV generation in its April months as our utility pursues its zero-carbon goal by adding more PV each year, our utility begins to look at how it could address this situation. It begins by looking at three possible approaches that are available to our utility at the time this book is written.[6]

The first approach is simply to "curtail" the PV output that would result in our utility's generation output exceeding the hourly load. Our utility's system operators would do this by not allowing this amount of PV output to flow into the electric grid. This can be achieved by the system operators remotely controlling the DC-to-AC inverters at the PV site. By doing so, the operator can "dial down" the amount of ac energy that is "released" into the electric grid.[7] This might be the simplest approach, but it would mean that all of the potential "excess" PV-generated MWh would be unused.

A second approach would be to try to sell these excess PV MWh to another utility or third party. Because there will always be uncertainty in both the hourly load and the PV output, such a sale would likely not be a "firm" sale in which both the MWh amount and time period of the sale are specified in advance. Instead, such a sale would more likely be a non-firm sale with the amount of energy and time of sale being on an as-available basis. As a result, the other utility or third party could not count on definite delivery of any excess PV-based MWh sales from our utility.[8]

A third approach would be for our utility to build batteries (or other energy storage facilities) in which the excess PV MWh could be stored for its later use. This approach has particular appeal to our utility because, as we saw in Chapter 10, it will need a lot of battery MW to serve night-time load if it is to meet its zero-carbon goal in Current Year + 29.

---

[5] These MWh values are derived by summing the values for each bar in Figures 13.5 and 13.6, respectively.

[6] A fourth potential approach, the use of alternate zero-carbon, or carbon-neutral, fuels, will be discussed in Chapter 14.

[7] A rough analogy is a dimmer switch for a lamp that can change the amount of electricity that flows into the lamp.

[8] In addition, if another nearby utility system is also pursuing a low-to-zero-carbon goal, it may be experiencing a similar excess solar output situation at the same time. If so, it would be an unlikely candidate to be a buyer of our utility's excess solar output.

So, let's assume that, in addition to adding 1,387 MW/year of PV starting in Current Year +5, our utility also adds a constant amount of battery storage MW per year that would be needed to meet its energy storage needs in Current Year + 29 when its zero-carbon goal is to be met. Continuing to use the results of the second estimation approach for projecting the MW amount of new resources needed to supply 100% of the Current Year + 29 GWh with zero-carbon emissions that was discussed in Chapter 10, the projection was that our utility would need 17,835 MW of 4-hour duration batteries by Current Year + 29.

If our utility plans to begin adding these new batteries in Current Year + 5, and to continue adding a constant MW amount each year for 25 years, it will plan to add approximately 713 MW (= 17,835 MW/25 years) of 4-hour batteries each year. Let's see where this would leave our utility in Current Year + 10 and Current Year + 15 in regard to being able to store (for later use) all of the projected excess PV output on an average April day.

- *In Current Year + 10*: Our utility will have added 713 MW of batteries in each of the 6 years from Current Year + 5 through Current Year +10, or 4,278 MW (= 713 MW/year × 6 years) by that year. With these batteries having a 4-hour duration, this total battery addition will result in the ability to effectively store approximately 17,112 MWh (= 4,278 MW × 4 hours) of energy. Because this amount of energy storage is larger than the projected 5,261 MWh/day of excess PV generation projected for this year, these battery additions should easily address our utility's excess PV generation situation on average April days in Current Year + 10.
- *In Current Year + 15*: Our utility will now have added another 5 years of 713 MW of batteries per year, or 3,565 MW (= 713 MW/year × 5 years) of additional 4-hour batteries. Therefore, the total amount of batteries on our utility's system in Current Year + 15 is 7,843 MW (= 4,278 + 3,565). With the 4-hour battery duration assumption, the 7,843 MW of batteries can store 31,372 MWh of energy (= 7,843 MW × 4 hours). However, this amount of energy storage falls far short of the 40,781 MWh of daily April excess PV generation in Current Year + 15.

Therefore, at least by Current Year + 15 (and likely before then), our utility cannot address its projected excess PV generation situation during April days solely by adding 4-hour batteries at a pace that will get them to the estimated MW amount of batteries needed to meet its zero-carbon goal by Current Year + 29.[9] Considering that the only battery resource option our utility has contemplated so far is a 4-hour duration battery, our utility will need to add more batteries of 4-hour duration, add batteries of greater duration, accept some level of solar curtailment, sell excess PV MWh to another utility or third party, and/or find another method by which to deal with the excess PV output.

---

[9] The astute reader may notice that this result is another suggestion why the second estimation approach for determining the new resource MW needed to serve 100% of our utility's GWh in Current Year + 29 may underestimate the new battery MW that would actually be needed.

As previously mentioned, we will return to our utility's excess PV generation situation/challenge in Chapter 14 where we will examine another potential approach to deal with this issue. We now take a different look at our utility's April load, and the projected PV additions, to see what impact (if any) the PV additions may have on the operation of our utility's existing fleet of generating units. This is the second challenge mentioned earlier that is associated with installing ever-increasing amounts of PV on our utility system.[10]

OPERATION OF EXISTING CONVENTIONAL GENERATION WITH EVER-INCREASING AMOUNTS OF PV

In this section, we put on a system operator's hat[11] and first take a look at how our utility's existing fleet of generating units might be operated on an average day in April in Current Year + 5. In that year, we assume our utility will only have 240 MW of PV installed on its system.[12] We then take a similar look at how our utility's generators might be operated on an average day in April in Current Year + 10. For that year, we assume that our utility is pursuing its zero-carbon goal and, therefore, will have installed 8,322 MW of PV on its system. The objective of this exercise is to see how a very large amount of PV on our system may influence the way that all of the generating units on the system are dispatched.

We begin by noting that the examples we are providing regarding how our utility's existing generating units might be dispatched on an average April day in either of these 2 years represent only one of many ways an actual system operator might dispatch the generating units on these days. Keeping in mind that the end objective is to see how the addition of significant amounts of PV would likely alter the operation of our utility's existing non-nuclear generating units, the approach we are taking is a simplified one. This approach takes into account the following operating rules and/or guidelines that system operators might typically use to dispatch their generating units:

1. Nuclear units are operated continually and are only brought off-line for planned refueling or if unplanned maintenance is required. Once nuclear units are in operation, they are rarely ramped up or down on an hourly basis.
2. In somewhat similar fashion, coal units are seldom taken off-line absent planned or unplanned maintenance. This is due to the relatively low operating cost of these units (as we saw in Part I of this book) and to the many hours it takes these units to regain full steam pressure once they

---

[10] For the remainder of Chapter 13, we will use the term solar "curtailment" to refer to our utility either blocking PV output from entering the transmission grid, and/or selling excess PV output, as the way in which our utility will address excess PV MWh. Although storing the excess PV output is definitely another option, this unnecessarily complicates the narrative at this point. The key points to focus on now are the amounts of excess PV MWh and how they impact the operation of the utility's conventional generating units.

[11] We do so reluctantly, and with a fair amount of trepidation.

[12] We assume that our utility selected Supply Only Resource Plan 3 (PV) to meet its resource needs in Current Year + 5.

are restarted. However, once they are being operated, these units can be ramped up or down.

3. All else equal, generators will be dispatched beginning with the generators with the lowest operating cost, followed in turn by the generators with the next lowest operating costs. We are assuming the same $ per MWh dispatch cost of the various generating units that we discussed previously in Part I of this book, specifically in Table 2.3. Therefore, after dispatching the existing nuclear capacity, our utility's existing coal units have the lowest dispatch cost (roughly $20/MWh assuming a $2/mmBTU coal price), followed by our utility's existing CC units (roughly $42/MWh assuming a $6/mmBTU natural gas price). Therefore, assuming that no special circumstances are in play on these two average April days, our utility will dispatch its existing coal units, then its existing CC units, after nuclear.[13]

4. Obviously, the total MW amount of generating capacity of each type provides an upper limit of how much of each type of generation can be dispatched. We will retain the total capacity MW amounts for each type of generating unit used in the Part I analyses of this book (and shown previously in Figure 2.4). Those total MW values are 3,500 MW for coal and 3,000 MW for CCs.

There are three other bits of information not previously discussed that we need in order to dispatch our utility system's existing generating units. These three other bits of information are (i) the numbers of individual coal units and CC units, (ii) the "ramp rate" (how quickly a generating unit's output can be ramped up or down) in terms of MW/minute, and (iii) the "low limit" for each generating unit, which is the lowest MW level at which the unit can be operated.

Table 13.1 provides our assumptions regarding these operating parameters for our utility's existing nuclear, coal, and CC generating units. In addition, Table 13.1 presents a simplified set of "ground rules" that we will use to dispatch the generating units on our system.[14]

---

[13] Because the electric load in April is much smaller than the summer or winter peak loads, our utility typically will not have to dispatch its steam (gas) units or its CTs. Assuming no generator outages on this April day, the load can be met with a combination of nuclear, coal CC, and PV output.

[14] With these assumptions, we continue to keep our hypothetical utility system simple in an attempt to make the discussion as easy to follow as possible.

TABLE 13.1

Assumptions and Ground Rules for Operation of Our Utility's Generating Units

Assumptions:

Type of Generating Unit	Total Capacity per Type of Generating Unit (MW)	Assumed Size of Each Generating Unit (MW)	Assumed Number of Each Type of Generating Unit	Assumed Low Operating Limit for Each Type of Generating Unit (MW)	Assumed Ramp Rate for Each Type of Generating Unit (MW per minute)
Nuclear	1,000	1,000	1	Not Applicable	Not Applicable
Steam (coal)	3,500	700	5	200	3
CC	3,000	600	5	200	30
Total =	7,500				

Ground rules: (As part of our continuing effort to keep the discussion simple, our ground rules are very basic.)

1. Nuclear is not cycled on/off or ramped up/down
2. Coal units are not cycled on/off.
3. Assuming $6 per mmBTU cost of natural gas, the coal units will be more economic to operate than the CC units. Consequently, coal units will be dispatched next (after nuclear and solar), followed by CC units.
4. CC units can be cycled on/off once per day
5. CC ramp rates and startup costs are ignored as the CC unit moves through its stages of operating its CT components. For example, if a CC unit is a 2 × 1 unit (meaning it has two CTs), the time and costs to switch from using one CT to using both CTs is not accounted for.

Regarding the assumptions, this table shows that our utility system has, in addition to its nuclear capability, five coal generating units, and five CC generating units. The MW value for each generating unit within a type of generating unit is identical; for example, each of the five 5 coal units have a maximum capacity of 700 MW, which results in a total coal generating capacity of 3,500 MW (= 700 MW/unit × 5 units). Each of the five CC units has a maximum capacity of 600 MW, which results in a total CC generating capacity of 3,000 MW (= 600 MW/unit × 5 units).

Table 13.1 completes our assumptions by providing ramp rates for our utility's generating units. Coal units typically have relatively slow ramp rates, and we have chosen a ramp rate of 3 MW/minute for these units. Conversely, CC units have relatively fast ramp rates, and we have chosen 30 MW/minute for these units. Finally, we have assumed that all 10 of our utility's coal and/or CC units have a low operating limit of 200 MW each.

Turning now to five ground rules presented in the lower half of Table 13.1, we will use these rules to dispatch the generators. These rules are both simple and based on the four general guidelines discussed above. The five ground rules are as follows:

1. The nuclear capacity will not be cycled on or off nor will it be ramped up and down during the 24-hour period. It will continue to operate at its maximum output of 1,000 MW.
2. The coal units will also not be cycled on/off, but the units, if they are being operated, can be ramped up or down within limits (*i.e.*, respecting the ramp rate and minimum operating limits).
3. By assuming coal prices at approximately \$2/mmBTU and natural gas prices at approximately \$6/mmBTU, the variable cost (\$ per MWh) to produce electricity from coal will be lower than the variable cost to produce electricity from the CC units. Therefore, our utility's coal units will usually be dispatched before the CC units.
4. The CC units can be cycled on/off once per day, and these units can also be ramped up/down to follow the load within limits (*i.e.*, respecting the ramp rate and minimum operating limits).
5. And, to keep things very simple, we will ignore both the time it takes, and the costs incurred, when switching a CC unit from one CT stage to another. (For example, a 2 × 1 CC has two CT components. As the CC unit ramps up, it goes from utilizing only one CT to utilizing both CTs. To do this requires some time and incurs some start-up costs.)

We will now use these ground rules to dispatch our utility's generation fleet on an average day in April in each of 2 years: Current Year + 5 (with 240 MW of PV) and Current Year + 10 (with 8,322 MW of PV). We will do so in a step-by-step approach in which each additional step adds new information.

Table 13.2 sets the stage for this examination by presenting only information regarding the hourly load, the nuclear contribution, and the PV contribution.

TABLE 13.2

Current Year + 5 Average April Day: Hourly Load and Output of Nuclear and PV (240 MW of PV Installed)

(1)	(2)	(3)	(4)	(5a) = (2) − (3) − (4)	(5b) (See Text Explanation)	(5c) = (5a) − (5b)
Hour	Current Year + 5 Hourly Load (MW)	Nuclear Output (MW)	PV Output (MW)	Remaining Hourly Load to Serve (MW)	Curtailed PV (MW)	Adjusted Remaining Load to Serve (MW)
1	4,046	1,000	0	3,046	0	3,046
2	3,742	1,000	0	2,742	0	2,742
3	3,575	1,000	0	2,575	0	2,575
4	3,507	1,000	0	2,507	0	2,507
5	3,561	1,000	0	2,561	0	2,561
6	3,837	1,000	0	2,837	0	2,837
7	4,349	1,000	0	3,349	0	3,349
8	4,641	1,000	30	3,611	0	3,611
9	4,904	1,000	115	3,789	0	3,789
10	5,248	1,000	163	4,085	0	4,085
11	5,607	1,000	175	4,432	0	4,432
12	5,882	1,000	185	4,697	0	4,697
13	6,063	1,000	187	4,877	0	4,877
14	6,238	1,000	178	5,060	0	5,060
15	6,398	1,000	162	5,236	0	5,236
16	6,545	1,000	144	5,400	0	5,400
17	6,697	1,000	122	5,574	0	5,574
18	6,745	1,000	98	5,647	0	5,647
19	6,602	1,000	63	5,539	0	5,539
20	6,431	1,000	16	5,416	0	5,416
21	6,447	1,000	0	5,447	0	5,447
22	5,980	1,000	0	4,980	0	4,980
23	5,388	1,000	0	4,388	0	4,388
24	4,723	1,000	0	3,723	0	3,723

Total Solar Output (MWh) = **1,638**

PV MWh Curtailed in Excess of Load = **0**

Percent PV Curtailed in Excess of Load = **0%**

Total PV MWh Curtailed = **0**

Percent PV Curtailed in Total = **0%**

In Table 13.2, columns (2) through (4), respectively, present the hourly load, the hourly output of 1,000 MW of nuclear, and the hourly output of the 240 MW of PV. Column (5a) then subtracts the hourly outputs of nuclear and PV from the hourly load to derive the remaining load to be served by the other generating units on the utility system.

Note that by comparing the values in columns (2) and (4), we see that the PV hourly output values are small compared to the hourly load. Therefore, there is still a relatively large remaining hourly load to serve after accounting for the hourly MW output of nuclear and PV (column [5a]). As a result, there is no need to curtail any of the PV output in Current Year + 5.[15] This is shown by the zero-MW values in column (5b). As a result, the remaining load values in column (5c) are identical to the values in column (5a).

At the bottom of the table, we see that the total PV energy output for this average April day is 1,638 MWh.[16] Because there was no need to curtail the PV output in any hour, the remaining values at the bottom of the page that relate to curtailed PV output are also zero.

Column (5c), the remaining load to be served after accounting for nuclear, PV, and PV curtailment (if applicable), is then carried over to Table 13.3[17].

---

[15] As we shall see, this will not be the case in Current Year + 10.

[16] The value of 1,638 MWh is the output of 240 MW of PV for 1 day in April. Therefore, it represents something entirely different than the 1,752 MWh value used in prior chapters that was assumed to be the annual output of 1 MW of PV with a 20% annual capacity factor.

[17] Columns (2) through (4) of Table 13.2 are not carried over to Table 13.3 in order to save space.

TABLE 13.3

Current Year + 5 Average April Day: Remaining Load to Serve After Nuclear and PV, with Potential Coal Dispatch (240 MW of PV Installed)

(1)	(5c) (from Table 13.2)	(6)	(7)	(8)	(9)	(10)	(11)	(12) = (5) − (11)
Hour	Remaining Hourly Load to Serve (MW)	Coal Unit 1 Output (MW)	Coal Unit 2 Output (MW)	Coal Unit 3 Output (MW)	Coal Unit 4 Output (MW)	Coal Unit 5 Output (MW)	Coal Total Output (MW)	Load Unserved by Coal (MW)
1	3,046	700	700	700	700	246	3,046	0
2	2,742	700	700	622	520	200	2,742	0
3	2,575	700	700	622	353	200	2,575	0
4	2,507	700	700	622	285	200	2,507	0
5	2,561	700	700	676	285	200	2,561	0
6	2,837	700	700	700	465	272	2,837	0
7	3,349	700	700	700	645	404	3,149	200
8	3,611	700	700	700	700	584	3,384	227
9	3,789	700	700	700	700	700	3,500	289
10	4,085	700	700	700	700	700	3,500	585
11	4,432	700	700	700	700	700	3,500	932
12	4,697	700	700	700	700	700	3,500	1,197
13	4,877	700	700	700	700	700	3,500	1,377
14	5,060	700	700	700	700	700	3,500	1,560
15	5,236	700	700	700	700	700	3,500	1,736
16	5,400	700	700	700	700	700	3,500	1,900
17	5,574	700	700	700	700	700	3,500	2,074
18	5,647	700	700	700	700	700	3,500	2,147
19	5,539	700	700	700	700	700	3,500	2,039
20	5,416	700	700	700	700	700	3,500	1,916
21	5,447	700	700	700	700	700	3,500	1,947
22	4,980	700	700	700	700	700	3,500	1,480
23	4,388	700	700	700	700	700	3,500	888
24	3,723	700	700	700	700	700	3,500	223

Because coal units have the next lowest dispatch cost of conventional generating units (after nuclear), the coal units are the next generators our utility would normally dispatch. Table 13.3 presents a dispatch scenario for our utility's five coal units, each with a capacity of 700 MW, after accounting for the output of nuclear and PV.

Recalling that the total amount of coal capacity is 3,500 MW (= 700 MW/ generator × five generators), and the remaining load to be served in Hour 1 is 3,046 MW, all five of our utility's coal units are projected to be operated in that hour, with only coal unit 5 not being operated at its maximum 700 MW output.

As shown in column (5c), the remaining load (after accounting for nuclear and PV) then drops in Hours 2 through 4. Our utility handles this by first ramping down coal unit 5 to its low operating limit of 200 MW. (By keeping coal unit 5 operating, we preserve its ability to ramp up as the load increases later in the morning. If coal unit 5 had been turned off instead, the process to restart that unit would take many hours and our utility may have been forced to use other, more expensive-to-operate generators instead.) In Hours 2 through 4, coal units 3 and 4 are also ramped down as needed to match the declining load. Coal units 1 and 2 remain at full output, where they are running most economically, during all 24 hours of the day.

Starting in Hour 5, the remaining load begins to increase and the coal units 3, 4, and/or 5 ramp up accordingly. However, by Hour 9, the coal units are fully dispatched at their maximum total capacity of 3,500 MW, but our utility's remaining load to be served in Hour 9 is greater than 3,500 MW. This is shown in column (12) by the non-zero value of 289 MW. Therefore, the next-up-to-bat generators, our utility's CC units, must come into play no later than Hour 9.

However, our utility has anticipated the need for operating the CC units and has chosen to start one of the five CC units (CC unit 1) at its minimum operating load of 200 MW in Hour 7 and then ramp that CC unit up in Hour 8 at the same time it is ramping up the coal units to their maximum output. This can be seen in in Hour 7 in column (12). The 200 MW load that remains in Hour 7 will be served by one of our utility's CC units, as we shall see in the next table.

Table 13.4 is an extension of Table 13.3 in which we now show the potential operation of our utility's five CC units by adding columns (13) through (19).[18]

---

[18] Recall the discussion in Part I of this book that the primary marginal, or "swing" generating units on our utility system were the CC units.

TABLE 13.4

Current Year + 5 Average April Day: Remaining Load to Serve After Nuclear and PV, with Potential Coal and CC Dispatch (240 MW of PV Installed)

(1)	(5c)	(6)	(7)	(8)	(9)	(10)	(11)	(12)	(13)	(14)	(15)	(16)	(17)	(18)	(19)
	(From Table 13.3)	Coal Unit 1	Coal Unit 2	Coal Unit 3	Coal Unit 4	Coal Unit 5	Coal Total		= (5c) – (11)	CC Unit 1	CC Unit 2	CC Unit 3	CC Unit 4	CC Unit 5	CC Total	= (12) – (18)
	Remaining Hourly Load to Serve	Output	Output	Output	Output	Output	Output	Load Unserved by Coal	Output	Output	Output	Output	Output	Output	Remaining Unserved Load
Hour	(MW)	(MW)	(MW)	(MW)	(MW)	(MW)	(MW)	(MW)	(MW)	(MW)	(MW)	(MW)	(MW)	(MW)	(MW)
1	3,046	700	700	700	700	246	3,046	0	0	0	0	0	0	0	0
2	2,742	700	700	622	520	200	2,742	0	0	0	0	0	0	0	0
3	2,575	700	700	622	353	200	2,575	0	0	0	0	0	0	0	0
4	2,507	700	700	622	285	200	2,507	0	0	0	0	0	0	0	0
5	2,561	700	700	676	285	200	2,561	0	0	0	0	0	0	0	0
6	2,837	700	700	700	465	272	2,837	0	0	0	0	0	0	0	0
7	3,349	700	700	700	645	404	3,149	200	200	0	0	0	0	200	0
8	3,611	700	700	700	700	584	3,384	227	227	0	0	0	0	227	0
9	3,789	700	700	700	700	700	3,500	289	289	0	0	0	0	289	0
10	4,085	700	700	700	700	700	3,500	585	385	200	0	0	0	585	0
11	4,432	700	700	700	700	700	3,500	932	600	332	0	0	0	932	0
12	4,697	700	700	700	700	700	3,500	1,197	600	397	200	0	0	1,197	0
13	4,877	700	700	700	700	700	3,500	1,377	600	537	240	0	0	1,377	0
14	5,060	700	700	700	700	700	3,500	1,560	600	600	360	0	0	1,560	0
15	5,236	700	700	700	700	700	3,500	1,736	600	600	536	0	0	1,736	0

Row															
16	0	1,900	0	200	500	600	600	1,900	3,500	700	700	700	700	700	5,400
17	0	2,074	0	274	600	600	600	2,074	3,500	700	700	700	700	700	5,574
18	0	2,147	0	347	600	600	600	2,147	3,500	700	700	700	700	700	5,647
19	0	2,039	0	239	600	600	600	2,039	3,500	700	700	700	700	700	5,539
20	0	1,916	0	200	516	600	600	1,916	3,500	700	700	700	700	700	5,416
21	0	1,947	0	200	547	600	600	1,947	3,500	700	700	700	700	700	5,447
22	0	1,480	0	200	200	480	600	1,480	3,500	700	700	700	700	700	4,980
23	0	888	0	0	0	288	600	888	3,500	700	700	700	700	700	4,388
24	0	223	0	0	0	0	223	223	3,500	700	700	700	700	700	3,723

Columns (13) through (17) show the hourly operation of the five individual CC units. CC unit 1 starts operation in Hour 7 as was just discussed. That CC unit then ramps up to its full 600 MW capability by Hour 11. CC unit 2 begins operation at its minimum operating limit in Hour 10 and quickly ramps up to 600 MW. As the remaining load continues to increase, CC unit 3 begins operating in Hour 12 and then also ramps up to its maximum 600 MW output. CC unit 4 only begins operation in Hour 16 and never ramps up to its full capability. The last CC unit, CC unit 5, is never called upon to operate on this April day.

Column (18) sums the total CC MW generated for each hour. As shown in this column, the CC units' total MW output increases continually beginning in Hours 7 through 18. The CC output begins to ramp down in Hour 19 as our utility's remaining load begins to decrease. By Hour 24, only CC unit 1 is still being operated, and then only near its low operating limit. (And, if the next day follows the same load pattern as shown beginning in Hour 1 for this day, even CC unit 1 will have ceased operation by Hour 1 of the next day.)

The key takeaways from Tables 13.2, 13.3, and 13.4 that address an average April day in Current Year + 5 are as follows:

- There is no need to curtail output of the modest 240 MW of PV in any hour on this April day.
- All five of the most economical-to-operate non-nuclear generating units (the coal units) are dispatched first and operated primarily as baseload units.
- Four of the five next most economical-to-operate other generating units (the CC units) are dispatched next and ramped up and down, and cycled on and off, as needed to follow the load. One of the CC units is not used on this April day.

Therefore, the dispatch of our utility's generating units in Current Year + 5 (which has only 240 MW of installed PV) is about what one would expect as our utility tries to make maximum use of its least expensive-to-operate generating units. The dispatch pattern is probably very similar to what our utility was used to seeing in prior years on April days before the 240 MW of PV was added.

Now we look at what the dispatch pattern might be in Current Year + 10 in which 8,322 MW of PV are installed on our utility's system. We take the same approach as before with Table 13.5 (presented on the next page) beginning our look at individual generator dispatch on an average April day in Current Year + 10 by focusing only on nuclear, PV, and PV curtailment.

TABLE 13.5

Current Year + 10 Average April Day: Remaining Load to Serve After Nuclear and PV (8,322 MW of PV Installed)

(1)	(2)	(3)	(4)	(5a) = (2) − (3) − (4)	(5b) (See Text Explanation)	(5c) = (5a) − (5b)
Hour	Current Year + 10 Hourly Load (MW)	Nuclear Output (MW)	PV Output (MW)	Remaining Hourly Load to Serve (MW)	Curtailed PV (MW)	Adjusted Remaining Load to Serve (MW)
1	4,252	1,000	0	3,252	0	3,252
2	3,933	1,000	0	2,933	0	2,933
3	3,758	1,000	0	2,758	0	2,758
4	3,686	1,000	0	2,686	0	2,686
5	3,742	1,000	0	2,742	0	2,742
6	4,032	1,000	0	3,032	0	3,032
7	4,571	1,000	0	3,571	0	3,571
8	4,878	1,000	1,049	2,829	0	2,829
9	5,154	1,000	3,978	176	1,624	1,800
10	5,516	1,000	5,660	(1,144)	2,944	1,800
11	5,893	1,000	6,076	(1,183)	2,983	1,800
12	6,182	1,000	6,406	(1,224)	3,024	1,800
13	6,373	1,000	6,470	(1,097)	2,897	1,800
14	6,556	1,000	6,169	(613)	2,413	1,800
15	6,724	1,000	5,616	107	1,693	1,800
16	6,878	1,000	4,998	880	920	1,800
17	7,038	1,000	4,246	1,793	7	1,800
18	7,089	1,000	3,415	2,674	0	2,674
19	6,939	1,000	2,178	3,761	0	3,761
20	6,759	1,000	544	5,216	0	5,216
21	6,776	1,000	0	5,776	0	5,776
22	6,285	1,000	0	5,285	0	5,285
23	5,663	1,000	0	4,663	0	4,663
24	4,963	1,000	0	3,963	0	3,963

Total Solar Output (MWh) = **56,804**

PV MWh Curtailed in Excess of Load = **5,261**

Percent PV Curtailed in Excess of Load = **9**

Total PV MWh Curtailed = **18,505**

Percent PV Curtailed in Total = **33**

This table uses the same format as Table 13.2 (which began our look at dispatch for an average April day in Current Year + 5). However, one quickly notes a few differences in the values in Table 13.5. First, our utility's hourly load values in column (2) are higher due to 5 years of load growth, but the nuclear hourly output values in column (3) have not changed because no new nuclear capacity has been added.

The most significant changes are found in regard to the projected PV hourly output that is presented in column (4). At the bottom of column (4), we see that the total PV output on this day is 56,804 MWh, which is much larger than the 1,638 MWh of PV output shown for Current Year + 5 in Table 13.2. In fact, starting in Hour 10 and continuing through Hour 13, the projected hourly PV output alone shown in column (4) is larger than our utility's entire hourly load shown in column (2).

When the PV output and the nuclear output are combined, then compared to the hourly load in column (5a), we see negative values for the remaining load in Hours 10 through 14. These negative values indicate that excess generation is being produced in each of those hours. (Note that these same excess PV output values were previously presented graphically in Figure 13.5.)

As a result, our utility must curtail at least the amount of PV output in Hours 10 through 14 that shows up as negative numbers in column (5a). The sum of those negative values is 5,261 MWh, which is shown at the bottom of column (5a). (We note that we saw the same 5,261 MWh value of excess PV output earlier.) This amount of "have-to-curtail" PV output represents 9% of the total PV output on this day (= 5,261 MWh curtailed/56,804 MWh of total PV output).

However, if only the amounts of PV that match the negative values in column (5a) are curtailed, that would leave our utility with 0 MW of load for those 5 hours. This would mean that our utility would have to shut down the operation of all its existing non-nuclear generating units for these 5 hours and then restart them. Restarting large coal and CC units takes a number of hours and increases the possibility of encountering problems with one or more units during the restart. In addition, this is only one average day in April. Our utility can expect to face similar situations on multiple days in April and other Spring/Fall months.

Therefore, our utility decides to continue to run at least some of its existing other generators and to curtail more PV output than is needed strictly due to the 5-hour excess generation period we just discussed. Our utility has to decide two things: how many of its existing other generators to continue to run, and how much solar to curtail above the necessary 5,261 MWh curtailment in Hours 10 through 14.

Our utility first looks at the peak remaining load it will have to serve during the 24 hours. What it sees is that peak remaining load is projected to be 5,776 MW as seen in column (5a) in Hour 21. In that evening hour, there is no PV output, so all of the 5,776 MW of remaining load will need to be served by coal and CC units. The total capability of these units is 3,500 MW (coal) and 3,000 MW (CC), totaling 6,500 MW. In order to serve that peak remaining load in Hour 21, our utility will need to have most of these ten units operating at or near their maximum capacity.

For example, our utility could meet its 5,776 MW remaining peak load in Hour 21 by running all five coal units to produce 3,500 MW (= 700 MW/unit × 5 units), plus running four CC units to produce 2,400 MW (= 600 MW/unit × 4 units). This would result in a total of 5,900 MW which is more than the 5,776 MW remaining peak load (so full output of all five CC units is not needed). Another way our utility could meet this peak load would be to run four coal units to produce 2,800 MW and run all five CC units to produce 3,000 MW. This would result in a total of 5,800 MW which also exceeds the 5,776 MW remaining load in Hour 21.

After considering the ground rules (particularly the ramp rates) for the existing generators, and wanting to maintain flexibility for the system operators (recognizing the minute-to-minute uncertainty they face in real-life situations), our utility decides to use something similar to the second approach. Therefore, it will be running nine conventional generating units at Hour 21: four coal units and five CC units. And, due to potential problems in restarting units that have been switched off, our utility will operate these nine units during all 24 hours.

What this means is that, with a 200 MW low operating limit for each of these nine generators, there will be <u>at least</u> an 1,800 MW hourly output in all hours (= 200 MW/ unit × 9 units). In turn, this means our utility must have a remaining load, after PV curtailment, of at least 1,800 MW.

Importantly, this includes Hours 10 through 14 in which our utility already has to curtail PV output just to get to 0 MW remaining load. In each of those hours, our utility will now be curtailing an additional 1,800 MW of PV to ensure the remaining load is 1,800 MW. Our utility also extends the PV curtailment in other "adjacent" hours both prior to, and after, the Hour 10 through 14 period to ensure that there is at least an 1,800 MW load in these hours as well. As a result, PV curtailment at some level is performed starting in Hour 9 and continuing through Hour 17.

The resulting total PV curtailment is 18,505 MWh shown at the bottom of column (5b). This represents approximately 33% of the total PV output of 56,804 MW that would be expected absent curtailment (= 18,505 MWh curtailed/56,804 MWh of total PV output).[19]

The "adjusted" remaining load to be served is presented in column (5c). That column carried over to Table 13.6 which now takes a look at how our utility could dispatch its five coal units to partially meet the remaining load.

---

[19] Note that this significant amount of PV curtailment, if it occurs on enough days of the year, can result in a substantial reduction in PV's PTC benefits that were discussed in Chapter 11.

TABLE 13.6

Current Year + 10 Average April Day: Remaining Load to Serve After Nuclear and PV, with Potential Coal Dispatch (8,322 MW of PV Installed)

(1)	(5c) (from Table 13.5)	(6)	(7)	(8)	(9)	(10)	(11)	(12) = (5c) − (11)
Hour	Adjusted Remaining Load to Serve (MW)	Coal Unit 1 Output (MW)	Coal Unit 2 Output (MW)	Coal Unit 3 Output (MW)	Coal Unit 4 Output (MW)	Coal Unit 5 Output (MW)	Coal Total Output (MW)	Load Unserved by Coal (MW)
1	3,252	400	400	400	300	0	1,500	1,752
2	2,933	300	300	300	300	0	1,200	1,733
3	2,758	300	300	300	300	0	1,200	1,558
4	2,686	300	300	300	300	0	1,200	1,486
5	2,742	300	300	300	300	0	1,200	1,542
6	3,032	300	300	300	300	0	1,200	1,832
7	3,571	300	300	300	300	0	1,200	2,371
8	2,829	300	300	300	300	0	1,200	1,629
9	1,800	200	200	200	200	0	800	1,000
10	1,800	200	200	200	200	0	800	1,000
11	1,800	200	200	200	200	0	800	1,000
12	1,800	200	200	200	200	0	800	1,000
13	1,800	200	200	200	200	0	800	1,000
14	1,800	200	200	200	200	0	800	1,000
15	1,800	200	200	200	200	0	800	1,000
16	1,800	200	200	200	200	0	800	1,000
17	1,800	200	200	200	200	0	800	1,000
18	2,674	350	300	300	300	0	1,250	1,424
19	3,761	480	400	350	300	0	1,530	2,231
20	5,216	660	546	530	480	0	2,216	3,000
21	5,776	700	700	700	700	0	2,800	2,976
22	5,285	700	550	550	550	0	2,350	2,935
23	4,663	520	500	500	400	0	1,920	2,743
24	3,963	400	400	400	300	0	1,500	2,463

We see that the dispatch of the coal units in Current Year + 10 is much different than in Current Year + 5 in which four of the five coal units were dispatched at their maximum 700 MW capacity starting in Hour 1 and continuing through most of the 24 hours. In addition, in Current Year + 5, all five coal units were being operated in each of the 24 hours.

Conversely, in Current Year + 10, only four coal units are operated during the 24-hour period, and these four coal units only are dispatched at their maximum in Hour 21 when the remaining load is at its highest. For a number of hours, the coal units are operated at their low limit of 200 MW. This helps allow our utility to better utilize its five CC units that have faster hourly ramping capability that will be useful in real-life situations to accommodate the minute-to-minute changes in PV output.

Column (12) presents the load that remains to be served after the dispatch of the coal units. That unserved-by-coal load now appears in all 24 hours, and that load will be met by use of the CC units. We note that the remaining load to be served in each hour is at least 1,000 MW. This will allow all five CC units to be operated at least at their 200 MW low operating limits.

Table 13.7 (presented on the next page) now shows how our utility might dispatch its CC units to serve this remaining load.

TABLE 13.7

Current Year + 10 Average April Day: Remaining Load to Serve After Nuclear and PV, with Potential Coal and CC Dispatch (8,322 MW of PV Installed)

(1)	(5c)	(6)	(7)	(8)	(9)	(10)	(11)	(12)	(13)	(14)	(15)	(16)	(17)	(18)	(19)
	(From Table 13.6)							= (5c) − (11)							= (12) − (18)
	Adjusted Remaining Load to Serve (MW)	Coal Unit 1 Output (MW)	Coal Unit 2 Output (MW)	Coal Unit 3 Output (MW)	Coal Unit 4 Output (MW)	Coal Unit 5 Output (MW)	Coal Total Output (MW)	Load Unserved by Coal (MW)	CC Unit 1 Output (MW)	CC Unit 2 Output (MW)	CC Unit 3 Output (MW)	CC Unit 4 Output (MW)	CC Unit 5 Output (MW)	CC Total Output (MW)	Remaining Unserved Load (MW)
Hour															
1	3,252	400	400	400	300	0	1,500	1,752	500	434	410	208	200	1,752	0
2	2,933	300	300	300	300	0	1,200	1,733	500	433	400	200	200	1,733	0
3	2,758	300	300	300	300	0	1,200	1,558	400	358	400	200	200	1,558	0
4	2,686	300	300	300	300	0	1,200	1,486	386	300	400	200	200	1,486	0
5	2,742	300	300	300	300	0	1,200	1,542	442	300	400	200	200	1,542	0
6	3,032	300	300	300	300	0	1,200	1,832	500	450	450	232	200	1,832	0
7	3,571	300	300	300	300	0	1,200	2,371	500	500	500	500	371	2,371	0
8	2,829	300	300	300	300	0	1,200	1,629	400	329	300	300	300	1,629	0
9	1,800	200	200	200	200	0	800	1,000	200	200	200	200	200	1,000	0
10	1,800	200	200	200	200	0	800	1,000	200	200	200	200	200	1,000	0
11	1,800	200	200	200	200	0	800	1,000	200	200	200	200	200	1,000	0
12	1,800	200	200	200	200	0	800	1,000	200	200	200	200	200	1,000	0
13	1,800	200	200	200	200	0	800	1,000	200	200	200	200	200	1,000	0
14	1,800	200	200	200	200	0	800	1,000	200	200	200	200	200	1,000	0

15	1,800	200	200	200	200	0	800	1,000	200	200	200	200	200	1,000	0
16	1,800	200	200	200	200	0	800	1,000	200	200	200	200	200	1,000	0
17	1,800	200	200	200	200	0	800	1,000	200	200	200	200	200	1,000	0
18	2,674	350	300	300	300	0	1,250	1,424	300	275	275	300	274	1,424	0
19	3,761	480	400	350	300	0	1,530	2,231	475	450	475	450	381	2,231	0
20	5,216	660	546	530	480	0	2,216	3,000	600	600	600	600	600	3,000	0
21	5,776	700	700	700	700	0	2,800	2,976	600	600	600	600	576	2,976	0
22	5,285	700	550	550	550	0	2,350	2,935	600	600	600	600	535	2,935	0
23	4,663	520	500	500	400	0	1,920	2,743	600	600	550	500	493	2,743	0
24	3,963	400	400	400	300	0	1,500	2,463	600	600	463	400	400	2,463	0

In Current Year + 5, only four of the five CC units were operated on this April day, and there were many hours in which no CC unit was running. In Current Year + 10, all five of the CC units are in operation for all 24 hours due to the faster ramping capabilities of the CC units.

As Table 13.7 shows, the CC units are operated for many hours at their low operating limit of 200 MW. In the late afternoon/early evening, as PV output declines and the remaining load increases, the output of the CC units increases. All the CC units are dispatched at their maximum 600 MW in Hour 20 and near their maximum in Hour 21. After that, the CC units ramp down to follow the declining load.

Column (19) shows that our utility's hourly load is met in all hours by nuclear, PV, coal, and CC capability in all hours. However, in order to accomplish this, our utility has curtailed a significant amount of PV potential energy.

The key takeaways from comparing generation operation in Current Year + 10, compared to operation in Current Year + 5, are as follows:

- Due to the PV MWh output now being larger than the load that remains after accounting for nuclear, a considerable amount of PV output will have to be curtailed during 5 hours. In our example, this "have-to-curtail" amount of energy is 5,261 MWh, which equates to 9% of the projected total PV output for the day.
- Furthermore, additional amounts of PV will need to be curtailed due to the real-life constraints on the operation of a utility's existing generating units that we have addressed in developing our utility's ground rules. This additional amount of PV curtailment is driven by how many conventional generating units a utility chooses to keep in operation, ramping limits, and the low operating limits of those generating units. In our example for Current Year + 10, our utility curtails more MWh of PV output in order to have enough load to allow the conventional generating units to operate at their low operating limits. The total PV output curtailment for this average April day was 18,505 MWh or 33% of the PV output that otherwise would have been projected on this average April day.
- Another outcome from having a total of 8,322 MW of PV on the system is that our utility now will operate its existing coal and CC units much differently than it would have absent this large amount of PV. In Current Year + 5, our utility basically maximized the output of all five coal units in most hours and used its CC units as the swing/marginal units to follow the changing hourly load. However, in Current Year + 10, only four of the five coal units were used, and they were operated at their low limits for many hours. This change in the operation of the coal units is, in large part, due to the PV output combined with the slow ramp rates of the coal units.
- In Current Year + 5, our utility did not use one of the five CC units at all, maximized the use of three CC units for a relatively few hours, and did not operate any of the CC units for a number of hours. In Current Year + 10, all five CC units were used in all hours. Their usage is at the low limit for many hours, but this usage ramps up to/near their maximum 600 MW

output for a couple of hours in the evening. This change in the operation of the CC units is again due, in large part, to the PV output and to the faster ramp rates of the CC units.

This simplified look at how the operation of our utility's fleet of conventional generating units is likely to change as our utility moves into/through the transition period helps to point out one important point: the value of the more flexible (due to faster ramp rates) existing natural gas-fueled generating units. Although the installation of batteries could reduce the amount of solar that our utility must curtail in the absence of the storage option, our utility recognizes that there may be limits on how many battery MW can be installed annually. Therefore, the existence of these fast-ramping, gas-fueled units not only continues to provide firm capacity but helps to lower the amount of PV MWh curtailment during at least the earlier portion of the transition period. In addition, these CC units have significantly lower carbon emission rates than the coal units that our utility would otherwise have dispatched.

In summary, the dispatch of the coal and CC units for our utility in Current Year + 10 (once approximately 8,300 MW of additional PV MW are installed) is significantly different than the dispatch in Current Year + 5. In addition, not all of the PV output can be immediately used by our utility in the hours in which those MWh are generated. Unless storage, sales, or other options are available to utilize the excess PV MWh, a significant amount of PV output that might be available for our utility will be curtailed by blocking those MWh from entering the transmission grid.

INVERTER-BASED RESOURCES (IBRs) AND SYSTEM STABILITY

In Part I of this book, we stated that our primary focus would be on how a utility conducts (or should conduct) its resource planning work to determine what the best generation and DSM resource options were for its specific utility system. As our discussion of that topic progressed, certain aspects of transmission planning and system operations were directly and/or indirectly incorporated, particularly into the analyses that have been presented in both Parts I and II of this book. In addition, we just discussed how our hypothetical utility system would need to be operated very differently once very large amounts of PV MW had been installed.

The utility functions of transmission planning and system operations are separate subjects about which individual books have been written and will continue to be written. This was another reason that the primary focus of this book is on resource planning in regard to generation and DSM resource options.

However, these three utility functions—resource planning, transmission planning, and system operations—are interconnected. And, as suggested near the beginning of this chapter, these interconnections will need to expand as utilities utilize "new" resource options, especially if the utility seeks to achieve a low-to-zero carbon goal.

For that reason, we will briefly mention aspects of moving toward zero carbon for which a utility's transmission planning and system operation functions will have primary responsibility. The actions and changes these other utility functions decide are needed from their perspectives will also impact a utility's resource planning work.

At the beginning of this chapter, we mentioned that, for many years, the coordination activities between the functional areas of resource planning, transmission planning, and system operations were well understood and only occasionally resulted in novel challenges. In large part, this is because the generation options the utility might select in the future shared a number of similar characteristics. Three of those similar characteristics are listed below:

1. Assuming no unplanned maintenance, existing generators and new generation options could operate as many hours as needed for as long as there was fuel with which to run the generators.
2. During the operation of these generators, their projected output and firm capacity value did not vary significantly as day turned to night, from month-to-month, or as one season turned to another.
3. All these generators are basically machines with spinning rotors.

However, as the second edition of this book is written, solar, wind, and batteries (plus other generation sources) are now clearly cost-competitive resource options and have become even more cost-competitive due to the new federal tax credits offered by the 2022 IRA. This obviously increases the number of types of generation resources with which resource planning, transmission planning, and system operations are working with. More importantly, these (relatively) new resource options pose different challenges for these utility functional groups because these new resource options do not share any of the three characteristics just described.

In regard to the first characteristic listed above—"can operate as many hours as needed," this obviously does not apply to PV (which does not operate at night), batteries (which must cease providing electricity when they need to be recharged), or to wind turbines (which cannot operate if the wind isn't blowing with sufficient strength). This has implications for both the economics of these resource options and for system reliability.

The second characteristic—"projected output and firm capacity values do not significantly vary"—also does not apply to PV and batteries (or to wind turbines). PV's output changes constantly, eventually dropping to zero during night-time hours. Battery output can be constant (if operated in that manner) for a period of time before operation must cease because the battery must be recharged (and the battery becomes an electric load during the recharging period). In addition, as discussed in Chapters 11 and 12, the firm capacity values of additional increments of PV and/or batteries do vary significantly as more and more of these increments are added to a utility system.

Therefore, these two characteristics that "conventional" generating units (fossil-fueled and nuclear) share, but which PV (and certain other zero-emission generation options) does not share, have been addressed either directly or indirectly in the previous chapters of this book. However, the third characteristic that PV and batteries do not have—*"All of these generators are basically machines with spinning rotors."*—has not previously been addressed in this book.

PV and batteries do not rely upon a spinning rotor for their operation. Instead, their operation is based on power electronics, including an inverter. In simple terms, an inverter converts direct current (DC) electricity to alternating current (AC) electricity and, in so doing, regulates the flow of electrical power. Because PV and batteries require an inverter, they (and some other resource options) are often referred to as inverter-based resources or IBRs.

Existing utility systems have been built over many decades to produce electricity from generators that utilize spinning rotors and then to carry the electricity to customers over transmission and distribution lines. An important part of the overall generation, transmission, and distribution system is to keep the overall electrical grid stable in terms of (i) matching the amount of electricity generated and consumed, (ii) voltage, and (iii) frequency. Stability in the electrical grid is accomplished through sophisticated system protection systems that were also developed based on a grid that relied on spinning rotor-based generators.

The introduction of IBRs in relatively modest MW amounts to-date has not had a major impact in regard to system stability on a daily basis. However, the state-wide utility system in California has already experienced a few moments in which very high percentages, approaching 100%, of their instantaneous electrical load was served by IBR resources such as PV and wind. These brief moments will undoubtedly stretch into longer and longer time periods as increasing amounts of IBR resources are added to utility systems. Ultimately, if zero-carbon operation becomes reality in the future, a very high percentage of the electricity produced in the United States is likely to come from IBRs.[20]

A couple of the characteristics inherent in current IBR technologies that can be problematic in regard to system stability are listed below:

- Current IBR technologies have inherently low fault current and also lack the inertia that is supplied by conventional generation.
- Fault current is the electrical current that flows through an electrical circuit during an electrical fault condition. System protection systems work on fault current that protective relays use to detect and clear faults on the transmission system. Because current IBRs do not have a spinning mass component that produces fault current, these IBRs produce fault current levels that are much smaller (typically less than 25%) than the amount of fault current produced from conventional generation.
- When a conventional generator experiences a problem, the inertia of its spinning rotor continues to provide support to the electrical grid for some discrete amount of time, thus allowing more time for the system protection system to react. Current IBR technologies lack inertia.

These, and other IBR-related issues, have been recognized as challenges in maintaining a stable electric grid as ever-increasing amounts of IBRs are installed. At the time the second edition of this book is written, various utilities in the United States and abroad have considered a variety of potential approaches to addressing these issues. Some of these potential approaches include the following:

- Awarding contracts for providing inertia (United Kingdom).
- Installing batteries specifically to provide fault current and inertia. (Australia). Batteries can currently discharge their maximum amount of

---

[20] That percentage value may not reach 100% due to the continued existence of nuclear generation plus other possibilities that will be discussed in the next chapter.

power in approximately 1 second. In the future, their discharge time will need to improve to less than 1 second, thus discharging in the realm of electricity cycles.[21]

- Installing synchronous condensers that are synchronous generators/motors that serve no load. The purpose of synchronous condensers is to generate or absorb reactive power as needed to adjust grid voltage (various places in the United States).
- Limiting IBR output to 80% of installed capacity to mitigate voltage instability (Texas).[22]

As these, and other, approaches evolve to maintain system stability with ever-increasing installations of IBR resources, at least four things are likely to emerge regarding resource planning. First, the three utility functions of resource planning, transmission planning, and system operations will have to work together even more closely than has been the case in the past. Second, the resource planning function may have new constraints placed on one or more of the IBR resource options, at least in the short term. Third, additional costs, which are likely to be very large, will be incurred by utilities to bolster their system protection systems. Fourth, modifications of a utility's existing generating units may be needed to help maintain system stability (as discussed below).

SUMMARY: KEY TAKEAWAYS REGARDING SYSTEM OPERATIONS, TRANSMISSION PLANNING, AND RESOURCE PLANNING

The key takeaways from the discussions in this chapter are as follows:

1. The addition of very large amounts of renewable energy generators, such as PV, will have significant impacts on the operation of the new renewable generators themselves, as well on the operation of the existing generating units that are currently on the system.
2. These impacts will likely include the need to curtail, sell, store, or otherwise use (for other purposes) the output of ever-increasing amounts of PV MWh output at various times during the year, particularly in periods of relatively low load and high PV output such as the Spring and Fall months.
3. These impacts will also include the need to operate the utility's existing generators in ways they have not been operated before. Utilities are likely to find that their existing generators will need to cycle on/off, and ramp up/ down, more frequently than they have in the past. This can create "wear and tear" problems because few, if any, of the existing generators were designed to operate in this manner.

---

[21] In the United States, the electric grid operates on a 60-cycles-per-second mode (or 60 hertz). Elsewhere, as in Australia, the grid operates in a 50-cycles-per-second mode.

[22] Limiting IBR output is obviously not compatible with a long-term objective of getting to low-to-zero carbon operation, but it may be useful (and even necessary) as an interim solution during the transition period for at least some utility systems.

As a consequence, utilities may need to modify their existing conventional generating units to lower the current low operating limits and/or increase the ramp rates. If so, there will be increased capital (and probably variable) costs as well as challenges to coordinate this retrofit work with regularly scheduled generator maintenance and overhaul activities stretching over a number of years.

4. The addition of very large amounts of renewable energy generators, such as PV, will also have significant impacts on a utility's transmission planning function. These impacts will largely be in the area of system stability. This is because the current IBR technologies do not provide the same fault current and inertia that are supplied by conventional generators that rely on spinning rotors.

5. As a result, the three utility functions of resource planning, transmission planning, and system operations will need to increase their already high levels of collaboration.

In Chapter 14, we will wrap up this book's discussion of electric utilities and resource planning starting with a summary of the information presented in the 13 previous chapters. In addition, opinions of the author regarding approximately 20 topics will be offered in question-and-answer (Q&A) format. Finally, a few words will be offered regarding what to look for in the future regarding electric utility resource planning.

14 Final Thoughts (Including Some Opinions)

A QUICK LOOK BACK

We have covered a lot of material in the preceding 13 chapters. Therefore, we will start this chapter by taking a quick look back to summarize the information that has been discussed.

In Part I of this book, we addressed the fundamentals of electric utility resource planning. In the course of presenting these fundamentals, we:

- Created a hypothetical utility system ("our utility") with its fleet of generating units and then showed how those generating units would typically be operated to meet the utility's load on both the utility's peak day and on an annual basis.
- Explained the three questions a utility must continually answer in its resource planning work.
- Discussed the two basic types of resource options, Supply options and demand side management (DSM) options, that our utility can choose from to meet its resource needs.
- Performed reliability analyses for our utility that showed it needed to add resources in 5 years, *i.e.*, in Current Year + 5.
- Performed economic and non-economic analyses for our utility for both Supply and DSM options in order to select the best overall resource option for our specific utility system to meet the resource need in Current Year + 5.
- Discussed a number of factors that can (and will) complicate utility resource planning.
- Introduced my four Fundamental Principles of electric utility resource planning.

In Part II of this book, we addressed a move toward low-to-zero carbon operation by our hypothetical utility. In addressing such a move, we:

- Examined the magnitude (MW) and types of new resources that our hypothetical utility would need to add in order to supply 100% of our utility's annual GWh in the target year of Current Year + 29.
- Looked at recent and/or projected future installed costs for PV and batteries.
- Discussed the economic impacts of the federal tax credits for PV and batteries that were included in the 2022 Inflation Reduction Act (IRA).
- Examined the firm capacity values of PV for both winter and summer peak hours, plus saw how the firm capacity value of new increments of PV decline as increasing amounts of PV are added to our utility's system.

DOI: 10.1201/9781003301509-16

- Introduced a way to calculate the duration needed for batteries to have their firm capacity value be 100% of the battery's nameplate rating, plus saw how the duration for new increments of batteries will need to increase as more and more batteries are added to our utility system.
- Discussed why the currently used reliability criteria of reserve margin and loss-of-load probability (LOLP) will likely need to be modified, and/or augmented by a new reliability criterion, to be able to address the new reality of intermittent generators and time-limited storage.
- Assuming that our utility has moved into the transition period on its way to lower carbon, examined how it will be unable to instantaneously utilize all solar MWh in certain hours in which those MWh are produced.
- Looked at how even greater amounts of solar MWh output will need to be "curtailed" (blocked, sold, or stored) during the transition period due to operating constraints of existing conventional generators.
- Examined how our utility's existing conventional generating units will need to be operated differently during the transition period.
- Discussed how the different electrical characteristics of inverter-based resources (IBRs), compared to conventional generators, will pose system stability challenges for transmission planners and system operators.

It is now time to begin wrapping up our discussions. We do so first by presenting summaries of key points presented in Parts I and II of this book. Then we will switch gears a bit. Up to this point, I have attempted to focus the discussion on an explanation of fundamental concepts, analytical approaches, and principles in regard to utility systems and resource planning. In so doing, I have tried to minimize presenting personal opinions (and attempted to identify those opinions when they showed up). However, later in this chapter, I shall offer some personal opinions regarding some of these issues we have discussed in a question-and-answer (Q&A) format. We will then close this chapter with a brief discussion of key factors that will help to shape future resource planning efforts.

We now head for the summaries of key points presented in Parts I and II of this book. You may surprise yourself with how much you now know about those topics.

SUMMARY OF THE KEY POINTS WE HAVE LEARNED ABOUT UTILITY SYSTEMS IN GENERAL

For many of you, it is likely that before you started reading this book, the extent of your knowledge regarding electric utility systems was "drive by" knowledge, *i.e.*, you could recognize a power plant when you drove by it.[1] Hopefully, your knowledge of utility systems has expanded considerably. In regard to electric utility systems, together we have actually created a hypothetical electric utility system and have examined how it operates.

---

[1] Or at least you could identify the facility as a power plant when you drove close enough to the plant property so that you could read the sign which listed the name of the plant.

This has enabled us to learn a number of things about utility systems. A list of the key points we have learned about utility systems in general includes the following:

- The demand for electricity that a utility system's customers will have varies considerably from hour-to-hour, and from one season to another. We saw that from examining both peak day load shapes and annual load duration curves for our utility.
- A utility system will typically have different types of generating units with which it will meet the hourly demand for electricity. These generating units are usually operated (or "dispatched") according to their operating or variable costs. All else equal, the least expensive-to-operate generating units will be dispatched first, followed in turn by increasingly expensive-to-operate generating units.
- Each utility will typically have one or more different types of generating units that regularly operate "at the margin" of the hourly demand for electricity (*i.e.*, their operation ramps up or down, respectively, as the demand for electricity increases or decreases).
- The types of fuels these "at the margin" generating units use as they ramp up or down are often termed the utility's "marginal" fuels. These are typically fossil fuels: primarily natural gas and, for some utility systems, coal. These "at the margin" generating units are important not only in respect to the day-to-day (and hour-to-hour) costs and air emissions of the current utility system, but they also play a pivotal role in the utility's analyses regarding which resource option(s) is the best choice when new resources are needed in the future. This is because, when a utility needs to add new resources, each of the resource options being considered will primarily impact the future operation of these "at the margin" generating units. Also, each new resource option the utility may consider will impact the future operation of these "at the margin" generating units differently.
- Finally, and perhaps most importantly, we have learned that, because of inherent differences in electrical demand patterns and the number and type of generating units, each utility system is unique. This often-overlooked fact is so important that it has been expressed as my Fundamental Principle #1 of Electric Utility Resource Planning.

We then turned our attention to how a utility actually performs its resource planning function to determine what type of resources it should add to meet the future needs of its customers.

SUMMARY OF THE KEY LESSONS WE HAVE LEARNED REGARDING UTILITY RESOURCE PLANNING

We utilized our hypothetical utility to demonstrate how a utility may proceed to make decisions regarding future resource options. Then we used an integrated resource planning (IRP) approach to perform economic and non-economic analyses of a variety of types of resource options. In so doing, we learned the following:

- There are three questions that must be answered in resource planning analyses: (i) *"When (in what year) does the utility need to add new resources?"*; (ii) *"What is the magnitude (MW) of the new resources that are needed?"*; and (iii) *"What is the best resource option(s) with which to meet this need?"*
- There are three basic types of analyses that utilities typically undertake in an IRP approach in order to answer these questions: (i) reliability analyses; (ii) economic analyses; and (iii) non-economic analyses. All of these three types of analyses in an IRP approach are designed to place all resource options on a level playing field so that the options can compete fairly against each other, *i.e.*, to perform analyses of resource options with no bias and/or pre-determined outcome for any type of resource option. The objective is to determine the best overall selection of new resource for our utility's customers.
- Reliability analyses are used to determine the answers to the first two questions listed above: *"When (in what year) does the utility need to add new resources?"*, and *"What is the magnitude (MW) of the new resources that are needed?"* At the time the second edition of this book is written, a reserve margin criterion, and/or an LOLP criterion, is most often used to answer these questions.
- Economic and non-economic analyses are then used to determine the answer to the third question: *"What is the best resource option(s) with which to meet this need?"*
- Economic analyses may be conducted in two stages. In the first stage, preliminary economic screening analyses may be used to "screen out" the least economically attractive resource options. However, these preliminary economic screening analyses can only provide meaningful results in very limited circumstances (which are explained in Part I of this book).
- These preliminary economic analyses are often performed using an individual-option-versus-individual-option approach that ignores a number of impacts to the entire utility system. While such preliminary screening analyses *may* be useful in very limited circumstances, these preliminary analyses cannot, and do not, account for all of the system impacts a resource option will have on the entire utility system. Therefore, the results of these preliminary economic screening analyses cannot be used to make a meaningful final decision about resource option selection. This fact is so important that it has been expressed as my Fundamental Principle #2 of Electric Utility Resource Planning.
- In the second stage of economic analyses, final (or system) economic analyses are conducted to determine which resource option is in the best economic interests of the utility's customers. The final economic analyses fully account for all of the system cost impacts a resource option will have on electric rates if it is selected. These final (or system) economic analyses

are typically conducted by developing, then comparing, different multi-year resource plans in which each resource plan includes at least one of the competing resource options in the first year in which the utility needs to add new resources.

- Furthermore, when both DSM and Supply options are competing for a role in a utility's resource plan, final (or system) economic analysis must be performed using an electric rate perspective. If only a total cost perspective is used, then important DSM-related impacts on electric rates that apply to all customers are overlooked. Utilizing an electric rate perspective is so important when performing economic analyses that it is expressed as my Fundamental Principle #3 of Electric Utility Resource Planning.

- Non-economic analyses generally address non-cost attributes associated with resource options' impacts on the utility system as a whole. (However, the costs of those impacts are typically also addressed in the IRP final economic analyses.) Examples of these non-cost attributes can include, but are not limited to, the following: (i) the number of years it takes for a resource option to emerge as the best economic option; (ii) the types and amounts of fuel the utility system uses; and (iii) the amount and types of system air emissions for the utility. Such attributes of a resource plan are meaningful, particularly regarding issues involving utility system reliability and risk.

- Non-economic analyses are particularly useful in helping to avoid making snap decisions about certain types of resource options based on commonly held (but often incorrect) beliefs about whether those types of resource options really reduce system fuel usage and system air emissions. The non-economic portion of an IRP analysis ensures that decisions are made only after a thorough comparison of these impacts has been made for all the resource options being considered. In other words, IRP-based non-economic analyses ensure that one answers the important question "compared to what?" regarding any preconceived beliefs that might exist regarding certain types of resource options. The "compared to what?" consideration is so important that it forms the basis of my Fundamental Principle #4 of Electric Utility Resource Planning which applies to both economic and non-economic analyses.

- A utility typically uses the results from both the economic and non-economic analyses to ensure that the utility has a form of risk assessment in regard to potential changes in key assumptions and/or regulation/legislation. This helps ensure that the utility is making a fully informed decision as to which resource option is the best selection for its customers.

- There are a number of other factors, which we have labeled as "constraints," that, if applicable to a utility system and/or regulatory jurisdiction, must be taken into account in that utility's resource planning efforts. These constraints can be grouped into three categories: "absolute," legislative/regulatory-imposed, or utility-imposed constraints. Such constraints can significantly alter the results of IRP analyses.

SUMMARY OF THE KEY LESSONS WE HAVE LEARNED REGARDING MOVING TOWARD ZERO CARBON

We next looked at what a move in the direction toward zero carbon for our hypothetical utility would entail. We examined both the magnitude (MW) of new resources needed to serve 100% of our utility's GWh in the target year (Current Year + 29) and the impacts the required new resources will have on the operation of a utility's transmission system, existing generating units, and new generating units. The following items summarize the key takeaways from this examination.

- The magnitude (MW) of new resources needed to achieve a low-to-zero carbon goal can only be known with accuracy by using a resource planning computer model. There is no other way to accurately account for the hourly changes in load, hourly changes in solar (and other renewable resource) output, when energy storage resources will need to be discharged and recharged, and in how the existing generating units currently on the system will need to operate in the transition period years before the carbon goal is to be achieved. However, we have used an estimation approach (the second estimation approach) in Chapter 10 that appears able to provide an "in the ballpark" estimate of the needed new resource MW, but which likely underestimates the needed new MW.
- Such analyses show that the magnitude of new resources needed to meet to serve 100% of the utility's annual GWh in the target year with zero-carbon emissions is enormous. The needed MW of new resources will likely be a multiple (approximately four-to-five times) greater than a utility's existing generation fleet total MW. In addition, the needed MW of new resources to serve 100% of the utility's annual GWh in the target year with zero-carbon emissions will likely be more than an order of magnitude greater (*i.e.*, more than 10× greater) than the firm capacity MW additions that would be needed to simply continue to meet the utility's reliability criterion absent a carbon goal.
- The federal tax credits from the 2022 IRA provide (as they were intended to do) a significant overall economic savings to a utility that chooses new resource options, such as PV, batteries, and wind turbines, which are eligible for the tax credits. These tax credits make it highly likely that more of these resources will be selected in IRP analyses, and that these resources will be selected earlier, than would be the case without these federal tax credits.
- However, the 2022 IRA's federal tax credits for PV have a bigger impact than the corresponding 2022 IRA federal tax credits for batteries. This leads to the possibility of PV additions outpacing battery additions, which can have negative impacts regarding system reliability, operation, and stability during the transition period.
- Certain characteristics of solar can pose challenges that must be addressed in resource planning analyses. One such challenge is that PV's firm capacity on winter peak days may be zero MW, or near-zero MW, if the utility's winter peak hour occurs in the early morning. In such a case, although ever-increasing PV additions *may* continue to address the utility's summer

reliability needs during the transition period, these PV additions will not address such a utility's winter reliability needs.

- Another challenge with PV is that the summer firm capacity value of future increments of PV decreases as more and more PV MW are added to the utility system. In addition to potentially creating summer reliability problems, this characteristic of solar may make it difficult for the utility to continue to economically justify the addition of new increments of PV absent a federal or state mandate to get to low-to-zero carbon goal.

- Battery firm capacity value is directly tied to a battery's duration. The best way to determine what duration a battery needs to have its firm capacity value be 100% of its nameplate capacity is to examine the utility system from an hourly perspective that accounts for both MW and MWh. A way to perform this examination was introduced in Chapter 12.

- For new increments of batteries to be given credit for 100% firm capacity value, the durations of those batteries will need to incrementally increase as more and more batteries are added to the system. As one might expect, this has cost implications for new increments of batteries.

- Regarding system operations, the dispatch of a utility's existing conventional generating units, during the transition period (and before these generating units are ultimately retired and/or converted to burn zero-carbon-emitting, or carbon-neutral, fuel), will have to change as more and more PV and batteries are added to the system. Existing conventional generating units were not designed to operate in some of the ways (such as significantly increased ramping and/or cycling) they will have to operate as ever-increasing amounts of PV are added to the system.

- As more and more PV is added to a system, the utility will increasingly face periods when solar output exceeds the instantaneous hourly load. This will be especially prevalent in relatively low-load months/seasons (such as Spring and Fall) which also have high levels of solar insolation. These "excess" solar MWh will need to be "curtailed," *i.e.*, blocked from entering the transmission grid, sold to other parties, stored, or used in other ways.[2]

- PV and batteries (and other renewable resources such as wind turbines) utilize inverters and, for that reason, are commonly referred to as IBRs. Current IBR technologies do not possess certain characteristics that existing conventional generators have such as a spinning rotor. As a result, current IBR technologies have much lower fault currents and inertia compared to existing conventional generating units. For this reason, these IBRs pose system stability challenges for utility transmission planners and system operators.

- As a result of all of these points, the already significant levels of coordination that exist between a utility's resource planning function, transmission planning function, and system operating function will need to increase because the challenges (and opportunities) inherent with current IBR technologies will require a more encompassing, holistic approach.

---

[2] We will discuss potential other uses later in this chapter.

A FEW OPINIONS ON VARIOUS TOPICS

It is now time for me to switch hats. As mentioned at the beginning of this chapter, the focus of the book to this point has been on explaining fundamental concepts, analytical approaches, and principles regarding utilities and utility resource planning. These concepts, analytical approaches, and principles were demonstrated using our hypothetical utility system.

In doing so, I have tried to keep my opinions to a minimum (and have attempted to identify them when they have shown up). Stated another way, I have been wearing an "instructor" hat. But now, I will remove the instructor hat and will put on a "yes-I-have-an-opinion" hat. With this new hat in place, I will discuss a variety of issues using a question-and-answer (Q&A) format. I hope that this change of pace proves interesting.

Let me make two points about the Q&A text that follows. First, in testifying before regulatory bodies as an expert witness in dozens of hearings, my attorneys have always directed to me to begin by answering the question as succinctly as possible and then expand upon that answer to the extent needed/advisable. I will attempt to utilize that approach here as well.

Second, let me state that the opinions expressed as solely my own personal opinions. As such, they may or may not be in agreement with opinions or positions my former employer FPL, or its parent company NextEra Energy, have taken or may take in the future.

Now let's begin the Q&A portion of our program.

Question (1): You have been performing and/or directing resource planning analyses for one of the largest electric utilities in the United States for more than 30 years. Based on your experience, do you believe that an IRP approach is the best way to select resource options to meet a utility's resource needs?

Answer: Yes. Before I explain the reason for this belief, let's review the definition of IRP that we introduced in Chapter 3. That definition is as follows: IRP is an analytical approach in which both types of resource options, Supply and DSM, are analyzed on a level playing field. For each resource option, an IRP analysis accounts for all known cost impacts on the utility system that are passed on to its customers through the utility's electric rates. In addition, non-economic impacts to the utility system from the resource options are also evaluated. In this way, IRP analyses result in a comprehensive competition among resource options.

From this definition, we see that an IRP approach is designed to foster a competition among different types of resource options in which all costs to the utility system from each resource option, which are passed on the utility's customers through electric rates, are accounted for. Other non-economic impacts to the utility system from each resource option, including system reliability impacts, are also accounted for.

An IRP approach requires that resource options *compete* to earn a place in a utility's resource plan specifically suited to the individual utility. Therefore, the competition uses information specific to that particular utility, including, but not limited to, the following: electric load patterns, fuel costs, and types of generating units.

In other words, an IRP approach allows one to make a fully informed decision, both from an economic and a non-economic perspective, regarding which resource option is the best selection for a specific utility based upon costs that are typically the purview of the utility and its regulators. Therefore, an IRP approach is the best way to select resource options to meet a specific utility's future resource needs.

Question (2): Over this more-than-30-year period, have you noticed any efforts to limit the head-to-head competition of resource options that is the foundation of the IRP approach, or to otherwise influence the outcome of IRP analyses?

Answer: Yes. The basic role of a resource planner is to perform analyses of all available resource options and to determine which resource option(s) is projected to be the best selection for the utility's customers. In performing that role, it is obvious when the analysis of resource options is influenced by legislative, regulatory, or other means.

There have been numerous occurrences, by a number of states, to move away from a full competition between resource options. One example is the implementation of prescriptive constraints that we discussed in Part I of the book in Chapter 8. Both the third constraint ("standards"/quotas for specific resource options) and the fourth constraint (prohibition of other specific resource options) limit competition. The "standards"/quotas constraint mandates that all utilities in the regulatory body's jurisdiction select some level of "favored" resource options, while the prohibition of specific resource options constraint prohibits these utilities from even considering certain "unfavored" resource options. These two constraints do not allow a true competition between resource options (the objective of an IRP approach) to occur.

Another example of attempting to limit competition between resource options, and/or influence the outcome of IRP analyses, is when an intervening party seeks to influence a regulatory body decision about resource options through arguments that use incomplete information. For example, the incomplete information may come from an intervening party using only the results of a preliminary economic screening analysis such as a screening curve analysis ($/MWh) or a particular DSM cost-effectiveness screening test.[3]

The most recent, and what I believe will ultimately be the most "successful", effort to influence the outcome of utility IRP analyses is the inclusion of significant federal tax credits in the 2022 IRA. These tax credits "tilt the scale" significantly in favor of solar, wind, batteries, and other resources so that the result is that IRP analyses will select more of these zero-carbon resources and will do so earlier than would otherwise be the case.

In summary, there have been numerous attempts to influence the outcome of utility IRP analyses for decades. And it would be surprising if there are not more attempts to influence IRP analyses in the future. However, I believe it is important that utilities and their regulators adhere as closely as possible to fundamental

---

[3] To make the situation even worse, these organizations frequently use preliminary economic screening analysis results from states other than the state the utility in question is located in. Therefore, this incorrect approach violates, at a minimum, both Fundamental Principles of Electric Utility Resource Planning #1 and #2.

IRP principles and objectives, particularly in regard to comparisons of competing resource options and resource plans.

Question (3): As part of your answer to the previous question, you included the statement *"… earlier than would otherwise be the case …"* in regard to utilities selecting resource options that would move the utility in the direction of lower carbon. Do you believe a movement in the direction of low-carbon emission utility generation fleets is inevitable?

Answer: Yes. The steady decline in installed costs for solar, wind, and batteries seen over the last decade has already resulted in these resources being selected in utility resource plans in increasing numbers even before the passage of the 2022 IRA. If, as projected, these costs continue to decline, even more of these resources will be selected and installed as a result of the normal course of IRP analyses, regardless of whether a utility has set a specific low-to-zero carbon goal for itself.

However, it is uncertain how far toward zero carbon the normal IRP analyses, absent a carbon goal, will take a utility.[4] It certainly seems possible that many, if not most, utilities will have achieved at least a 50% reduction in their carbon emissions in the next two decades or so even without a specific carbon goal. Some utilities may achieve a 75% to 80% reduction by that time without setting a specific carbon goal. Much, if not most, of the discussion in the Q&As that follow will apply to these utilities. However, the primary focus in many of the following Q&As is for utilities that have chosen to set a specific low-to-zero carbon goal beyond what would be the result of normal IRP analyses.

Thus, the direction of movement of utility resource plans is not the question. The question, for me, is how far and fast this movement will be for each individual utility. All utilities will face a significant amount of uncertainty in a number of areas as they move inevitably in the direction of lower carbon.[5]

Question (4): What do you believe are the most significant uncertainties in this inevitable movement of utilities toward lower carbon?

Answer: I believe that much of the uncertainty will fall into three categories: (i) cost-effectiveness, (ii) system reliability, and (iii) regulatory approval. All three categories are interrelated, but each category has its own type of uncertainty.

Question (5): Regarding the first category of uncertainty, do you believe that a resource plan that is designed to achieve a specific low-to-zero carbon goal can be cost-effective?

Answer: In order to answer this question, I return to my Fundamental Principle #4 of Electric Utility Resource Planning: "Always Ask Yourself: Compared to What?"

---

[4] And, of course, it will vary from one utility to another as dictated by Fundamental Principle #1 of Electric Utility Resource Planning.

[5] Even NextEra Energy's bold announcement of achieving a zero-carbon goal by 2045, along with a blueprint of its plan with which to accomplish this goal, contained a number of prudent caveats regarding the assumptions it used in its analyses in recognition of uncertainty regarding these assumptions.

Then we remind ourselves that resource planning analyses are typically based on comparing one resource plan to another resource plan(s). Then we can address the question after first considering two possible premises for the question.

If the premise of the question is that the utility definitely has a state (or future federal) mandate to get to zero carbon, then the question can be restated as follows: "*Is there a resource plan that achieves the mandated carbon goal that is the most economic resource plan for all customers; i.e., a plan that achieves the carbon goal with the lowest possible electric rates for all customers?*" The answer to this specific question is definitely: "Yes." There will always be a resource plan that can achieve an objective for a given set of assumptions regarding load, fuel costs, resource option costs, tax law, etc., which results in the lowest electric rates for customers given this premise.

However, there is another possible premise to the question that is the more interesting (to me) premise and arguably the more important premise. With that premise, the question can be restated as follows: "*Can a resource plan that achieves a specific low-to-zero-carbon goal be as economic (have electric rates as low) as a resource plan without a carbon goal?*" Note that, just as in all IRP analyses, we are discussing a comparison between two competing resource plans. And, for this discussion, we assume that the utility's carbon goal is something more "aggressive" than what the utility's IRP analyses says will be achieved anyway without a specific carbon goal.

So, based on this premise to the question, I have a two-part answer to the question of: "*Can a resource plan with a specific carbon goal be as economical as the resource plan a utility would develop absent a carbon goal?*" If one only considers technology that is currently (at the time the second edition of this book is written) available at utility-scale, then I believe it is unlikely that such the resource plan with the carbon goal can be economic compared to the best resource plan without a carbon goal.

However, if one assumes that certain technologies, which are only concepts today, successfully come to fruition at utility-scale in the future, then my answer is that it may be possible for a resource plan with a specific carbon goal to be as economic as a resource plan without a carbon goal.[6]

Question (6): Based on your decades of IRP analyses, and considering only current PV and battery technology, would lowering the projected costs of PV and batteries be likely to result in the CPVRR cost of a resource plan with a specific carbon goal being as low as the CPVRR cost a resource plan without a carbon goal?

Answer: I do not believe this is likely. One might think that if one were to use a lower projection of installed costs for PV and batteries (or wind), this would automatically make the CPVRR cost of the resource plan with a specific carbon goal

---

[6] Note that NextEra Energy's Blueprint stated that part of its plan to achieve zero-carbon operation by 2045 included two promising new technology concepts that are not currently proven at utility-scale: use of renewable natural gas at utility-scale and use of hydrogen as a fuel in converted existing CC units.

lower than the CPVRR cost of a resource plan without a carbon goal. The reality is that it is not that simple.

When developing and comparing two competing resource plans on a level playing field, the same basic assumptions and inputs must apply to both resource plans. In the case of a resource plan with a specific carbon goal and a resource plan without a carbon goal, the projected $ per kW installed cost inputs for new PV, batteries, CC units, etc. must apply to both plans.

Using our hypothetical utility as an example, it has to add 52,515 new MW of PV and batteries in order to supply 100% of our utility's GWh with zero-carbon emissions in Current Year + 29. Now let's assume for this discussion that, for a given projection of installed costs for PV and batteries, a competing resource plan without the carbon goal is projected to (i) have a lower CPVRR cost, and (ii) install fewer total new MW of PV and batteries.

The question is what happens if the projected costs for PV and batteries are lowered? The resource plan with the carbon goal likely does not change significantly (because it has to add the same MW amount of PV and batteries to achieve its carbon goal), but its CPVRR cost gets lower due to the lower costs of PV and batteries it must add. Conversely, one would expect the resource plan without the carbon goal to change to include more MW of the lower cost PV and batteries that are now selected based on economics. Consequently, the CPVRR cost of this second resource plan is also lowered.

Lowering the PV and battery cost inputs in this comparative analysis drives two results. First, the resource plan with the specific carbon goal is now chasing a moving (downward) target of CPVRR costs for the resource plan without a carbon goal. Second, the resource plan without a carbon goal increasingly begins to look like the resource plan with the carbon goal because it adds more PV and batteries due to their projected lower costs.

Unless the new projection of cost inputs is substantially lower than the original cost projection, the CPVRR cost of the resource plan with the specific carbon goal is not likely to be lowered enough to match the CPVRR cost of the resource plan without the carbon goal.

Therefore, all else equal, the higher MW amount of needed new zero-carbon resources to meet a carbon goal will likely result in a fixed cost disadvantage for a resource plan with a specific carbon goal, regardless of the installed cost inputs used for PV and batteries.

Question (7): Beyond the variable cost savings that will come from using less fossil fuel, is there any other fuel cost-related factor that may work as a cost advantage for a resource plan with a specific carbon goal versus a resource plan without a carbon goal?

Answer: Yes. Because a resource plan with a specific carbon goal should require less natural gas than a resource plan without a carbon goal, the costs associated with firm transportation of natural gas should be lower, and perhaps significantly lower, for the resource plan with a specific carbon target.

Utilities who do not own their own natural gas pipelines typically contract with natural gas pipeline owners for transportation rights on the pipelines. These contracts essentially guarantee that enough space is reserved on the pipelines to deliver a minimum volume of natural gas per day. These contracts may cover a dozen years or more. Near the end of the contracted time period, the contracts are typically reviewed to see if changes in the volume of natural gas transportation are needed.

In certain utility resource planning analyses, the annual costs for these existing firm gas transportation contracts may not be accounted for because the contract has been signed and because this amount of natural gas transportation will be needed by the utility in all of the resource plans it considers for a time period well beyond the "decision year" in question (*i.e.*, Current Year + 5 for our hypothetical utility system). A utility might only consider the costs for any additional firm gas transportation that is projected to be needed. The analyses for our hypothetical utility in Part I of this book took that approach when evaluating Supply Only Resource Plan 1 (CC) due to the addition of a new natural gas-fueled CC unit in Current Year + 5. The costs for the additional firm gas transportation needed were shown in column (5) of Table 5.11.

However, the costs associated with firm gas transportation need to be accounted for when comparing long-term resource plans with and without carbon goals. As the percentage of annual load that is served by renewable energy sources grows, the volume of natural gas that will be used to serve customers in the future will decrease. Utilities will need to secure less and less firm gas transportation over time. Therefore, as existing contracts approach the end of the contractual period, utilities will likely begin contracting for lower volumes of firm gas transportation, thus lowering these costs.

Resource plans with a specific low-to-zero carbon goal will eventually have greatly reduced-to-no firm gas transportation costs. Conversely, resource plans without a carbon goal may have reduced firm gas transportation costs, compared to the historic level of those costs (due to addition of more renewable energy sources in the future), but they will likely have some remaining costs for firm gas transportation.

Assuming all else equal, this would represent one fixed cost advantage for resource plans with a specific carbon goal.[7]

Question (8): What are your thoughts regarding how the modifications/additions to address transmission system stability, and the potential need for generating unit modifications, discussed in Chapter 13 will comparatively impact costs for resource plans with and without a specific carbon target?

Answer: As increasing amounts of inverter-based resources (IBRs) and other renewable resources are added to utility systems based on economics, these modifications/additions will likely be needed in resource plans both with and without a carbon goal. Although there is a lot of uncertainty regarding the magnitude, timing, and cost associated with these modifications/additions, it seems reasonable to assume that there will be higher CPVRR costs for the modifications/additions for a resource plan with a specific carbon goal than for a resource plan without a carbon goal.

---

[7] However, this firm gas transportation advantage may be minimized, or even eliminated, if a utility uses the existing natural gas pipelines to transport alternate zero-carbon, or carbon-neutral fuels, to its existing generating units (as will be discussed in a subsequent Q&A). In such a case, one would expect the pipeline owners to again charge a transportation fee.

In Chapter 13, we discussed the need utilities will likely face to address transmission and generation stability issues as increasing amounts of IBRs are added to utility systems. This will involve modifications/additions to the utility's existing transmission system. We also discussed the potential need to alter existing generating units to allow faster ramping times and/or lower low operating limits as a utility enters the transition period. Assuming that a significant amount of PV and batteries (or wind) would be installed in resource plans both with and without a carbon goal, a certain amount of these needed modifications/additions will be needed in both resource plans.

Discussion of this topic takes us beyond what has typically been addressed to-date by most utilities in their IRP analyses. From conversations I have had with planners at a number of utilities, this appears to be the one area associated with moving toward lower carbon that is perceived to have the greatest level of uncertainty.

There are (at least) several reasons for this high level of uncertainty. First, transmission planning models often do not attempt to address, or cannot address, much more than 10 years in the future. Second, one of the key inputs for transmission planning models is where new generation will be sited in relation to the existing transmission grid. Using our hypothetical utility as an example, it projects needing 34,680 MW of new PV by its target year. It is highly unlikely that our utility can currently know precisely where it will site that many MW of new resources. Third, transmission planning models will also need inputs regarding when existing generating units will be retired and/or converted to burn zero-carbon, or carbon-neutral, fuels. And, for any unit that will be converted, will the mode of operation of that unit change, perhaps to serve only at night-time in lieu of the utility installing more battery storage? As is the case with sites for new PV resources, these retirement/conversion/operation inputs will likely not be known with a high level of confidence for a number of years.

Combined with this high level of uncertainty about when and where modifications and additions are needed is the potential for high costs for these modifications/additions. Although relatively little work appears to have been made publicly available in the industry regarding these potential costs at the time the second edition of this book is written, 30 years of performing/directing IRP analyses have shown that modifications/additions to transmission systems and generating units are usually large ticket items. Therefore, it seems reasonable to assume these new types of modifications/additions will entail large costs as well.

In summary, we are left with the expectation that, regardless of whether a utility has a specific carbon goal, significant modifications/additions will be needed to a utility's transmission system and/or generating fleet once the utility begins adding a significant amount of IBRs and other renewable resources. However, the timing and magnitude of these changes is not only unknowable now with a high level of confidence, the timing and magnitude of these changes may not be fully knowable for years to come due to modeling limitations and/or limited knowledge about sites for the new resources and potential retirements/changes to existing generators. On top of that, the costs for such modifications/additions are not now known, but potentially large. In short, there is a lot of uncertainty regarding these modifications/additions.

However, it seems reasonable to assume that a resource plan without a carbon goal will need fewer and/or later modifications/additions than a resource plan with a specific carbon goal. For that reason, and assuming all else equal, it also seems reasonable to assume that a resource plan with a specific carbon goal would be at a disadvantage regarding the Fixed Costs of these modifications/additions.

Question (9): The last several questions were based on an assumption of existing regulations and technology that is currently available at utility-scale. What are examples of changes in regulations and/or emerging technologies that could result in making it more likely that a resource plan with a specific carbon goal could be as economic (*i.e.,* with electric rates as low) as a resource plan without a carbon goal?

Answer: This is a two-part question that deals with both regulations and technologies. We'll address each of these in turn.

Let's start with regulations that could change. First, the state or federal government could mandate a particular carbon goal. This would essentially make this question moot and shift the discussion to the first of the two possible premises we mentioned before: *"What is the most economic resource plan that would achieve this mandated goal?"*

Second, we have already discussed how the new federal tax credits for renewables and batteries in the 2022 IRA have significantly lowered a utility's overall costs if the utility selects resource options that are eligible for these credits. One could assume that these federal tax credits will be changed in the future to provide even more of an overall economic benefit to utilities that select the eligible resources. In terms of "carrot and stick" approaches, this would represent a bigger "carrot."

Third, taking a "stick" approach, one could assume higher taxes and/or more stringent regulations could be placed on fossil fuels, primarily coal and natural gas, to discourage their use and improve the relative economics of renewables.[8] Fourth, using perhaps the ultimate "stick" approach, the federal government could prohibit the use of fossil fuels in new and/or existing generating units.[9]

Steps, such as these could be taken in the future, and/or resource planning efforts could simply assume that one or more of these steps will be taken in the future. However, all of these steps would have political consequences. So, rather than make such assumptions, utilities will likely be looking for technical approaches that might narrow the gap regarding the economics of a resource plan with a specific carbon goal compared to resource plans without a carbon goal.

We will briefly discuss two technical concepts that have been discussed in the electric utility industry for at least several years. Both of these concepts are based on continuing to use a utility's existing generating units, but with these generators

---

[8] The 2022 IRA included the Methane Emissions Reduction Program (Sec. 60113) that imposes a fee/tax on energy producers that exceed a certain level of methane emissions. The fee/tax is set to rise to $1,500 per metric ton of methane by 2026-on. This marks the first time the federal government has directly imposed a fee/tax on GHG emissions.

[9] There is a precedent for such an action. In 1978, Congress passed the Powerplant and Industrial Fuel Use Act which prohibited the use of natural gas for electricity generation. The reason for this legislation was a perceived shortage of natural gas. The intended solution was to use more coal which was (and is) abundant. This legislation was repealed in 1987.

running on either a zero-carbon fuel or a carbon-neutral fuel.[10] The first of these approaches uses renewable-generated MWh (such as from PV or wind) to power electrolyzers that separate water (H_2O) molecules into hydrogen (H_2) and oxygen (O_2) molecules. The H_2 molecules can then be used as a zero-carbon fuel in conventional natural gas-fueled generating units, such as CCs and CTs, after modifications are made to the generators. The O_2 molecules would be released into the air.[11]

The second of these approaches involves creation and use of renewable natural gas (RNG) or synthetic natural gas (SNG). RNG can be produced in a variety of ways, including from animal/human/crop waste materials. SNG can be produced by at least a couple of approaches, including from processes that produce CO_2, or from CO_2 that is already in the air. Both of these SNG approaches would result in no net increase in carbon dioxide, making these carbon-neutral fuels.

Both RNG and SNG are being produced at the time the second edition of this book is written. However, quantities being produced are very low relative to the volumes of natural gas that utilities currently use. The industry consensus at the time this book is written appears to be that SNG has greater potential in regard to producing utility-scale volumes of carbon-neutral fuel in the future. For that reason, we will briefly discuss one way in which SNG could be produced: direct air capture.

Using renewable-generated MWh as an energy source to create hydrogen, already existing CO_2 in the air would be combined with hydrogen to create SNG. With this method, CO_2 could be captured from the air in one location to produce SNG, then the SNG could be used as a fuel in a power plant, perhaps in another geographic location using existing pipelines (with needed modifications), where the CO_2 would be released back into the air. The concept is that there would be no net gain in total CO_2, i.e., SNG would be a carbon-neutral fuel.

However, one can envision a situation in which CO_2 is extracted from the air in one state, converted to SNG that is transported to another state using pipelines where it is used as fuel in a generator, thus releasing CO_2 in the second state. The first state could claim a net reduction in CO_2, while the second state would show a net increase in CO_2 emissions. This could create problems, especially in the case of GHG emissions reporting on a state level.

Question (10): Other than potentially lowering any projected CPVRR cost increase that IRP analyses might show for a resource plan with a specific carbon goal compared to a resource plan without the carbon goal, are there any other advantages to using these potential new fuels?

Answer: Yes. These two alternate fuel-based approaches have at least four potential advantages and several of these have not always been addressed in IRP analyses.

---

[10] The 2022 IRA provides federal tax credits for the use of such fuels.

[11] The concept is also listed in NextEra Energy's Blueprint as one approach FPL plans to take to achieve zero-carbon operations. FPL has also received approval from the Florida Public Service Commission for cost recovery of an approximate 20 MW electrolyzer at its existing Okeechobee CC unit. The electrolyzer will produce a relatively small amount of hydrogen which will be used, in conjunction with natural gas, to power one of the 3 CTs at this 3×1 CC unit. The electrolyzer is projected to be put into operation in late 2023.

One such advantage is that a utility's existing generators could continue to be used, thus minimizing the potential early writing off of the investments made in these generators (*i.e.*, a stranded investment). Second, transportation of these alternate fuels through existing natural gas pipelines may be possible, after pipeline modifications, thus allowing existing pipelines to also remain useful as well.[12] Third, the use of PV- (or wind-) based energy to power the production of hydrogen or other alternate fuels could at least minimize the curtailment of PV-generated MWh when those MWh are produced in hours in which the instantaneous electric load is lower than PV-generated MWh as discussed in Chapter 13. Fourth, the continued use of these spinning rotor-based generators would also minimize the system stability problems, and the cost of modifications/additions needed to address the problems, that were also discussed in Chapter 13 and above.

There are certainly a number of technical and cost issues to be resolved before alternate fuel approaches such as the ones we have discussed can become viable at utility-scale. However, I believe that such approaches have a reasonable chance of minimizing any CPVRR cost differential that might exist between a resource plan with a specific carbon goal and a resource plan without a carbon goal.

Question (11): You mentioned earlier that the second of the three main areas of uncertainty inherent in a movement toward zero carbon was system reliability. You discussed this topic earlier in Chapter 11. Is there anything else you would like to say about this topic?

Answer: Yes. There are two additional things I would like to mention. First, I believe that utilities' and regulators' perspective of system reliability will eventually need to change. Second, scenario analyses will become more valuable in system reliability work.

In Chapter 11, I stepped aside briefly from the instructor role to offer my opinion that I did not believe the currently used reliability criteria of reserve margin and LOLP, as now structured, would adequately address system reliability as a utility moved closer to a specific carbon goal. My rationale was that the structure of both of these criteria is based on a focus on a single hour: the highest load hour in either summer or winter with a reserve margin criterion, and the highest daily load hour with an LOLP criterion.

This focus has worked well for more than half a century with conventional generators whose output each hour is controllable, but which can be a constant MW value (absent loss of fuel or an unplanned outage, both of which are relatively rare). Therefore, there is less concern with whether sufficient generator output will be available hour-to-hour than there is concern over what the load will be hour-to-hour. This lends itself to a focusing only on the highest daily and/or seasonal load hour.

However, as a utility adds increasing amounts of PV and/or wind, it becomes more reliant on these highly intermittent generation sources, and on time-limited

---

[12] However, as previously mentioned, the use of existing pipelines to transport these alternate fuels to generating units would likely minimize/eliminate any firm gas transportation savings that might otherwise occur.

batteries, and less reliant on the existing constant output generators. As a result, there will be much more concern regarding the availability of generation on many, if not all, hours each day. This concern will be heightened as a utility with a zero-carbon goal moves further into the transition period and begins to either retire these existing generators and/or converts them to be able to use zero-carbon/carbon-neutral fuels (the production of which also depends on processes powered by solar and/or wind).

Therefore, I believe that the structure of these currently used reliability criteria will eventually need to be modified to address more hours, and/or that a new reliability criterion will need to be developed. However, these developments do not need to occur at the beginning of a utility's transition period. This is because the utility's total generation fleet will still be comprised mostly of conventional generators in the early stages of the transition period. But these developments will be needed in advance of the point in the transition period when the intermittent resources begin supplying a majority of the utility's energy and/or fuel (through the production of hydrogen and/or SNG).

I also believe that system reliability analyses will need to make greater use of scenario analyses than is common practice today. For an example of such a scenario analysis, consider a utility that plans to be heavily dependent upon PV. What is the probability that this utility will be able to continue to serve all of its load, both day-time and night-time, if it experiences 5 or 6 days of heavy cloud cover or precipitation? A similar question can be asked of a utility that plans to be heavily dependent upon wind. Can this utility be confident it can continue to serve all of its load if it experiences a prolonged period of less-than-expected wind? Finally, a similar question can also be asked of a utility that plans to rely on renewable energy-generated renewable or synthetic fuels. A lengthy period of cloud cover, precipitation, and/or low wind at a variety of sites may challenge the ability to continually produce these fuels that are needed by the utility to serve all of its load around the clock.

Therefore, scenario analyses such as these will need to be incorporated more and more in utility resource planning work if a utility is going to achieve a specific carbon goal and still maintain the current high level of system reliability that customers enjoy and, more importantly, expect.

Question (12): The last of the three major areas of uncertainty inherent in a movement toward zero carbon that you mentioned was regulatory approval. Please discuss this topic.

Answer: At the time the second edition of this book is written, there is no federal mandate for electric utilities to achieve a certain carbon emission level by a target year. Recognizing this, we will discuss the topic of regulatory approval for two categories of states: those that have a state mandate to achieve a carbon emission level by a target year, and those that do not have a state mandate. We will frame the discussion using the following question: *"Will utility regulators in both categories of states be likely to approve cost recovery of the large capital expenditures needed to achieve a specific carbon goal?"*

We will start with states that do have a mandate to achieve a carbon emission level by a target year. In such a case, the state is requiring utilities to achieve a particular goal. Utilities will have to make capital expenditures in order to achieve the goal. Therefore, it is logical that utility regulators will approve the needed expenditures as long as those expenditures are demonstrated to be prudent.

A simpler way to state this is to that this state will be looking for the most economical resource plan with which the state's mandate can be met (and which satisfies whatever system reliability criteria is applicable). (Note that this is the first of two premises we discussed previously in Question [5].) Therefore, the regulatory situation in which a state has a carbon goal mandate is fairly straightforward.

However, the situation may not be as straightforward if the state does not have a carbon goal mandate, and a utility sets its own specific carbon goal that would result in many more new resource MW being added over time than are called for to simply meet the existing system reliability criteria.

Looking at this situation from solely an economic perspective, the utility would typically have to be able to show that each next increment of renewable resources is cost-effective to add on its own, or that its long-term resource plan that is needed to achieve its carbon goal is cost-effective, *i.e.*, results in electric rates at least as low as what the utility's best resource plan without a carbon goal would be. As stated before, my opinion is that, given only current technology, this will be hard to demonstrate and achieve. Although utilities to-date have been successful in demonstrating that incremental renewable resource additions are cost-effective, this may be increasingly difficult to do in the future as the magnitude of these additions ramps up as needed to meet the utility's goal, and the firm capacity value of the new increments decreases.

One approach that has been successfully used to-date is to analyze the next increment of renewable energy (a PV or wind facility, for example) as if it were the last renewable energy increment that would ever be added. In other words, a resource plan with the next increment of renewable energy, but no more renewable energy after that increment, was compared to a resource plan with no new renewable energy at all.

However, this premise would appear to be harder to defend once a utility states it has a specific carbon goal with a target year that requires the utility to add ever-increasing amounts of renewable energy annually. Given such a publicly stated goal, it would be illogical to try to assume that the next increment of renewable energy was the last renewable energy increment the utility would ever add.

The case for regulatory approval of a near-term increment of renewable energy that the utility needs to meet its carbon goal, but for which it cannot show is cost-effective, is further weakened if the utility's projections show steadily declining costs for this type of renewable energy option over time. That raises the question: *"If there is no federal or state mandate to get to a carbon goal, and the utility does not need a large amount of renewable energy this year for reliability purposes, and renewable energy costs will be lower next year, why should a regulatory body approve a large expenditure for renewable energy for the current year?"*

However, an economic-only perspective is not the only perspective that utility regulators may use in evaluating a utility request for approval of new resources. Under traditional rate regulation by state commissions, new resource additions typically have to be in response to a projected resource need and/or show an economic advantage for customers. At its essence, this is the IRP paradigm at work. However, utility regulators have the discretion to evaluate resource additions from perspectives other than economics. Public and/or political pressure regarding carbon emission reduction may lower the economic bar for renewable energy additions and regulators may supplant IRP economic analyses with their own judgment. If that is the case, then utility regulators would be selecting resource additions based on their own view of "strategic" considerations.

This question will likely play out in a number of states in the next decade or so as utilities who have self-imposed a specific carbon goal, in a state that does not have a carbon goal mandate, move further into their transition periods. At these points in time, if the utility's resource plan to meet its carbon goal begins to significantly diverge from what the utility's best resource plan without a carbon goal would be, with significant incremental costs, it will be very interesting to see how utility regulators make their decisions. This will be particularly interesting as regulators broaden their review to account for the cost of modifications/additions of the transmission system and existing generators that will be needed to ensure that system stability issues do not become problems as IBRs are increasingly added.

Question (13): Assume that a utility has set a low-to-zero carbon goal with a specific target year. What are the likely economic impacts would occur if a utility decided to move up (*i.e.,* accelerate) the target date?

Answer: This is an interesting question. Furthermore, I believe utilities that have already set a specific carbon goal will almost certainly face pressure in the future to accelerate the target date for achieving their goal. For example, if a utility sets a carbon goal with a target date of 2050, they will likely face pressure to accelerate the target date to 2045, 2040, or even earlier.

This pressure may come from their own internal management (after learning of carbon goal statements from other utilities that set earlier targets), environmental advocacy groups, and/or the investment community (just to name a few potential sources). The rationale for asking for/demanding an accelerated target date is that GHG emissions would be reduced or eliminated more quickly. Fossil fuel-related costs would also be reduced or eliminated more quickly.

However, there will also be other costs that will be higher due to such an acceleration. First, capital or installed costs to achieve the carbon goal would be higher with an acceleration. No analysis is needed to reach this conclusion; one only has to consider the commonly accepted downward cost trends for PV and batteries. For example, let's look back at the projected installed costs for batteries that were presented in Figure 12.1 which shows a projected decline of battery installed costs from 2023 through 2030. Assume that a similar trend of downward costs for both batteries and PV continues out to a utility's original carbon goal target year.[13] What this means is that each year the installed cost of that year's batteries and PV will be lower than their costs in the preceding year.

---

[13] Most, if not all, current projections for future PV installed costs follow a similar downward trend.

Now assume that the utility in question decides to accelerate its carbon goal target year by 5 years, from 2050 to 2045. What this means is larger MW amounts of new resources will be needed years earlier than would have been the case if the utility had stuck to its original carbon goal target year. In other words, most of the new resources that had originally been planned for installation in the years 2046 through 2050 will now have to be installed by 2045 or earlier.[14] These greater numbers of new resources in the earlier years will be installed at the higher installed cost values that are applicable in those earlier years. In other words, if a utility accelerates its carbon goal target year, it is "moving back up the declining cost curve" and incurring higher installed costs for the needed new resources as a result.

Another negative economic impact of an accelerated target year is that it widens the CPVRR cost difference between the costs for a resource plan with a specific carbon goal and resource plans without a carbon goal. This occurs for three reasons. The first is the just discussed "moving back up the cost curve" effect which increases the annual nominal costs of each new resource that otherwise would have been added in later years when costs are lower.

Second, the utility now needs to add more MW of new resources in earlier years because it now has fewer years with which to meet its accelerated carbon goal. As an example, let's look back to our previous discussion with our hypothetical utility.

In Chapter 10, using the second estimation approach for determining the total combined MW of PV and batteries that would be needed by our hypothetical utility to meet its carbon goal in Current Year + 29, we saw that 52,515 MW of new PV and batteries combined would be needed. Our utility assumed it would begin adding new resources to meet its carbon goal starting in Current Year + 5. Therefore, it had 25 years to add 52,515 MW of new resources which equates to needing to add approximately 2,100 MW per year assuming a constant, steady installation pattern (= 52,515 MW/25 years).

If our utility now decides to accelerate its carbon goal by 5 years from Current Year + 29 to Current Year + 24, it now has only 20 years in which to achieve its carbon goal. After adjusting for the lower annual GWh load that will have to be served 5 years earlier, it finds it needs to add approximately 49,950 MW of new resources in the earlier Current Year + 24. This equates to needing to add approximately 2,500 MW per year (= 49,950 MW/20 years). Consequently, our utility will need to add about 400 MW more each year to meet its accelerated target date. Not only is our utility installing resources years earlier than originally planned (thus moving back up the cost curve), but our utility is also installing more MW of them in each of the earlier years than originally planned. Both factors drive installed costs for the new resources higher.[15]

The third reason that an accelerated carbon target year results in increasing the CPVRR cost difference between the resource plan which meets the carbon goal and

---

[14] The total MW of new resources needed in 2045 will be somewhat lower than the MW needed in 2050 because the GWh to be served in 2045 will be somewhat lower than the GWh to be served in 2050. See the example that follows in the text.

[15] The alert reader will note that this acceleration of the carbon target date will also save more fuel costs in the earlier years. These cost savings would, at least to a degree, offset the higher capital costs of acceleration.

resource plans that do not have a carbon goal is due to the effect of discounting. In addition to the two factors just mentioned that drive up <u>nominal</u> installed costs for new resources, acceleration of the carbon target date also reduces the number of years over which some of these expenditures would be discounted. Therefore, the higher total annual nominal costs for the new resource additions also result in higher <u>present value</u> costs. The total CPVRR costs of the revised resource plan that meets the accelerated carbon goal target year have now increased while the CPVRR costs for resource plans that do not have a carbon goal are unaffected and thus remain unchanged. Consequently, the CPVRR difference between the resource plans increases even more.

In summary, assuming all else equal, accelerating a carbon goal target date to an earlier year will likely result in significantly higher total net CPVRR costs and electric rates for customers when compared to a resource plan without a carbon goal (and which, therefore, does not have this acceleration).

Question (14): Are there also non-economic impacts that are likely to occur if a utility decides to accelerate its carbon goal target date?

Answer: Yes. There are at least three non-economic impacts that come readily to mind regarding such an acceleration (although it is likely that these will have indirect economic consequences as well). These three impacts are all based on the fact that the number of years in which the utility has to meet its specific carbon goal will have been reduced if the original carbon goal target year is accelerated to an earlier year.

The first of these is the increased pressure and workload that will be placed upon governmental entities whose job it is to review resource additions under an established permitting process. Using our previous example of our hypothetical utility, an acceleration of its carbon goal target date from Current Year + 29 to Current Year + 24 resulted in the MW amounts of new resources needed annually increasing from approximately 2,100 to 2,500 MW. Assuming all else equal, this will cause greater strain on regulatory permitting processes and increase the possibility of project delays. Such impacts are even more likely if our hypothetical utility is not the only utility in the state that is attempting to achieve a carbon goal.

A second, and related, impact is the increased demand an accelerated carbon goal target year will have on suppliers of the needed new resources. As our example showed, an acceleration increases the MW amount of new resources needed each year. This will place a strain on the suppliers and contractors, thus further increasing the possibility of project delays, but from a different source: supply chains in addition to permitting processes.

A third non-economic impact is that the transmission planning and system operation functions, in collaboration with the resource planning function, will have less time to resolve how to address the system stability and system operations issues previously discussed in Chapter 13. In addition to having fewer years to resolve these issues, the very resources that are causing these issues would be added to the utility system annually in larger amounts and at an increasing pace. This increases the possibility of system stability and reliability problems negatively impacting customers.

It would likely also increase the CPVRR costs of addressing these system stability issues for the reasons previously discussed (for example, fewer years of discounting) because any needed modifications/additions would need to be added earlier than would have been the case if the original target year had been retained.

Question (15): In regard to the actual mechanics of IRP analyses, assume that a utility has been using projected carbon compliance costs from a third party in its resource planning work and the utility has now set a low-to-zero specific carbon goal. Is it appropriate for the utility to continue to use these third-party projected carbon compliance costs in its resource planning work?

Answer: No. As I define the term "carbon compliance costs," such an action would equate to double counting which would be an error in economic analyses.

The term "carbon compliance costs" means (to me) a projection of costs that are forecast to occur to comply with potential future legislation and/or regulation that sets a carbon goal. The legislation/regulation's goal could be, for example, an 85% reduction in carbon emissions by a certain year. Another example could be zero-carbon emissions by 2050. The projected costs to comply with such a potential target are usually projected by a consulting firm or other third party. These projected "compliance costs" account for capital and fixed costs for new resources needed to reach the forecasted target, as well as reductions in fuel and other variable costs. In this way, a total net compliance cost projection is made.

These carbon compliance cost forecasts are typically provided in terms of $ per ton values that are projected, not for individual utilities, but for a country as a whole (or for large regions of a country). These values can be used by utilities in IRP analyses as assumed costs to be avoided by the installation of renewable resources and DSM. In other words, these values are used as assumed "avoided cost" benefits attributable to these types of resources.

The use of projected compliance costs made sense before utilities began to set their own carbon targets. (In fact, we used just this approach in Part I of this book in our analyses of Supply and DSM resources for our hypothetical utility.) That is because the possibility of federal legislative and/or regulatory action mandating carbon limits was (is) very real. In addition, the projected costs of complying with these potential carbon limits could be quantified through computer modeling that compared resource plans with and without the potential carbon limits. Furthermore, assuming the carbon limits were in place, the compliance costs would be recovered from utility customers through electric rates.

However, this approach of using compliance costs should no longer apply once a utility sets its own specific carbon goal. In its resource planning analyses, our hypothetical utility will determine the capital and fixed costs for resources needed to reach its own carbon goal and will also account for any fuel or other variable cost savings.

These additional costs and cost savings will be specific to each utility. As such, these utility-specific costs represent exactly what the costs will be for each utility to comply with its own specific carbon target. To also account for a consultant's/third party's projection of national/regional compliance costs for potential future legislative/regulatory-imposed mandates would be an attempt to account for compliance costs twice, *i.e.*, double-counting.

Therefore, once a utility sets its own specific carbon goal, it should no longer utilize someone else's projected compliance costs that are not specific to that individual utility.[16]

Question (16): What impacts will a more rapid consumer adoption of electric vehicles (EVs) have on a utility achieving a specific carbon goal?

Answer: There will be several impacts. Some of these impacts will pose further challenges for utilities. However, one impact could potentially benefit the utility's customers economically.

First, significant EV adoption will result in increased growth in electrical load for the utility. FPL provides a good example of this. At the time the second edition of this book is written, FPL's publicly available Ten Year Power Plant Site Plan for the years 2022 through 2031 projected a growth in summer peak (MW) load through this time period that was attributed solely to adoption of EVs of approximately 1,200 MW. Therefore, when compared to total projected summer peak load growth over the same period of approximately 3,600 MW, EV-driven load was projected to be about 33% of total peak MW load growth. In terms of projected energy (GWh) growth, FPL projected EV-driven energy increases of approximately 4,900 GWh through this period. When compared to projected total GWh growth of approximately 13,450 GWh, EV-driven load was projected to be approximately 36% of total energy growth.[17] Therefore, whether viewed from the perspective of peak load growth, or energy load growth, EV-driven load growth at FPL was forecast to be significant either from an absolute or percentage perspective.[18]

Second, as a result of this increased load growth, a strong, continued movement into EVs will require a utility to need even greater amounts of new resource MW to achieve its carbon goal than would be the case without the EV growth. In turn, the utility will have to acquire more land and install more transmission facilities for the larger amount of renewables, thus further increasing its capital and fixed cost expenditures. The utility's near-term emissions will likely also increase, at least in the early years of the transition period, as more of the utility's marginal fuel is used to produce additional energy to charge the EVs.

---

[16] Note that this discussion of the inappropriateness of continuing to use carbon compliance costs in resource planning work for a utility that has set its own carbon goal does not extend to the use of potential carbon taxes in resource planning work. The possibility of such a tax remains regardless of whether a utility has set a specific carbon goal.

[17] These values can be found in the text, and/or derived from the tables, found on pages 76, 77, 89, and 93 in FPL's Ten Year Power Plant Site Plan 2022–2031 which can be found on the Florida Public Service Commission website.

[18] Because each utility system is different, the magnitude of EV-driven growth in peak load projected by FPL may not be projected by other utilities due, in part, to the utility's peak load hour. However, it is likely that some growth in peak load will occur for most, if not all, utilities.

The third impact of a strong movement into EVs is that, despite having to pay for more new resources and accompanying land and transmission, the utility's electric rates may actually be lowered compared to what the electric rates would be absent the EV growth.

Question (17): Please explain how a utility's electric rates can be lowered by a strong movement into EVs if the EVs result in more utility expenditures.

Answer: The short answer is that EVs will also increase the number of GWh that the utility must serve and the number of GWh is one of two key factors in the calculations of electric rates. The other factor is the total cost, or revenue requirements, required to serve load and authorized by the utility's regulator(s).[19]

Although such a movement into EVs will increase a utility's capital/fixed cost expenditures, and with some increase in variable costs (more fuel will be used which may be partially offset due to running the utility's marginal generating units in off-peak hours at higher capacity factors which are more economical), serving the additional EV load will also increase the number of GWh that the utility serves. Recall that a utility's electric rate is basically a fraction in which total costs are the fraction's numerator and total GWh sales are the fraction's denominator. For example, assume that a (very tiny) utility had costs of $100 and sales of 1,000 kWh. This utility's electric rate would be $0.10 per kWh (= $100/1,000 kWh) or 10 cents/kWh.

The additional costs for serving EVs will certainly increase certain costs. Let's assume that these costs increase by one-half of 1% or by 0.5%. This tiny utility's costs (the numerator) would now be $100.50 (= $100 × 1.005). Let's also assume that the increase in annual sales from EVs is 1%. The utility's sales (the denominator) would now be 1,010 kWh (= 1,000 kWh × 1.01). This utility's electric rate would now be $0.0995 per kWh (= $100.50/1,010) or 9.95 cents/kWh, which is a reduction from the previous 10 cents/kWh.

Assuming all else equal, if the percentage increase in EV-driven GWh sales is greater than the percentage increase in EV-driven net costs, the utility's electric rate will be lowered by the utility serving a higher number of GWh due to EVs.

Question (18): Are there any other points you would like to make related to this discussion regarding EVs?

Answer: Yes. There are two other points I would like to make. The first is that this discussion again points out the importance of focusing economic analyses on electric rates, not on costs. Second, programs that encourage increased EV use should properly be thought of as DSM programs.

The first point is that one should always think of any impact to a utility system that affects the amount of GWh sales in terms of electric rates, not solely in terms

---

[19] Note that Appendix F provides another perspective on how changes in GWh sales may affect electric rates.

of costs.[20] In addition, the simple exercise described in the previous Q&A is basically an application of the Rate Impact Measure (RIM) test. In fact, one could formally apply the RIM test to this question and derive the same result. The only difference in applying the RIM test to a load-increasing DSM program (like an EV-promoting program) rather than to a load-reducing DSM program (like an efficient air conditioner-promoting program) is that the benefits and costs are reversed. With a load-reducing DSM program, the benefits are expenditures that are avoided. Part of the costs are revenue requirements that are unrecovered. However, with a load-increasing DSM program, the benefits are additional revenues that are recovered (from increased GWh sales), and the costs are the additional fixed and variable net cost expenditures that the utility has to make.

This example again points out that the RIM test is the only correct way to perform economic screening analyses of DSM programs, whether load-reducing or load-increasing, because this test accounts for changes in both costs and sales that affect electric rates that all customers are served under. Any other DSM screening test that does not account for the impact on revenue requirements recovered (*i.e.*, ignores electric rates) would view only the increased costs to serve the higher EV load and could never view EV-promoting programs as potentially cost-effective actions for a utility to take.

The second point is that, because EVs impact a utility's electrical load (or demand for electricity), any utility program that promotes or otherwise addresses EVs is, by definition, a DSM program. However, in this case the DSM program increases, not reduces, the customers' demand for electricity. Utilities and regulators should recognize this fact and view programs that increase electrical demand through EV-encouraging efforts, as well as other potential electrification efforts, as DSM programs.

Question (19): Speaking of DSM programs, have there been any recent trends in DSM that warrant a mention?

Answer: Yes. In the decade or so preceding the writing of the second edition of this book, many utilities across the country have seen a noticeable decline in the cost-effectiveness of DSM options.

This trend of steadily declining DSM cost-effectiveness is primarily driven by two factors. The first factor is a decline in a number of the costs that DSM programs are designed to avoid. This lowers the projected benefits of DSM which, in turn, lowers the projected benefit-to-cost ratios for DSM programs (regardless of which preliminary screening test for DSM is used).[21] At least three types of utility costs that DSM can potentially avoid have lower costs than they did previously.

The first of these are lower installed capital costs for new generation. DSM programs have traditionally (and correctly) been compared to the type of generating unit the utility would build absent the DSM program. For many utilities, these generating units have been new CC and/or CT units and the $ per kW installed costs for these

---

[20] In other words, we are once again faced with my Fundamental Principle #3 of Electric Utility Resource Planning.

[21] Other than the Participant Test which is unaffected by this trend of lower avoided costs.

units have decreased in recent years. The manufacturers of these technologies, recognizing that they are in competition with other technologies used by utilities to supply electricity (particularly solar and wind whose costs have dropped significantly), have continually worked to make the installed $ per kW costs of these generation options more competitive. In addition, the efficiency with which these types of generators (especially CC units) utilize natural gas has also improved dramatically, thus lowering the cost of operating these units. Therefore, the manufacturers have been successful in lowering both these building and operating costs which, in turn, lowers the potential DSM benefits of avoiding new CC and CT units.

In addition, the cost of natural gas has also declined from what it was a decade ago. Largely due to hydraulic fracturing drilling techniques, the prices for natural gas have generally been much lower in the last decade than they were in 2008 or so.[22] Lower fuel costs mean even lower potential DSM benefits of reducing fuel usage. A third type of cost that has declined over the last decade has been projected environmental compliance costs. These costs have declined as utilities have lowered their system emission rates. The lower emission rates have occurred as more and more utilities have retired coal units, replaced coal units with lower emission natural gas-fueled units, and installed more zero-emission generation such as solar and wind. Similar to lower fuel costs, lower projected environmental compliance costs also mean lower potential DSM benefits of reducing system emissions.

The second factor that is driving the decline in utility DSM program cost-effectiveness is the significant amount of energy efficiency that is being delivered by federal and state energy-efficiency codes and standards. Utilities account for the impacts of these energy-efficiency codes and standards in their load forecasting efforts. As a result, forecasts of both system peak loads (MW), and annual energy (GWh) loads, are lower than what would be in a case without these codes and standards. Thus, there is less load growth to address with any type of new resources (DSM or Supply) in IRP analyses.

In addition, the potential market for some utility DSM programs has been reduced by the mandated higher efficiency of appliances, equipment, and lighting. For example, without the codes and standards, customers who have home air conditioning (AC) units with a low seasonal energy efficiency rating (SEER) might consider purchasing a new AC with a slightly higher SEER when it came time to replace their AC. For example, a customer with an AC with a SEER of 10 might consider stepping up to a new AC with a SEER of 12. However, at the time the second edition of this book is written, the minimum SEER ratings for a residential AC unit are 14 for the northern United States, and 15 for the southern United States. As a result, the market potential for improving air conditioner SEER from 10 to 12 via utility DSM programs has vanished. That fact, combined with the fact that installed costs for air conditioners, become increasingly expensive at higher SEER ratings, lower the realistic achievable market potential for these types of utility DSM programs.

---

[22] However, as the second edition of this book is being written in 2022, natural gas prices initially increased to approximately $9/mmBTU. This was due to various factors (inflation, the current federal administration's actions to reduce fossil fuel drilling and use, the war in Ukraine, etc.). However, natural gas prices subsequently decreased to approximately $2.50/mmBTU at the time the second edition was sent to the publisher.

Absent draconian measures being implemented by government to make it more costly to use fossil fuels in the near-term to produce electricity, I believe this trend of lower DSM benefits, and lower realistic achievable market potential for some DSM programs, will continue. This is especially as utilities pursue specific carbon goals and begin installing solar and wind facilities almost exclusively.

Question (20): Do you believe that utility DSM programs continue to make sense as utilities pursue specific carbon goals?

Answer: Perhaps, but with some qualifications. Those qualifications are based on my belief that the case for utilities continuing to pursue types of DSM programs that reduce electric usage by customers is much weaker than it was back in the 1980s and early 1990s when interest in utilities pursuing such programs began in earnest.[23]

In those years, and for two or three decades thereafter, the interest in utility DSM programs was driven largely by concerns (in no particular order) regarding emissions, imported and domestic fuel supply and cost, and the costs of building and operating conventional generating units. However, around 2010 or so, many of these concerns lessened. Generating units became cleaner, less expensive to build, and less expensive to fuel/operate. Plentiful supplies of domestic fuel lowered the cost of natural gas and all but eliminated concerns over reliance on imported fuel supplies. The use of natural gas instead of coal greatly reduced emission levels. All of these factors not only contributed to the trend of declining DSM benefit-to-cost ratios that we just discussed, but also cut away at the underlying rationale for utility DSM programs (*i.e.*, the above-mentioned concerns).

In the last 10 years or so before the second edition of this book was written, concerns regarding climate change and GHG emissions have greatly increased and have largely replaced the previously mentioned concerns regarding other emissions, fuel cost/supply, and costs of conventional generating units. These new concerns would seem to breathe new life into utility DSM programs that reduce usage by providing a new rationale for these programs.

However, as utilities begin to build, almost exclusively, new solar and wind generation, this poses perhaps an even greater challenge for economically justifying continued utility DSM programs that reduce customer usage. The $ per kW installed costs of solar and wind facilities are currently approaching the $ per kW installed costs of CCs and CTs, and the costs of solar and wind facilities are projected to be lower than CCs and CTs in the not-too-distant future.

Because DSM programs that reduce energy usage are (logically) compared to the type of generating unit that would be added absent the DSM program, DSM programs that reduce energy usage will need to compete with new solar and wind options. Having DSM programs that avoid solar and wind facilities will result in even lower avoided capital cost benefits than these programs do now in avoiding CCs and CTs.

---

[23] As the reader considers this Q&A, and the one that follows, it may be informative to point out that I began my utility career with FPL in the DSM area. I spent more than 10 years designing and monitoring DSM programs. After I subsequently moved into the resource planning area, I spent about 30 years analyzing all competing resource options, including DSM options.

Furthermore, avoiding solar and wind facilities will likely increase utility system fuel usage, GHG emissions, and environmental compliance costs, at least in the first half of the transition period. This is because the existing conventional generating fleet will now have to supply the energy to customers that would have been supplied by the avoided solar and wind facilities, thus resulting in more natural gas usage, GHG emissions, and environmental compliance costs than would be the case if the solar and wind facilities had not been avoided. This not only eliminates/minimizes the potential variable cost savings that DSM programs have been projected to have when avoiding CCs and CTs, but it also undermines the new rationale for these types of DSM programs: the reduction of GHG emissions.

An attempt might be made to shift DSM's "target" away from avoiding new generating facilities that the utility would otherwise build, to now target early retirement of existing generating units. However, early retirement of these generating units during the transition period would only exacerbate the system stability and/or system reliability issues that utilities will be attempting to address as they move in the direction of zero carbon. Therefore, this would likely not be a fruitful approach to attempt to apply to justify DSM programs.

For these reasons, I believe that it will be increasingly difficult to develop utility DSM programs that reduce customer usage and are cost-effective. Then, in regard to the underlying rationale for such DSM programs, it would also seem to be difficult to justify avoiding zero-emission generating units such as solar and wind during the transition period if the ultimate objective is to significantly lower carbon emissions.

However, as previously discussed in regard to EV adoption, the possibility exists that utility programs that *increase* electricity use may be able to lower electric rates.

Question (21): Do you have any recommendations for utilities and their regulators regarding how DSM should be approached in the future?

Answer: Yes. Those recommendations are listed below.

1. Utilities and regulators should recognize that utility DSM programs that reduce customer usage have become increasingly hard to justify based on economic analyses that include all DSM costs and benefits that are accounted for in electric rates.
2. In addition, utilities and regulators should acknowledge that significant amounts of energy efficiency are now being delivered to customers annually through federal and state energy efficiency codes and standards. In other words, energy efficiency is still being delivered to customers, but the delivery mechanism has changed to codes and standards instead of utility DSM programs. (And, one might argue, that the codes and standards delivery process is more "efficient" than the utility DSM process because codes and standards are mandated, whereas utility DSM programs rely on voluntary customer participation.)
3. Utilities and regulators should acknowledge that, just as once very useful oil- and coal-fueled generators have been replaced by superior technologies such as CCs, solar, batteries, and wind, the usefulness of utility DSM programs that reduce customer usage has significantly declined.

4. Utilities and regulators should resist the temptation to continue such utility programs anyway (perhaps through the use of fundamentally flawed DSM cost-effectiveness tests) largely, or solely, because of how it would be perceived if these programs were scaled back/discontinued. Parties should instead look to new and better ways to guide and assist the ongoing restructuring of the electric utility industry towards zero-carbon.

5. However, because utility resource planning should be an open competition between all resource options applicable to the utility in question, including both Supply and DSM options, utilities and their regulators should continue to evaluate existing and new DSM options that are designed to reduce customer usage. But they should do so with analyses that fully account for all of the costs and benefits of such programs that are reflected in the utility's electric rates.

6. In addition, utilities and regulators should recognize that demand side management programs, by definition, also include programs that increase customer demand for electricity. Such programs have the potential to result in lower electric rates for all customers. If this is the projected outcome, these programs should be viewed as cost-effective and pursued. An example of such a DSM program is one that promotes the use of EVs and which results in more projected incremental revenues from increased sales than projected incremental costs, thus lowering electric rates. Other potential programs that fall in the realm of "electrification" and increase electricity demand should be considered DSM programs as well.

In summary, the utility industry does not necessarily need to eliminate utility DSM programs designed to reduce customer usage as it pursues lower carbon. But it must recognize that the economical and philosophical justifications for such programs have been significantly diminished with the emergence of lower cost, zero-carbon generation resources. Policy makers/regulators should not continue such programs for appearances sake. They should let go of these programs when their usefulness has ended and expand the perception of DSM to include potentially beneficial programs that increase electric usage and lower electric rates.

WHAT LIES AHEAD FOR ELECTRIC UTILITIES AND UTILITY RESOURCE PLANNING?

The only correct answer to this question, other than recognizing a movement toward lower carbon emissions, is "no one knows." Indeed, this has been the correct answer to this question ever since I first began my work in electric utility resource planning analyses several decades ago.

There are simply too many factors that can quickly, and significantly, change what one may have thought was a safe-to-predict direction or outcome for the electric utility industry as a whole, or for a specific electric utility. A listing of such potential "game changing" factors would certainly include, but not be limited to, those listed below. (Note that the factors listed below are presented in no particular order.)

- *The actual trajectory of installed costs for PV and batteries (plus other renewable energy technologies).* As we discussed in Chapter 11, installed costs for PV significantly declined from 2010 through 2021 but then leveled out in 2022 as supply chain disruptions and other issues arose. A key question going forward will be how much lower the installed costs for PV and batteries will go, and how quickly such cost reductions will occur.

  A historical note may be of interest. In the early 1980s, the U.S. government instituted a 40% federal Income tax credit for consumers of solar water heaters. A (likely unintended) consequence was that solar water heater manufacturers significantly raised the cost of the water heaters. The consumer still received some break in their net costs (water heater plus taxes) while the manufacturers substantially increased their profits. A few years later, the manufacturers lowered costs immediately once the federal tax credits were eliminated. It will be interesting to see whether the 2022 IRA tax credits will have a similar effect on the projected declining cost curve for PV.

- *The actual consumer adoption rate of EVs.* Current utility load forecasts generally project an upwards trend of EV adoption along with the associated higher electric load. However, as the second edition of this book is written, the purchase price of EVs remains significantly higher than gasoline-powered vehicles in the absence of subsidies/rebates, the used EV market is too new to provide much certainty regarding long-term value, and lack of charging infrastructure is still a concern. For these reasons, another of THE key questions going forward is what the actual adoption rate of EVs will be.

- *The resolution of system stability challenges inherent with increasing amounts of IBRs.* From discussions with planners at numerous utilities regarding this subject, it is apparent that at least most utilities are in the early stages of trying to determine how best to address transmission system stability challenges that will accompany a move toward lower carbon. One of the many issues regarding this is that a utility's transmission planners cannot possibly know now where the tens of thousands of MW of new resources that will be needed to reach a low-to-zero carbon goal are going to be located over the next 10-to-20 years. Transmission costs in general tend to be quite large so the costs for address the system stability issues will likely be very large as well. It is possible that this could be THE key question going forward in regard to the incremental net cost of the movement toward zero carbon.

- *The resolution of how to evaluate utility system reliability as increasing amounts of intermittent generating resources, and time-limited storage resources, are added to utility systems.* As mentioned previously in the text, it may not be crucial to resolve this question now as utilities are just entering their transition periods because it will take a while before these new resources actually begin to usurp the dominant role that conventional generators now play. However, the sooner this piece of the movement to lower carbon puzzle is solved, the sooner its ramifications on costs, system stability, etc. can be factored in.

- *The approach(es) that utility regulators, in states without a state carbon mandate, take toward approving costs for increasing amounts of zero-carbon resources for utilities who have a self-imposed specific carbon goal.* How the regulators will decide to adjudicate utility requests for cost recovery for new resource MW that may not meet traditional cost-justification requirements of a reliability need and/or economic advantages for all customers, but which are required by the utility to meet a self-imposed carbon goal, will be very interesting to see. It will not be surprising if state legislatures also play a role in such situations.
- *The rate of progress in developing zero-carbon or carbon-neutral fuels at a scale that will be meaningful for electric utilities.* Such fuels exist today, but only in very small volumes and at costs much higher than natural gas. If these types of fuels are to be a serious competitor/complement to solar, batteries, and wind in time to have a significant impact as utilities plan for the mid-2030s through the 2040s, then the development pace of these fuels, as well as their transportation/distribution networks, needs to accelerate rapidly.
- *Unforeseen events.* There will almost certainly be future events that affect the electric utility industry that were not foreseen, or if they were foreseen, were projected to have a very low probability of occurring. Examples of such events include the earthquake- and tsunami-induced damage to several Japanese nuclear reactors in 2011, and the COVID pandemic that began in late 2019-to-early 2020. As unforeseen events such as these occur, certain resource options will be looked upon more favorably and/or supply chains will again be disrupted. Such events are likely to have unexpected impacts on many aspects of the utility industry, including utility resource planning.
- *Federal and state legislation/regulation.* Enactment of far-reaching legislation or regulation, such as potentially setting federal carbon-level mandates, imposing federal carbon taxes, and/or making further changes in federal tax policy, could be game-changing events. This is especially true because such legislation/regulation would almost certainly have unexpected consequences.
- *A potential re-thinking of the need to reach zero carbon.* As the second edition of this book is written, there is pressure from many directions for electric utilities to get to low-to-zero carbon and to do so quickly. But as we have discussed in Part II of the book, there are a number of significant issues that utilities will have to resolve as they continue to move to lower carbon. As these issues are more thoroughly analyzed, and all of the cost, reliability, and system stability ramifications are better understood, it is possible that many will reach a conclusion that something less than zero carbon is not only sufficient, but desirable, at least during the next few decades.
- *Changing economic conditions.* The timing and duration of economic cycles, including the effects of inflation, will affect all aspects of electric utility planning and operation from capital investment in new/more efficient technologies to the levels of demand for electricity.

- *New nuclear-generating units.* The first "wave" of new nuclear-generating units in the United States can be said to have officially commenced in 2013 with the start of construction of Plant Vogtle units 3 and 4 in Georgia. Unfortunately, the construction of these units has not gone well. At the time the second edition of this book is written, the plants are approximately 6 years behind the original schedule and approximately $16 billion over the original budget. Consequently, no other new nuclear units of the same type are projected to be built in the United States. However, a lot of development work has gone into new nuclear generating unit designs that are smaller and modular in concept. (In addition, in late 2022, a breakthrough in nuclear fusion occurred in which, for the first time, more energy was created than was used in the process.) It is far too early to know if any of these efforts will yield new nuclear generation options that can be cost-competitive with other resource options, particularly other zero-carbon-emitting options. But because new nuclear technologies offer the potential for around-the-clock operation with zero emissions (without any concerns regarding cloud cover, wind speeds, etc.) and can also address system stability issues, the potential for new nuclear technology will continue to hold interest.
- *The emergence of other new technologies.* New technologies at a utility-scale (such as efficient, relatively low-cost CO_2 capture/removal methods), plus technologies that increase electrical use (such as "must have" electronic devices for the home and office) can be expected to emerge. Such technologies could have widespread (and, once again, unexpected) impacts on utility planning and operation.

Therefore, in regard to electric utility planning and operation, the only certainty is that there will be an enormous amount of *uncertainty* that utility resource planners will face. Both the challenges to, and the complexity of, electric utility resource planning will likely continue to increase due to this uncertainty.

CLOSING THOUGHTS

With this uncertainty in mind, I close this book with two final thoughts. These thoughts represent hopes that I have in regard to electric utility resource planning in the future.

First, the inherent uncertainty in electric utility planning and operation is the best reason to undertake utility resource planning using an approach that allows one to examine all resource options from the perspective of the utility system as a whole. An IRP approach, such as the one discussed at length throughout this book, is the best utility resource planning approach with which to address this future uncertainty. Therefore, I hope that future generations of legislators, regulators, utility executives, and utility resource planners will continue to recognize the inherent advantages of using an IRP approach in helping to make fully informed decisions.

Second, and in conclusion, I hope that I have been able to convey through this book both the continual challenges that face anyone whose job is to conduct resource planning for electric utilities, plus the enjoyment I have had in trying to meet those challenges.

Appendix A: Fundamental Principles of Electric Utility Resource Planning

Throughout the book, we have introduced a number of concepts that are important to utilities in their ongoing work of planning for future resources that are needed to maintain a reliable and cost-effective utility system. Of these concepts, I believe there are a handful of concepts that can be viewed as truly "fundamental principles" of electric utility integrated resource planning (IRP).

These fundamental principles, discussed previously in the book, are listed again in this appendix in order to provide a handy reference. In addition, a few comments about how to "apply" these fundamental principles are provided.

FUNDAMENTAL PRINCIPLE #1 OF ELECTRIC UTILITY RESOURCE PLANNING: "ALL ELECTRIC UTILITIES ARE DIFFERENT"

Each electric utility is different in regard to (at least) its electrical load characteristics and its existing generating units. Therefore, when faced with a particular problem or issue, such as "which resource option is the best selection?", the correct answer for one electric utility may not be the correct answer for another electric utility.

Because each utility system is different, any assumption that what the best resource option selection for Utility A is will automatically be the best choice for another utility (or, even worse, for all other utilities) is very likely to be wrong, and, in some cases, spectacularly wrong.

Understanding this fundamental principle is of particular importance to legislative and/or regulatory bodies that may be tempted to impose certain resource option selections, or a certain levels of resource options, on all utilities within its jurisdiction. Although such an action may be tempting due to expediency, political and otherwise, a "one-size-fits-all" approach will almost certainly guarantee that the best choices for individual utilities, and their customers, will not be made.

In regard to how best to apply this fundamental principle, the answer should be obvious to one who has read through the book to this point. Each utility needs to perform a complete set of analyses (as is the case when an IRP resource planning approach is used), utilizing utility-specific assumptions and inputs, to ensure that the best choices for the specific individual utility are made.

FUNDAMENTAL PRINCIPLE #2 OF ELECTRIC UTILITY RESOURCE PLANNING: "SYSTEM COST IMPACTS OF PRODUCING OR CONSERVING ELECTRICITY ARE OF UPMOST IMPORTANCE. INDIVIDUAL RESOURCE OPTION COSTS OF PRODUCING OR CONSERVING ELECTRICITY ARE OF LITTLE OR NO IMPORTANCE WHEN CONSIDERED SEPARATE FROM THE UTILITY SYSTEM AS A WHOLE"

The projected costs of producing/conserving electricity for any individual resource option by itself, often expressed in terms of cents per kWh, or $ per MWh, regardless of whether the resource option is a Supply option or a demand side management (DSM) option, are of little/no consequence when performing economic evaluations whose objective is to select the most economic resource option for the utility system as a whole. When selecting a resource option for a particular utility, the objective is to identify the resource option that results in the lowest electric rates that are charged to customers. Only an analysis that accounts for the costs of the resource option itself, plus the cost impacts on the system that will occur if the resource option is added to the system, can determine the most economic resource option.

The application of this fundamental principle is most important when one runs across an argument that Resource Option A should be selected because "Resource Option A can produce/save energy at a cost of only 'X' cents per kWh." Alarm bells should go off immediately when one reads or hears such a statement because the party issuing such a statement either: (i) doesn't understand the significant limitations of such cents per kWh comparisons of individual resource options, or (ii) understands these significant limitations, but is trying to pull a fast one on his/her audience (who may not realize these limitations).

In order to obtain a complete answer to what the cost of a particular resource option would be for a specific utility, analyses of the utility system as a whole must be conducted that evaluate different resource plans that address not only Resource Option A, but also all other possible resource options as well. Only in this way can one obtain a complete and, therefore, meaningful picture of the system cost impacts of competing resource options.

FUNDAMENTAL PRINCIPLE #3 OF ELECTRIC UTILITY RESOURCE PLANNING: "ELECTRIC RATE IMPACTS ARE THE MOST IMPORTANT CONSIDERATION WHEN ANALYZING DSM AND SUPPLY OPTIONS; TOTAL COST IMPACTS ARE LESS IMPORTANT"

In regard to economic analyses, projections of only total utility system costs for competing Supply options can be used to correctly select the most economic Supply resource option. As previously discussed, this is because the Supply option which results in the lowest total costs will also result in the lowest electric rates. That is because: (i) an electric rate is basically total system costs ($) divided by total system sales (GWh), and (ii) Supply options do not change the total system sales (GWh) value (the denominator in the electric rate calculation). Therefore, the Supply option with the lowest system costs will also be the Supply option with the lowest electric rate.

However, when evaluating DSM options versus Supply options, economic analyses must be carried out one step further. Analyses of DSM versus Supply options must account for the fact that DSM options change the number of GWh of sales over which a utility's costs are recovered (and most DSM options reduce the number of GWh of sales). This unique characteristic of DSM options makes it necessary to conduct an electric rate calculation in order to really determine which option, DSM or Supply, is the best economic choice from a customer's perspective. The importance of total costs is lessened in these calculations because total costs are merely one of two inputs into the calculation of electric rates.

Any economic analyses of competing DSM and Supply options must apply this principle to get a complete and meaningful answer to the question of which resource option is the most economic option for the utility's customers. However, there is no need to perform this electric rate analysis when evaluating only Supply options because none of the competing Supply options will directly change the number of GWh of sales over which the utility's costs are recovered. Therefore, once one knows the Supply option that is projected to result in the lowest system costs, one also knows that this Supply option is projected to result in the lowest electric rates. One could perform an additional step of calculating electric rates for all of the Supply options, but it is not needed in such cases because the system cost analysis has already determined the Supply option which will result in the lowest electric rates.

This is obviously not the case when DSM options are compared to Supply options because DSM options, by definition, either decrease or increase the number of sales GWh over which the utility's costs are recovered. Consequently, the additional step of calculating the projected electric rates that will result from each of the competing resource options is absolutely necessary to determine which resource option will result in the lowest electric rates for the utility's customers, *i.e.*, which resource option is the economic choice.

FUNDAMENTAL PRINCIPLE #4 OF ELECTRIC UTILITY RESOURCE PLANNING: "ALWAYS ASK YOURSELF: 'COMPARED TO WHAT?'"

In all aspects of resource planning, one must always ask the question "compared to what?" when one is analyzing a particular resource option.

For example, in Chapter 14, we posed the question of whether a resource plan that is designed to achieve a zero-carbon goal could be cost-effective for the utility's customers. With this system economics-based question, the way the question is stated makes it unclear what this resource plan would be compared to. If all of the competing resource plans must achieve the same zero-carbon target, and in the same target year, then it is certain that one of the resource plans will be more economic (*i.e.*, will result in lower electric rates for customers) than the other resource plans that are considered.

On the other hand, if the question is whether a resource plan that achieves the zero-carbon target can be cost-effective versus another resource plan(s) that does not have to achieve a zero-carbon target, then we have a completely different question. Analyses of both types of resource plans, those that must achieve a carbon target and those that do not have to achieve this target, will have to be performed to answer this question.

Similar situations can apply regarding non-economic analyses. An example might be a question regarding air emissions. The fact that Resource Option A is projected to either lower the number of kWh served by a utility (such as with DSM options), or to produce kWh without burning fossil fuel (such as with a renewable energy option), is no guarantee that Resource Option A will actually result in lower system air emissions for a specific utility. The key point is "what is Resource Option A being compared to?"

Although Resource Option A may reduce air emissions from what the emissions otherwise would be *if no other resource option were added or considered*, there may be a Resource Option B that would result in even lower system air emissions. (In such a case, the selection of Resource Option A would actually increase system emissions compared to the selection of Resource Option B.) Therefore, in order to know which resource option will really result in lower system air emissions, all resource options which are applicable for a specific utility must be evaluated/compared.

Furthermore, in regard to the application of Fundamental Principle #4 in non-economic analyses, this principle can be thought of as the counterpart of Fundamental Principle #2 that addresses economic analysis. Both principles point out that a comparison of the characteristics of individual resource options can only provide an incomplete and, therefore, meaningless "answer" to the question of which resource option is the best selection.

Appendix B: Glossary of Terms

A number of terms commonly used in discussing electric utilities and resource planning for electric utilities are used throughout this book. An attempt has been made to explain those terms when the term is introduced. However, for handy reference purposes, a number of the terms that individuals outside of the utility industry may find unfamiliar are listed and explained in the compilation below.

Air Emissions: As used in the book, this term refers to certain gases that are emitted into the air as a result of burning fossil fuels in electric utility generating units. Three air emissions are used in the book when discussing utility system operation and resource planning: sulfur dioxide (SO_2), nitrogen oxides (NO_x), and carbon dioxide (CO_2).

Availability: Availability refers to the percentage of the annual hours per year that a generating unit is projected to be able to produce electricity. In simplistic terms, the projected availability value for a specific generating unit is developed by accounting for two factors: (i) the projected number of hours each year that the unit will be out of service for planned or scheduled maintenance and (ii) the projected number of hours each year that the unit is expected to be out of service for unplanned maintenance (*i.e.*, the unit is "broken"). As a consequence of these two factors, the projected availability for a generating unit is typically less than 100%.

Baseload Generating Unit: A baseload generating unit is a generator that is projected to operate much of the time during a year. For example, a generating unit that is projected to operate more than 70% (or so) of the hours in a year is generally considered a baseload unit. Such a generating unit operates many hours of the year because it is among the least expensive-to-operate generating units on the utility system. (Please refer to the explanations/definitions of "Intermediate Generating Unit" and "Peaking Generating Unit.")

Battery Duration: In regard to system reliability analyses, battery duration is the amount of time, usually referred to in terms of hours, during which the full MW output of a battery can be delivered. For example, a 100 MW, 4-hour duration battery can be expected to be able to produce 100 MW of output for 4 hours before the battery needs to be recharged. Note that this battery could also deliver many other output combinations such as 50 MW for 8 hours or 80 MW for 5 hours. These potential output combinations are limited only by the total energy (MWh) output capability of the battery. In our example, a 100 MW, 4-hour battery can produce 400 MWh (= 100 MW × 4 hours) of energy in any combination of MW and hours before the battery needs to be recharged.

British Thermal Unit (BTU): This term refers to the amount of heat required to raise the temperature of 1 pound of water 1 degree Fahrenheit at sea level.

The term is used frequently in discussing the cost of fuel in which the cost is frequently referred to as "dollars per million BTU" ($ per mmBTU, or $/mmBTU). The BTU term is also used in regard to the fuel efficiency (or heat rate) of generating units in which the efficiency is referred to in terms of BTU/kWh. The BTU term is again used in regard to the amount of air emissions that result from burning fuel in which the amount of emissions is referred to in terms of pounds or tons of emission per BTU of fuel burned. (Please refer to the explanations/definitions of "mmBTU" and "Heat Rate.")

Capacity: In this book, the term "capacity" is being used to denote the projected maximum MW output of a generating unit based on a given set of conditions (such as air temperature, humidity, and type of fuel being used). This projected maximum output value is also referred to as "nameplate capacity."

Capacity Factor: Similar to the term "Availability," the term "Capacity Factor" is a value representing a percentage of the total hours in a year in regard to a generating unit. While the "Availability" value tells us the projected percentage of the annual hours that a generating unit is *capable* of operating (after accounting for both planned and unplanned maintenance), the "Capacity Factor" value tells us the percentage of the annual hours that a generating unit is projected to actually *operate* on a particular utility system. The Capacity Factor value is typically less than, but could be equal to, the Availability value. All else equal, the lower the operating cost of a generating unit, the higher its capacity factor will be.

Conventional Generating Unit: As used in this book, the term refers to electricity-producing generating units (or power plants) that use coal, natural gas, oil, or uranium as fuel. The term is used to distinguish these generating units from renewable energy-based generating units whose sources of energy are solar or wind.

Cumulative Present Value of Revenue Requirements (CPVRR): The term "Cumulative Present Value of Revenue Requirements (CPVRR)" refers to the sum of the annual present values of revenue requirements for a multi-year period. (Please see the explanation/definition of "Revenue Requirements" in this appendix and the discussion of CPVRR in Appendix C.)

Curtailment: As used in this book, the term refers to a utility that cannot, or does not, use the MWh output of a PV facility to serve its customers' load at the same time the facility generates the energy. Instead, the utility either (i) does not allow the energy from the PV facility to enter its transmission system, (ii) stores the energy for use at a later time, (iii) sells the energy to another party, and/or (iv) uses the MWh for purposes other than directly serving customer load (for example, those MWh might be used to create a zero-carbon or carbon-neutral alternative fuel).

Decision Year: In this book, the term refers to the year in which additional resources are needed based on reliability analyses and, therefore, the year for which the utility must make a decision for regarding which resource option(s) to add. (Please refer to the explanation of "Reliability Analysis".)

Demand Response Programs: (Please refer to the explanation/definition of "Load Management Program.")

Demand Side Management (DSM): This term refers to one of two types of resource options a utility can select to meet a future resource need. These two types of resource options are: Supply options (options that generate electricity) and DSM options (options that change customers' demand for, and use of, electricity). In regard to DSM programs, there are three types of DSM programs. Two of these program types reduce the usage of electricity by customers and, therefore, reduce the number of GWh of sales by a utility. These two types of programs are referred to by various terms. In this book, they will be referred to by the terms: energy conservation programs and load management programs. (Please see the explanations/definitions of "Energy Conservation Programs" and "Load Management Programs.") The third type of DSM programs increases customers' use of electricity and, therefore, increases the number of GWh of sales by a utility. An example of this third type of DSM program is one that encourages the use of electric vehicles.

Discount Rate: A discount rate is a value, expressed as a percentage (such as 8%), which one uses to calculate the time value of an amount of money that will be spent or received in one specific year relative to its worth in another year. (Please see Appendix C for a further explanation/discussion of the concept of discounting.)

Electrical Demand: This term refers to the demand for electricity by a utility's customers at any one point in time. The concept of electrical demand is critical when utilities evaluate the reliability of their utility system at the time (hour) of customers' highest electricity usage during a year. This value is also critical when a utility evaluates how large a resource is needed in the future to ensure that the utility system remains reliable. Electrical demand is usually referred to in terms of megawatts (MW) when discussing the utility system and Supply options, and in terms of kilowatts (kW) when discussing individual DSM options. (Please refer to the explanations/ definitions of "Megawatt" and "Kilowatt.")

Energy: Energy is the amount of electrical demand or usage over a given period of time. The period of time most frequently used in discussing an amount of energy is 1 hour and the terms most often used in discussing energy are: kilowatt-hour (kWh), megawatt-hour (MWh), or gigawatt-hour (GWh). (Please refer to the explanations/definitions of "Kilowatt-hour," "Megawatt-hour," and "Gigawatt-hour.")

Energy Conservation Programs: An energy conservation program refers to one of two basic types of demand side management (DSM) options or programs that reduce participating customers' use of electricity at the utility's peak load hour and/or over the course of a year. (Please see the explanation/ definition of "Demand Side Management.") A utility's energy conservation programs typically offer one-time rebates or incentives to utility customers who voluntarily install energy-efficient appliances, materials, or equipment in their homes or buildings. (Energy conservation programs are also referred to as "energy efficiency" programs.)

Energy Efficiency Program: (Please refer to the explanation/description of "Energy Conservation Program.")

Environmental Compliance Costs: For the purposes of the book, this term is used to denote the projected cost complying with potential future legislation or regulation which sets a limit on one or more air emissions, either on a "per generating unit" basis or in total for an electric utility system. Certain air emissions (for example, SO_2) are currently regulated in the United States, while other air emissions (for example, CO_2) are not regulated in the same manner at the federal level (at the time the second edition of this book is written). The environmental compliance cost value for air emissions is typically expressed in terms of $ per pound or $ per ton. For a given year, the annual number of pounds or tons of emission is first calculated and then multiplied by the $ per pound or $ per ton cost value applicable for that year, to derive the utility's total environmental compliance cost for that year for that specific emission type. The total cost over a year (or over multiple years) is also referred to as the environmental compliance cost. Note that environmental compliance costs may be projected costs only, not actual costs that the utility is incurring today.

Equivalent Capacity Factor: This term represents a type of "Capacity Factor" value for DSM programs. Just as the term Capacity Factor refers to the percentage of the hours in a year in which a generating unit is projected to operate, an "Equivalent Capacity Factor" represents the projected percentage of hours in the year that DSM programs will "operate," *i.e.*, will impact the electricity demand and supply of a utility system. This percentage value is calculated by dividing the projected annual energy (kWh) change from the DSM program by the projected peak hour demand (kW) change from the DSM program, then dividing the resulting hourly value by the 8,760 hours in a year. This value is typically used for illustrative or comparison purposes when discussing two or more resource options.

Firm Capacity Value: As used in this book, the term means the expected projected output for a generating unit at the time of the utility's system peak hour. This value is used in system reliability analyses such as reserve margin calculations. For conventional generating units, such as those that are fueled by fossil fuels and/or uranium, the firm capacity value usually matches, or closely approximates, the nameplate MW rating of the generator. For renewable energy-sourced generating units, such as solar and wind, the firm capacity is typically significantly lower than the nameplate MW rating of the generator due to the fact that the energy sources for these generators, sunshine or wind are intermittent and can vary in intensity from minute-to-minute and over the course of the day.

Fossil Fuels: As used in this book, this term refers collectively to three different types of fuel: coal, oil, and natural gas.

Gigawatt (GW): A gigawatt (GW) is a measure of electrical demand and represents 1,000 MW or 1,000,000 kW.

Gigawatt-hour (GWh): A gigawatt-hour (GWh) is a measure of electrical energy and represents 1,000 MWh or 1,000,000 kWh. The amount of electricity that an electric utility produces or sells over the course of a year is typically referred to in terms of GWh.

Green Hydrogen: This term refers to hydrogen that is produced through an electrical/chemical process using a zero-carbon energy source to power the process. For example, solar- or wind-generated electricity could power an electrolysis process in which water (H_2O) is separated into oxygen (O_2) and hydrogen (H_2).

Heat Rate: This term refers to the efficiency with which a conventional electrical generating unit produces electricity by utilizing fuel. Heat rate is measured by the ratio of how much fuel must be used to produce 1 kWh of electricity. The heat rate ratio is presented as "BTU per kWh" in which the amount of fuel used is measured in British thermal units (BTUs). The lower the value of this ratio (*i.e.*, the lower the amount of BTUs of fuel needed to produce 1 kWh of electricity), the more efficient the generating unit is. (Please refer to the explanation/definition of "British Thermal Unit.")

Integrated Resource Planning (IRP): Integrated resource planning (IRP) is an analytical approach in which both types of resource options, Supply and DSM, are analyzed on a level playing field (*i.e.*, there is no bias or predisposition for or against any type of resource option). For each resource option, an IRP analysis accounts for all known cost impacts on the utility system that are passed on to its customers through the utility's electric rates. In addition, non-economic impacts to the utility system from the resource options are typically evaluated as well. In this way, IRP analyses foster an open competition among resource options.

Intermediate Generating Unit: An intermediate generating unit is a unit that is projected to operate on a utility system less frequently than a baseload unit, but more frequently than a peaking unit. For example, a generating unit that is projected to operate in a range between 15% and 70% of the hours in a year might be referred to as an intermediate unit. The cost of operating an intermediate generating unit falls roughly in the middle of a range of operating costs for all of the generating units on the utility system. (Please see the explanations/definitions of "Baseload Generating Unit" and "Peaking Generating Unit.")

Kilowatt (kW): A kilowatt (kW) is a measure of electrical demand and 1 kW equals 1,000 watts (W).

Kilowatt-hour (kWh): A kilowatt-hour (kWh) is a measure of electrical energy and represents the equivalent of 1 kW of electricity being produced or consumed over 1 hour of time. For example, 1 kWh represents the amount of electricity used to light a 100-Watt light bulb for 10 hours (= 100 Watts × 10 hours = 1,000 Watt-hours or 1 kWh). A kWh is also a standard unit of electricity usage upon which electricity rates and bills are based.

Levelized Cost: This term refers to a proxy cost value. A levelized (or constant) cost value is the value that, when present valued each year and then summed, will result in the same cumulative present value as the cumulative present value of the sum of another group of non-constant annual cost values. (Please see Appendix C for a further explanation/discussion of this concept.)

Load Center (or Load Pocket): A geographical region of a utility's service territory or area in which a large portion of the utility's entire electric load exists.

Load Duration Curve: As used in this book, the term refers to a graph of the electric load for each of the 8,760 annual hours in a year. The hourly load values are presented either from the highest hourly load sequentially down to the lowest hourly load, or vice versa. (Please refer to the explanation/description of "Load Shape.")

Load Management Program: A load management program refers to one of two types of demand side management (DSM) options or programs that reduce the electricity usage of participating customers. (Please see the explanation/definition of "Demand Side Management.") A utility's load management programs typically fall into two categories. One category of load management program, typically referred to as a "load control" program, offers monthly incentives or credits on the electric bill to utility customers who voluntarily allow the utility to remotely control the operation of selected appliances/electrical equipment during times of high electrical demand on the utility system. The other category of load management programs typically offers electric rates in which the rate values differ either by the time of day or by the timing of the electrical peak load of the utility system. Electric rates such as these are designed to encourage customers to alter their electrical usage so that their usage is lower during times of high load on the utility system. (Load Management programs are also referred to as "demand response" programs.)

Load Shape: A load shape is a graph or chart that shows the total electrical demand from customers on a utility system for a selected period of time. In this book, a peak day load shape is used to show the hourly electrical demand for a 24-hour period. In addition, an annual load shape (or annual load duration curve) is also used to show the electrical demand for all hours over the course of a year ranging from highest load to lowest load (or vice versa).

Loss of Load Probability (LOLP): The term "Loss of Load Probability (LOLP)" refers to a reliability analysis approach that is designed to determine the probability that a utility will not have enough generating resources to meet all of it projected demand for electricity at any point during a year. This type of analysis is one of two basic types of "reliability analyses" that a utility may utilize as part of its resource planning process. (Please see the explanations/definitions of "Reliability Analysis" and "Reserve Margin.")

Megawatt (MW): A megawatt (MW) is a measure of electrical demand and represents 1,000 kW or 1,000,000 watts. This measurement is commonly used in a variety of references to utility systems, including the peak output of an electric generating unit and the amount of resources a utility needs to add in the future to maintain the reliability of its system.

Megawatt-hour (MWh): A megawatt-hour (MWh) is a measure of electrical energy and represents 1,000 kWh or 1,000,000 Watt-hours.

mmBTU: This term is a short-hand abbreviation for 1 million BTUs. (Please refer to the explanation/definition of "British Thermal Unit (BTU).")

Nameplate Capacity: In this book, the term "nameplate capacity" is being used to denote the projected maximum MW output of a generating unit based on a given set of conditions (such as air temperature, humidity, and type of fuel being used). This projected maximum output value is also referred in the book simply as "capacity."

Peaking Generating Unit: A peaking generating unit is a unit that is projected to operate only infrequently on a utility system. For example, a generating unit that is projected to operate 15% or less of the hours in a year is often referred to as a peaking unit. Peaking units are typically the generating units on a utility system with the highest operating costs. (Please see the explanations/definitions of "Baseload Generating Unit" and "Intermediate Generating Unit.")

Present Value: This term refers to the concept of the time value of money, *i.e.*, the relative value of money that is spent or received in a year (or years) different than the present year or some other year. Present valuing refers to a calculation method by which one can determine the relative value of money at different points in time. (Please see Appendix C for a further explanation/ discussion of this concept.)

Reliability Analysis: As used in this book, the term refers to a type of analysis that is used to answer two questions in regard to the continued ability of a utility system to reliably deliver electric service to its customers. The two questions are: (i) "When does the utility need to add new resources?", and (ii) "What is the magnitude (MW) of the resources the utility needs to add?" Reliability analyses are typically conducted using one or both of two criteria: a deterministic criterion such as a reserve margin criterion and/or a probabilistic criterion such as a LOLP criterion. (Please see the explanations/ definitions of "Reserve Margin" and "Loss of Load Probability.")

Reserves: The amount of generating capacity (MW) that a utility is projected to have above the utility's projected peak hour loads (MW) for summer and winter. The calculation of reserves is one of several steps in reserve margin analysis. (Please see the explanation/definition of "Reserve Margin.")

Reserve Margin: The term "Reserve Margin" refers to a reliability analysis calculation that is designed to show how much generating capacity a utility has compared to the projected peak summer and winter hourly loads for the utility. Reserve margin analyses can be of two types. The most common type is an analysis that fully accounts for the projected impacts from both Supply and DSM resources. This type of reserve margin analysis is also referred to as a "total" reserve margin analysis. The second type of reserve margin analysis is one in which the contribution of the utility's DSM resources is not accounted for. This type of reserve margin analysis is typically referred to as a "generation-only" reserve margin analysis and it is used to make it easier to see how much of a utility's reserves consist of Supply options. Overall, reserve margin analysis is one of two basic types of reliability analyses, along with LOLP analyses, that a utility may perform as part of its resource planning process. (Please see the explanations/ definitions of "Reliability Analysis" and "Loss of Load Probability.")

Resource Need: The term "Resource Need" refers to the magnitude (MW) of additional resources that are projected to be needed by a utility in a future year in order to maintain reliable electric service. The magnitude (and the timing) of a utility's resource need are determined through reliability analyses. (Please see the explanation/definition of "Reliability Analysis.")

Resource Option: This term refers to options that the utility system can choose from in order to meet a current or future resource need. There are two basic types of resource options: a Supply option (such as a new generating unit or a power purchase from another party) that allows the utility to provide more electricity, and a DSM option (such as an energy conservation program or a load management program) that changes the demand for, and usage of, electricity. (Please see the explanations/definitions of "Supply Option" and "Demand Side Management.")

Resource Planning: As used in this book, the term refers to a group of analyses that a utility performs to answer three fundamental questions regarding its projected resource need: (i) "When does the utility need to add new resources?", (ii) "What is the magnitude of the resources the utility needs to add?", and (iii) "What is the best resource option(s) with which to meet the projected resource need?"

Revenue Requirements: The term "Revenue Requirements" refers to the amount of revenues that are needed so that the utility can pay its bills and earn an authorized return for its investors. (Please refer to Appendix C for a further discussion of this concept.)

Round Trip Efficiency: The efficiency with which a battery is able to convert its stored energy into discharged electrical energy. The battery's round trip efficiency is usually discussed in terms of a percentage. For example, if a battery is able to store 100 units of electrical energy, then able to discharge 90 units of that energy, the battery is said to have a round trip efficiency of 90%.

Supply Options: This term refers to one of two types of resource options a utility can select to meet a future resource need. These two types of resource options are: Supply options (options that generate electricity) and DSM options (options that change the demand for, and usage of, electricity). Supply options can either be a generating unit owned by the utility or a purchase of power from another party's generating unit. Supply options can be conventional generating units (such as nuclear- or fossil-fueled generators) or renewable energy-sourced generators (such as solar- and wind-sourced generators).

Target Year: In this book, the term is used to denote the year in which a utility has committed to reaching its low- or zero-carbon goal.

Temperature (Dry Bulb): The temperature of the air measured by a thermometer freely exposed to the air, but which is shielded from moisture and radiation. Most utility load forecasts use dry-bulb temperature (DBT) as an important input.

Transition Period: The years from the time a utility makes a low- or zero-carbon goal announcement until the target year in which the goal is to be achieved.

Watt (W): This term refers to electrical demand. It is also the basic measure of the amount of electricity used in electrical equipment or devices (such as a 100-watt light bulb).

Appendix C: Mini-Lesson #1

Concepts of Revenue Requirements, Present Valuing of Costs and Discount Rates, Cumulative Present Value of Revenue Requirements, and Levelized Costs

When discussing resource planning for electric utilities, a number of economic and financial terms are invariably used. Among the most commonly used terms are the following:

1. Revenue requirements;
2. Present valuing of costs and discount rates;
3. Cumulative present value of revenue requirements (CPVRR); and
4. Levelized costs.

Each of these terms will be discussed in order to provide a basic understanding of the terms and how they are used throughout this book.

REVENUE REQUIREMENTS

The term "revenue requirements" is actually a very simple term. For a traditional regulated utility, it means the amount of revenues that are needed so that the utility can pay its bills and earn an authorized return for its investors. A list of the type of bills an electric utility has would include (but not be limited to) the following: purchase of fuels to use in power plants, purchase of new meters and transformers with which to serve new homes and new commercial buildings, replacement of old or damaged poles, salaries for its staff, maintenance for its existing power plants and service trucks, costs for new generation, transmission, and distribution facilities, payments to financial lending institutions for monies that have been borrowed to pay for large capital investments, tax payments, etc. In addition, utilities have financial obligations to their investors/shareholders. As a result, utilities strive to pay monies to these investors in order to give the investors a reasonable return on the money they have invested in the utility.

In other words, the term "revenue requirements" is simply the utility's total cost of doing business for a given time period. Another way to look at revenue requirements is that they are the total revenues that the utility requires to meet its financial obligations. In this book, we refer both to revenue requirements over the course of a year (annual revenue requirements) and over all of the years addressed in our analyses (cumulative revenue requirements).

These revenue requirements are "recovered" by a utility through the monthly bills its customers pay. These bills are based on electric rates that are set by a utility's regulatory body. These electric rates are basically set at a level(s) that, when applied to the utility's forecasted annual sales, results in the utility recovering the amount of money (*i.e.*, revenue requirements) that the regulatory body believes is necessary for the utility to pay its projected bills and its investors.

PRESENT VALUING OF COSTS AND DISCOUNT RATES

The term "present valuing" simply refers to a way to look at amounts of money received or spent in different years that allows one to consider these monies on a comparable basis. Most of us have heard the expression "the time value of money." This expression refers to the fact that one's view of a dollar available to us today is likely different than one's view of the same dollar that would be available to us in a future year.

One can spend a dollar today or invest it to receive more dollars in the future. The expected rate of return on this investment (after accounting for risk) is used to "present value" the future dollars so that a comparison to the dollar today can be made. In other words, the term, or concept, of "present valuing" is simply a way to compare the value of dollars that are available in different years on a common basis.

This comparison works by first selecting a specific point in time and then determining the values of monies in each different year relative to that specific point in time. This specific point in time is usually the current year (but it may be a prior or future year). In the example we will soon discuss, we will be present valuing monies back to the Current Year (just as we did in all of our analyses for our hypothetical utility system throughout this book).

The way that one then determines the values of monies in each different year (after accounting for risk) is by applying something called a "discount rate" to the monies spent or received in each different year. For purposes of discussing the example that follows, we will assume a discount rate of 8% per year.[1] (Note that this is the same discount rate we have used throughout this book in all of our discussions regarding economic analyses for our hypothetical utility system.)

Let us now look at a simple example of how the present value of an expenditure of $1 million, 5 years in the future is calculated. We do so in Table C.1.

---

[1] Discount rates can be developed in several ways. For utilities and other large companies, a discount rate is typically determined by its current after-tax cost of capital (which is developed using the percent of debt and equity instruments by which the utility raises capital, plus the respective costs of the debt and equity instruments after accounting for taxes). For individual utility customers, particularly residential customers, their discount rates would be developed in different ways. However, for purposes of this discussion, how one develops a discount rate is not as important as how a discount rate is utilized.

TABLE C.1

Present Valuing Example: Costs in one Year Only

(Discount Rate = 8.00%)

(1) Year	(2) Annual Discount Factor	(3) Expenditure (Nominal $)	(4) = (2) × (3) Expenditure (Present Value $)
Current Year	1.0000	$0	$0
Current Year + 1	0.9259	$0	$0
Current Year + 2	0.8573	$0	$0
Current Year + 3	0.7938	$0	$0
Current Year + 4	0.7350	$0	$0
Current Year + 5	0.6806	$1,000,000	$680,583

In this example, we see in column (1) that we are starting with the Current Year and then extending our view 5 years into the future. In Column (2), we have used an annual discount rate of 8% to calculate annual discount factors for each of the 6 years addressed in the table (*i.e.*, the Current Year plus 5 more years). Because we are present valuing back to the Current Year, there is no discounting of any expenditure that might have been incurred in the Current Year. Consequently, the annual discount factor for the Current Year is a value of 1.0000.

However, the annual discount factor for the next year (Current Year + 1) is 0.9259. This value is derived by dividing the previous year's annual discount factor (which is 1.0000 in this case) by a value of (1 + the discount rate), or 1.08 in this example. Therefore, the discount rate for Current Year + 1 is 0.9259 (= 1.0000/1.08).

Then, for Current Year + 2, the same calculation approach is used and the annual discount factor for Current Year + 2 is 0.8573 (= 0.9259/1.08). This calculation is then performed for all subsequent years through 5 years after the Current Year. In that last year, Current Year + 5, the annual discount factor is 0.6806 (= 0.7350/1.08).[2]

In column (3) of the table, we present an assumption of annual expenditures. This assumption for this simple example is that there will be no expenditures until 5 years after the Current Year. In Current Year + 5, an expenditure of $1,000,000 is made. (The term used in the subtitle of the column, "Nominal $," merely indicates the actual expenditures that are made in each year before making an attempt to account for the time value of money by present valuing the annual costs.)

In column (4), the annual discount factors in column (2) are multiplied by the annual expenditures in column (3). The resulting product of this multiplication is the present value of the annual expenditures back to the Current Year. We see that

---

[2] This example presents, and uses, an annual discount factor, shown in Column (2), in its calculation. Note that the use of present valuing formulae is also common in these types of calculations. In such cases, the annual discount factors are accounted for in formulaic calculations.

the present value of the $1,000,000 expenditure in Current Year + 5 is $680,583 in relation to the Current Year.[3]

The interpretation of this calculation is that, for whatever company or individual might be represented in this example, an amount of money equaling $1,000,000 spent 5 years from the Current Year is equivalent to $680,583 in the Current Year.

If the situation had been changed and, for example, there had been expenditures in each of the 6 years (from the Current Year through Current Year + 5) of $1,000,000, then the respective present value amount for each of the annual expenditures would be as presented in column (4) of Table C.2.

TABLE C.2
Present Valuing Example: Costs in All 6 Years
(Discount Rate = 8.00%)

(1)	(2)	(3)	(4) = (2) × (3)
Year	Annual Discount Factor	Annual Expenditure (Nominal $)	Annual Expenditure (Present Value $)
Current Year	1.0000	$1,000,000	$1,000,000
Current Year + 1	0.9259	$1,000,000	$925,926
Current Year + 2	0.8573	$1,000,000	$857,339
Current Year + 3	0.7938	$1,000,000	$793,832
Current Year + 4	0.7350	$1,000,000	$735,030
Current Year + 5	0.6806	$1,000,000	$680,583

As one might expect from our discussion so far, the present value amount of annual expenditure of $1,000,000 becomes less (by 8% which is the discount rate) in each year as one moves further away from the Current Year.

CUMULATIVE PRESENT VALUE OF REVENUE REQUIREMENTS

We now combine the two terms or concepts of revenue requirements and present valuing to discuss the term "cumulative present value of revenue requirements" or its abbreviation, "CPVRR." The term simply means the sum of the annual revenue requirements for a utility when viewed from a present value perspective. A simple example should help illustrate this.

We will use Table C.2 as a starting point and then make a couple of changes. First, we change the column titles for columns (3) and (4) from "Expenditures" to "Revenue Requirements" so that the example denotes the annual required revenues for a utility. Second, we sum the annual nominal and present value revenue requirements in these same two columns and show those sums at the bottom of each column. The result is presented in Table C.3.

---

[3] The fact that the $680,583 value represents a present value number is often abbreviated as $680,583 (NPV$) in which the "NPV" stands for net present value.

TABLE C.3
Cumulative Present Value of Revenue Requirements' (CPVRR) Example (Discount Rate = 8.00%)

(1)	(2)	(3)	(4) = (2) × (3)
Year	Annual Discount Factor	Annual Revenue Requirements (Nominal $)	Annual Revenue Requirements (Present Value $)
Current Year	1.0000	$1,000,000	$1,000,000
Current Year + 1	0.9259	$1,000,000	$925,926
Current Year + 2	0.8573	$1,000,000	$857,339
Current Year + 3	0.7938	$1,000,000	$793,832
Current Year + 4	0.7350	$1,000,000	$735,030
Current Year + 5	0.6806	$1,000,000	$680,583
Sums =		$6,000,000	$4,992,710
		CPVRR =	$4,992,710

Column (3) of this table shows that this utility has annual revenue requirements of $1,000,000.[4] When these nominal annual revenue requirements are summed over the 6-year period, the total nominal annual revenue requirements over the 6-year period is $6,000,000. Column (4) presents the associated present value amount of the required revenues for each year. The sum of these present valued amounts is $4,992,710. As shown in the last line of the table, this sum of the annual present valued revenue requirements is called the cumulative present value of revenue requirements or CPVRR.

At this point, before we move on to discuss our next financial/economic term; it may be useful to remind ourselves of how we use a projected CPVRR value for a utility in resource planning work for that utility. Let's assume that the utility is examining two Supply options. We have seen from the discussions in the book that the introduction of a particular Supply option onto a utility system will have numerous cost impacts on the system.

The cost impacts to the utility system from either of the Supply options are likely to be different from year-to-year. Consequently, the utility's projected annual revenue requirements if the first Supply option is selected are likely to be different from the projected annual revenue requirements if the second Supply option is selected. Converting the two different "streams" of annual revenue requirements into annual present valued revenue requirements, then summing the present valued costs, results in two projected CPVRR values for the utility system, one for each of the two Supply options.

---

[4] In reality, it is highly unlikely that any utility will have exactly the same amount of expenditures (or revenue requirements) from one year to the next. The annual revenue requirements are assumed to be identical in the example solely to simplify our discussion.

The Supply option that results in the utility's lowest projected CPVRR value is the more economic Supply option.[5]

LEVELIZED COSTS

The last term or concept we will discuss in this Appendix is "levelized cost." As discussed in various places in the book, levelized values can be used in utility resource planning in two ways. First, levelized costs can be used in preliminary economic screening analyses of Supply options (if the Supply options are identical, or at least very similar, in regard to four key characteristics) in an analytical approach called a "screening curve" approach.[6] This form of preliminary economic screening analysis produces a single levelized cost value, usually in terms of $ per MWh or cents per kWh. Second, levelized values can be used in the process of developing a levelized electric rate in the final (or system) economic analyses. The result is a single levelized electric rate value, usually in terms of cents per kWh.

Perhaps the best way to explain how a levelized cost value is developed is to show an example of the calculation. We will do so by using Table C.3 as a starting point. We will then change the values in column (3) and add two new columns, columns (5) and (6). The result is Table C.4.

TABLE C.4
Levelized Cost Example (Discount Rate = 8.00%)

(1)	(2)	(3)	(4) = (2) × (3)	(5)	(6) = (2) × (5)
Year	Annual Discount Factor	Actual Revenue Requirements (Nominal $)	Actual Revenue Requirements (Present Value $)	Levelized Revenue Requirements (Nominal $)	Levelized Revenue Requirements (Present Value $)
Current Year	1.0000	$1,000,000	$1,000,000	$1,258,782	$1,258,782
Current Year + 1	0.9259	$1,100,000	$1,018,519	$1,258,782	$1,165,539
Current Year + 2	0.8573	$1,210,000	$1,037,380	$1,258,782	$1,079,202
Current Year + 3	0.7938	$1,331,000	$1,056,591	$1,258,782	$999,261
Current Year + 4	0.7350	$1,464,100	$1,076,157	$1,258,782	$925,242
Current Year + 5	0.6806	$1,610,510	$1,096,086	$1,258,782	$856,706
		CPVRR =	6,284,732		6,284,732

[5] And we know from our discussions in the book that the Supply option with the lowest projected CPVRR value will also be the Supply option that provides the lowest present valued electric rates for the utility's customers. This is because the CPVRR costs will be divided by the same number of GWh of sales by the utility. We also know that if either of the Supply options is to be compared to a DSM option, the economic analysis must be carried out a step further to actually determine the projected electric rates for both options because the GWh sales values over which the costs will be recovered will differ between the Supply and DSM options.

[6] Please see Chapters 3 and 5, plus Appendix D, for a further discussion of the use of a screening curve approach in preliminary economic screening analyses, and of the severe limitations to using such an analytical approach.

From looking at the values in column (3) in this table, it is obvious that we have changed the projected nominal annual revenue requirements from the identical value of $1,000,000 each year that was assumed previously. We have kept the $1,000,000 value for the Current Year, then we have escalated the annual nominal revenue requirement by 10% each year as shown in column (3).

This automatically changes the present valued annual revenue requirements shown in column (4). The new projected CPVRR value for these changed annual revenue requirement stream is $6,284,732 as shown at the bottom of column (4).

We then determine a single, constant annual revenue requirement value that when applied, without escalation, for each year, results in the same $6,284,732 CPVRR value as determined in column (4). This value can be determined through a simple iterative process: (i) plug a number into column (5) and see whether the resulting CPVRR value in column (6) is too low or too high compared to the $6,284,732 value at the bottom of column (4); then (ii) continue to alter the original plugged-in number until the CPVRR values at the bottom of columns (4) and (6) match. (One can also more efficiently use the "goal seek" function that is commonly found on commercially available spreadsheets.)

The concept behind a levelized cost is to find a single cost value (*i.e.*, a cost value that remains constant or "level") that, when present valued and summed for all years, results in the same projected CPVRR value as the actual annual revenue requirements (which typically do change each year).[7]

Therefore, in our example, we wish to find a levelized annual cost value that is projected to result in the same $6,284,732 CPVRR value as the original annual revenue requirements. We do so through the use of two new columns in this table. Column (5) presents the levelized annual cost value and column (6) is used to calculate the associated annual present values, then to sum up these values to produce a CPVRR that corresponds to the levelized annual cost.

As shown in columns (5) and (6), the levelized annual cost value in this example is $1,258,782 because it produces a CPVRR value that is identical to that for the original annual revenue requirements, $6,284,732. The levelized value of $1,258,782 is shaded in the first row of Column (5).

In this way, or by using the more sophisticated goal seek formula that performs the same basic calculation, but does not require the use of the two additional columns that are shown here, one calculates levelized cost values that can be used in preliminary economic screening analyses of Supply options (in the relatively few cases where it is appropriate to do so, *i.e.*, where all four key characteristics, as explained elsewhere in the book, of the competing Supply options are identical or very similar).

In Appendix D, we further discuss the limitations of a screening curve approach when performing preliminary economic screening analyses.

---

[7] And, as demonstrated in Chapters 5 and 6 of this book, the same basic approach can also be used to compute a levelized electric rate.

Appendix D: Mini-Lesson #2

Further Discussion of the Limitations of a Screening Curve Analytic Approach

The topic of economic analyses of resource options is discussed in numerous places in this book. In those discussions, particularly in Chapters 3, 5, and 6, it was pointed out that there are two basic types or stages of economic analyses: preliminary economic screening analyses and final (or system) economic analyses.

For preliminary economic screening analysis of Supply options, the most commonly used approach is a "screening curve" approach. The screening curve approach is also referred to as a "levelized cost of electricity (LCOE)" approach. This analysis approach is simple to perform using a spreadsheet. In comparison, a final (or system) economic analysis uses a much more comprehensive and accurate approach using sophisticated computer models that can account for all of the economic and non-economic impacts on a utility system that the addition of a resource option will have.

Assuming the analysis is being applied to a Supply option (generating unit), the basic differences between these two types of economic analyses can be summarized as follows:

1. Preliminary economic screening analyses using a "screening curve" approach:
   a. Takes the perspective of a generating unit sitting alone in a field, completely unconnected to the utility system for which it will be a part of (if the generating unit in question is actually selected by the utility);
   b. Does <u>not</u> account for a number of key economic impacts on the utility system which will result if this generating unit is actually placed into the utility system;
   c. Can be useful, but only under *very limited* circumstances, to quickly eliminate or "screen out" the least economic Supply options from a list of Supply options that are identical, or very similar, in regard to four key characteristics, but do not provide meaningful results if the Supply options being compared are dissimilar in one or more of four key characteristics; and
   d. Should *never* be used to make a final decision regarding the selection of a resource option.

2. Final (or system) economic analyses:
   a. Takes the perspective of the entire utility system, including the generating unit in question (if the generating unit in question is actually selected by the utility);

b. <u>Does</u> account for all key economic impacts on the utility system which will result if this generating unit is actually placed into the utility system;

c. Always provides meaningful results for a resource option decision; and

d. Should *always* be used to make a final decision regarding the selection of a resource option.

Chapter 5 presented this summary level information, plus additional text that explained the severe limitations of a "screening curve" analysis approach in more detail.

We return to that discussion in this mini-lesson (which will actually turn into a bit longer than a mini-discussion). We shall first discuss the limitations of a screening curve analysis approach qualitatively. This qualitative discussion is a "lead in" to a quantitative discussion that uses an example of a screening curve analysis of a new CC generating unit. This quantitative example shows how misleading the results of a screening curve analysis can be.

QUALITATIVE DISCUSSION

As previously mentioned, the usefulness of a screening curve analysis approach is actually very limited. This approach can be used in a meaningful way to compare the economics of two competing resource options that are identical, or at least very similar, in regard to the following four key characteristics: (i) capacity (MW), (ii) annual capacity factors, (iii) the percentage of the option's capacity (MW) that can be considered as firm capacity at the utility's system peak hours, and (iv) the projected life of the option. If two resource options are identical, or very similar, in regard to each of these four key characteristics, then a screening curve analysis can be meaningful in performing preliminary analyses to "screen out" the less attractive of these two very similar options. (This leads to the common terminology of this type of analysis as a "screening curve" analysis.)

However, a screening curve analytical approach that attempts to compare resource options that are <u>not</u> identical, or very similar, in all of these four key characteristics will produce incomplete results that are of little value. Indeed, the less comparable any of these four characteristics are for the resource options being analyzed, the less meaningful are the results. For example, because a DSM measure and a CC generating unit are about as different in terms of resource options as one can get, a screening curve approach that attempts to compare these types of resource options provides meaningless results.

The reason that a screening curve approach can only provide meaningful results under very limited circumstances is because a typical screening curve analysis approach does not address numerous economic impacts that these resource options will have on the utility system as a whole. Instead, a screening curve approach merely looks at the cost of operating the individual option itself. One can think of a screening curve analysis as examining the costs of a resource option if it were placed out in an open field by itself and operated without its operation having any impact on the rest of the utility system (or if it were the only generating unit the utility has). The numerous impacts an individual resource option has on the utility system—for example, how it impacts the operation of all the other generating units on the system—are ignored in a typical screening curve approach.

However, the system impacts of any resource option are very large and can result in significant system costs, or system cost savings, that must be accounted for in order to have a complete picture of the total net system costs with any resource option. Any analytical approach, such as a screening curve approach, that ignores system cost impacts can only provide an incomplete, and, therefore incorrect, result.

This is true regardless of whether only one resource option is being evaluated, or more than one resource option is being evaluated and the evaluation results for each resource option are being compared to each other.

LIMITATIONS WHEN ANALYZING ONE RESOURCE OPTION

We shall first look at the limitations one runs into when using a screening curve approach to analyze a single resource option. Let's assume that the resource option in question is a CC unit. In a screening curve analysis, one assumes that this generating unit will operate at a particular capacity factor or within a range of capacity factors. For purposes of this discussion, we will assume the generating unit operates 90% of the hours in a year.[1] Then, using the generating unit's capacity and heat rate, plus the projected cost of the fuel the generating unit would burn, the annual fuel cost of operating the generating unit for 90% of the hours in a year is calculated. This calculation is then repeated for each year addressed in the screening curve analysis.

In a screening curve analysis, the unit's annual fuel costs—which will be very large for a generating unit with a 90% capacity factor—are added to all of the other costs (capital, O&M, etc.) of building and operating this individual generating unit. The present value total of these costs is then used to develop the levelized $ per MWh or cents per kWh cost of building and operating this CC unit.

However, the screening curve analysis approach does not take into account the fact that this new generating unit would not operate on a utility system at 90% of the hours in a year if it was not cheaper to operate this new unit than to operate other existing generating units on the system. In other words, for almost every hour the new generating unit operates, the MWh it produces displace more expensive MWh that would otherwise have been produced by the utility's existing generating units. Whatever the annual fuel cost is of operating this new generating unit 90% of the hours in a year, the utility will save an even greater amount of system fuel costs by reducing the operation of more expensive existing generating units during these hours.

For example, let's assume that the new generating unit's annual fuel cost would be $100 million, but that the operation of this new unit will also result in an annual savings of $110 million in fuel costs from reduced operation of the system's more expensive existing units. A typical screening curve analysis will include the $100 million cost for

---

[1] In this book, we used an 80% capacity factor for the CC options our hypothetical utility system was considering. We now use a 90% capacity factor assumption for this immediate discussion just to see if you are paying attention. However, for your viewing comfort, we shall return to an assumed 80% capacity factor when we get to the quantitative discussion.

operating the individual unit, but it <u>will ignore the $110 million in system fuel savings</u> that will also occur from not having to operate more expensive generating units.

For this reason, a typical screening curve analysis approach utilizes an incomplete set of information and, therefore, is an incorrect way to thoroughly analyze resource options. A complete analytical approach would take into account the total system fuel cost impact which is a net system fuel <u>savings</u> of $10 million (= $110 million in system fuel savings − $100 million in unit fuel cost) instead of only the $100 million fuel expense of operating only the individual CC unit. Consequently, a typical screening curve analysis will grossly overstate the actual net system fuel cost impact of the new CC unit.

In a similar fashion, other system cost impacts, such as system environmental compliance costs and system variable O&M, are not accounted for in typical screening curve analyses because this analytical approach does not take into account the fact that the new generating unit will reduce the operating hours, and the associated costs, of the utility's existing generating units.

Nor does a typical screening curve approach account for the impact the resource option will have in regard to meeting the utility's future resource needs. Therefore, the screening curve approach utilizes incomplete information for a number of cost categories, thus providing incomplete, and incorrect, results.

The preceding discussion discussed how a screening curve analytical approach utilizes incomplete information, thus leading to an incomplete accounting of all system cost impacts that would result from analyzing even a single new resource option. One might ask: *"Is a screening curve approach even more problematic when attempting to compare two or more different types of resource options?"*

The question is a good one and the answer is "yes." We next offer a qualitative discussion that looks at problems that occur when attempting to analyze several different types of resource options using a screening curve approach. However, we will first take a look at two resource options of the same type, but which differ in only one of the four key characteristics.

LIMITATIONS WHEN ANALYZING MORE THAN ONE RESOURCE OPTION

Now we shall assume that a screening curve approach is used in an attempt to compare the economics of three different CC-generating options:

1. Combined cycle Option A (1,000 MW);
2. Combined cycle Option B (1,000 MW);
3. Combined cycle Option C (500 MW).

Just as we did in Chapter 5, let's assume that the first comparison attempted is of two virtually identical CC units, CC Options A and B, in which the four key characteristics of these two CC units are essentially identical. In fact, let's assume that the two CC options are identical in all respects except one: the capital cost of CC Option A is lower by $1 million than the capital cost of CC Option B.

In this comparison, even though a screening curve analysis will not provide an accurate system net cost value (as per the just concluded discussion), a screening curve comparison can provide meaningful results in regard to a comparison of these two 1,000 MW CC options, CC Option A and CC Option B. This is because the impacts on the operation of existing generating units on the system will be identical from two CC units that are the same in regard to the following: (i) capacity (1,000 MW), (ii) capacity factor, (iii) the amount of firm capacity (1,000 MW) each unit will provide, and (iv) the life of the two units. For this reason, a screening curve analysis will provide a meaningful comparison of these two options.

In other words, these two Supply options are identical in regard to each of the four key characteristics that must be identical, or very similar, in order for a screening curve analysis approach to be able to provide meaningful results. What this means is that, even though the results will not be accurate from a system cost perspective for either of the two options, the results will be "off" by the same amount and in the same direction. This allows a screening curve approach to provide a meaningful comparison between these two very similar options.

As would be expected, a screening curve comparison would show that CC Option A results in a slightly lower $ per MWh (or cents per kWh) value for CC Option A compared to CC Option B due to its $1 million lower capital costs. As a result, CC Option B can be screened out and not analyzed further.

As this example shows, a screening curve analytical approach can produce meaningful results in a case in which each of the four key characteristics of resource options are identical or very similar. However, as the ongoing discussion will show, once any of these four characteristics for competing resource options are no longer comparable, a typical screening curve approach cannot produce meaningful results. We now look at the two remaining CC units: CC Option A (which survived its initial screening versus CC Option B) and CC Option C. These units are identical in regard to three of the four characteristics, but differ in regard to the other characteristic: capacity (MW).[1]

The capacities of these two remaining CC options are significantly different: 1,000 MW for CC Option A and 500 MW for Option C. Although the other three key characteristics for the two units are identical (capacity factor, percentage of capacity that is firm capacity, and life of the units), the significant difference in capacity offered by the two options would cause a screening curve approach to yield incomplete, and therefore incorrect, results.

The capacity difference between these options would result in at least two system impacts that are not captured by a typical screening curve approach. The first of these is the impact of each of the two CC options on the utility's future resource needs. The 1,000 MW of CC Option A will address the utility's future resource needs twice as much as will the 500 MW of CC Option C. In other words, CC Option A will avoid/defer future resource additions to a greater extent than will CC Option C.

This will show up in a system cost analysis in the form of different system capital, fuel, O&M, environmental compliance, etc. costs beginning at in the next future

---

[1] In order to keep the discussion as simple as possible, we will ignore any total installed cost differences between the two CC options.

year in which the utility would again need new resources if it selects the smaller CC Option. For example, assume that the selection of the smaller CC Option C would meet the utility's resource needs for 3 years, but the selection of the larger CC Option A would meet the utility's resource needs for 6 years. Therefore, starting 4 years in the future from the in-service date of either CC option, the utility's projected costs would be different depending upon whether the larger CC Option A, or the smaller CC Option C, had been selected. In the case of a selection of CC Option C, the utility would need to add another resource after 3 years. However, in the case of a selection of CC Option A, no other resource additions would be needed until after 6 years.

In addition, even prior to that point in the future when other new resources are needed, the 500 MW difference in capacity between the two CC options will result in different system total costs for fuel, variable O&M, and environmental compliance. In other words, the operation of the utility's existing generating units is impacted to a greater extent by the larger CC Option A than by the smaller CC Option C. This is because, assuming all else equal, the addition of a 1,000 MW generating unit will affect the operation of the utility's existing generating units more than the addition of a 500 MW unit will.

None of these system economic impacts that are driven by the difference in the capacity of two competing resource options are captured in a typical screening curve approach. The earlier discussion pointed out that a screening curve approach applied to even a single new resource option will omit a variety of significant system cost information that is necessary to develop a complete cost perspective of the one resource option. Now we see that an attempt to use a screening curve approach to compare the economics of two resource options that differ significantly even in one of the four key characteristics (*i.e.*, in their capacity) will omit an even greater amount of important system cost information. Therefore, the use of a screening curve approach is definitely flawed when used to compare two new resource options that differ in even one of the four key characteristics.

From this discussion, one can see these problems will also exist if two different types of resource options (for example, a 1,000 MW CC option and a 160 MW CT option) are evaluated using the fundamentally flawed approach of a screening curve.

LIMITATIONS WHEN ATTEMPTING TO ANALYZE FUNDAMENTALLY DIFFERENT TYPES OF RESOURCE OPTIONS (*i.e.*, SUPPLY AND DSM OPTIONS)

Because the previous examples discussed only Supply options, one might ask the following question: *"Do similar problems exist if one were to attempt to compare DSM options to Supply options using a screening curve approach?"* In other words, does the fact that Supply and DSM options are fundamentally different in various characteristics also lead to limitations in the usefulness of a screening curve approach to compare such options?

The answer is a resounding "yes." All of the problems inherent in using a screening curve approach that omits the system cost impacts discussed earlier are equally applicable whether Supply or DSM options are being addressed. For discussion purposes, let's assume that the utility is also considering a DSM option which will reduce peak load by 100 MW.

In this example, the system impacts of the lower amount of DSM (100 MW) on future resource needs compared to the 1,000 MW CC Option A would not be captured in typical screening curve analyses. This would lead to the same type of incomplete and incorrect analysis discussed previously regarding a comparison of a 1,000 MW CC unit and a 500 MW CC unit. Even if one were to adjust the 100 MW of demand reduction from DSM to account for the fact that 100 MW of DSM would be equivalent to 120 MW of Supply option capacity (assuming the utility had a 20% reserve margin criterion), 120 MW of the DSM option will have much different impacts on the utility system than will a 1,000 MW Supply option.

In addition, DSM options vary widely in terms of their actual contribution during system peak hours. In other words, would a DSM option(s) really contribute 100 MW of demand reduction at the utility's peak hour? The use of a screening curve approach that produces results in terms of $ per MWh or cents per kWh makes it easy to gloss over what the real contribution of specific DSM options may be in regard to reducing the utility's demand at its peak hour.

Many DSM programs are projected to reduce demand during a utility's summer and winter peak hours. Examples of such DSM programs include the following: load control, building envelope improvement, and heating/ventilation/air conditioning (HVAC) programs. However, other DSM programs may contribute little or no demand reduction at the utility's summer peak hour, at its winter peak hour, or at either peak hour. A streetlight program that addressed lighting systems that operate only at night would be an example of such a program because utility systems usually do not experience their system peak hours at night.

Furthermore, at the time this book is written, attempts by various parties to analyze DSM options using a screening curve analytical approach have lumped a wide variety of DSM options together regardless of the capability of these DSM options to lower demand at a utility's peak hour. No distinction is typically made regarding what impacts the various DSM options will really have in reducing demand at the utility's peak hour. This is simply another reason why attempting to apply a screening curve analysis approach to compare DSM and Supply options can only provide incomplete, and therefore meaningless, results.

QUANTITATIVE DISCUSSION

This discussion to this point has been a qualitative one. We shall now take a quantitative look at the limitations of a typical screening curve analysis approach. We first take a look at what a typical screening curve analysis might show for a new CC unit. As is true with all typical screening curve analyses, a number of system impacts that would occur if the new CC unit were to be placed on the utility system are simply not accounted for by this type of preliminary analysis approach for the new CC unit.

Second, we shall take a look at how significantly the results of a typical screening curve approach can change even if one were to account for only two of the many system impacts that are unaccounted for.

Table D.1 is an example of what the summary page of a typical screening curve analysis might be for a 1,000 MW CC unit that is going in-service 5 years from the Current Year. In this analysis, it is assumed that the CC unit will have an economic

(This page/space is intentionally left blank.)

TABLE D.1
Typical Screening Curve Results for a CC Unit: With No System Impacts

(1) Capacity Factor (%)	(2) Levelized Cost ($/kW)	(3) Levelized Cost ($/MWh)	(4) Year	(5) Capital $000	(6) Fixed O&M $000	(7) Capital Replacement $000	(8) Firm Gas Transport $000	(9) NO_x Emission $000	(10) SO_2 Emission $000	(11) CO_2 Emission $000	(12) Fuel Costs $000	(13) Variable O&M $000	(14) Total $000
0	223	0	Current Year (CY)	0	0	0	0	0	0	0	0	0	0
5	253	579	CY + 1	0	0	0	0	0	0	0	0	0	0
10	284	324	CY + 2	0	0	0	0	0	0	0	0	0	0
15	314	239	CY + 3	0	0	0	0	0	0	0	0	0	0
20	345	197	CY + 4	0	0	0	0	0	0	0	0	0	0
25	375	171	CY + 5	95,000	3,000	5,000	38,325	32	0.1	33,822	191,501	438	367,118
30	405	154	CY + 6	91,200	3,060	5,100	38,325	33	0.1	38,896	195,331	447	372,391
35	436	142	CY + 7	87,400	3,121	5,202	38,325	33	0.1	43,969	199,237	456	377,744
40	466	133	CY + 8	83,600	3,184	5,306	38,325	34	0.1	49,042	203,222	465	383,178
45	496	126	CY + 9	79,800	3,247	5,412	38,325	35	0.1	54,116	207,286	474	388,695
50	527	120	CY + 10	76,000	3,312	5,520	38,325	35	0.1	59,189	211,432	484	394,298
55	557	116	CY + 11	72,200	3,378	5,631	38,325	36	0.1	64,262	215,661	493	399,987
60	587	112	CY + 12	68,400	3,446	5,743	38,325	37	0.1	69,336	219,974	503	405,764
65	618	108	CY + 13	64,600	3,515	5,858	38,325	37	0.1	74,409	224,373	513	411,632
70	648	106	CY + 14	60,800	3,585	5,975	38,325	38	0.1	79,483	228,861	523	417,591
75	678	103	CY + 15	57,000	3,657	6,095	38,325	39	0.1	84,556	233,438	534	423,644
80	709	101	CY + 16	53,200	3,730	6,217	38,325	40	0.1	89,629	238,107	545	429,793

85	99	739								
90	98	769								
95	96	800								
100	95	830								

CY + 17	49,400	3,805	6,341	38,325	40	0.1	94,703	242,869	555	436,039
CY + 18	45,600	3,881	6,468	38,325	41	0.1	99,776	247,726	567	442,384
CY + 19	41,800	3,958	6,597	38,325	42	0.1	104,849	252,681	578	448,831
CY + 20	38,000	4,038	6,729	38,325	43	0.1	109,923	257,735	589	455,382
CY + 21	34,200	4,118	6,864	38,325	44	0.1	114,996	262,889	601	462,038
CY + 22	30,400	4,201	7,001	38,325	45	0.1	120,069	268,147	613	468,802
CY + 23	26,600	4,285	7,141	38,325	46	0.1	125,143	273,510	626	475,675
CY + 24	22,800	4,370	7,284	38,325	46	0.1	130,216	278,980	638	482,661
CY + 25	19,000	4,458	7,430	38,325	47	0.1	135,289	284,560	651	489,760
CY + 26	15,200	4,547	7,578	38,325	48	0.1	140,363	290,251	664	496,977
CY + 27	11,400	4,638	7,730	38,325	49	0.1	145,436	296,056	677	504,312
CY + 28	7,600	4,731	7,884	38,325	50	0.2	150,510	301,977	691	511,768
CY + 29	3,800	4,825	8,042	38,325	51	0.2	155,583	308,017	704	519,348
NPV (to in-service year) costs =	734,884	41,064	68,440	441,840	437	1	871,028	2,621,252	5,995	4,784,941
NPV $/kW at 100% Capacity Factor =	127	7	12	77	0	0	151	455	1	830

life of 25 years and that the discount rate is 8% (the same assumptions that have been used in the economic analyses throughout this book).

In Table D.1, the projected annual costs by the different cost categories, assuming the generating unit operates at a capacity factor of 100%, are shown in Columns (5) through (14). The net present value (NPV) total costs by category are shown on the next-to-last row of the table. The final row on the table shows the same NPV costs after converting them to a $ per kW value (again assuming a 100% capacity factor). These values are then used in the summary box that appears on the left-hand side in Columns (1) through (3).

In that box, Column (1) lists possible capacity factors the CC unit might operate at in increments of 5%. Column (2) offers a "demand" perspective by presenting the projected levelized costs over the 25-year period in terms of $ per kW values for the associated capacity factor level. In addition, Column (3) offers an "energy" perspective by presenting the projected levelized cost values in terms of $ per MWh values for the associated capacity factor level.

We see from the highlighted/shaded row in Columns (1) through (3) of the box that a typical screening curve analysis projects that the levelized cost for this CC unit at an 80% capacity factor (i.e., the same capacity factor that we assumed for the CC options in the economic analyses for our utility system throughout this book) is $101 per MWh. (Or, converting this value to a cents per kWh value, the projected cost is 10.1 cents/kWh.)[2]

The results of this typical screening curve analysis tell us that this CC unit is projected to produce electricity, assuming it operates at an 80% capacity factor, at $101 per MWh. However, recall that the premise of a typical screening curve analysis is that the CC unit is assumed to operate without any connection to the utility system as a whole, i.e., the generating unit is operating out in a field by itself with no connection to the utility system. Therefore, a typical screening curve calculation does not account for a number of system cost impacts that would actually occur if the CC unit were to be placed in-service on a utility system.

We shall now see how much this result from a typical screening curve analysis might change if only a couple of these system impacts were to be accounted for. We will account for only two of the many system impacts using the results from Table D.1 as a starting point.

The conservative assumption that we shall use is that both the system net fuel cost savings and the system net environmental compliance cost savings will be 10% of the CC unit's costs in those categories. For example, the fuel cost value for the new CC unit for the in-service year (Current Year + 5) shown in Column (12) in Table D.1

---

[2] This might be a good time to point out another problem with screening curve analyses: there are different ways to calculate the levelized number for a given capacity factor after one has calculated the annual costs. For example, one can present value back to the current year or to the in-service year if those two years are different. (Tables D.1 and D.2 utilized the in-service year.) That choice will give widely differing values that can make comparing results from one person's screening curve analysis to results from another person's screening curve analysis even less meaningful (if such a thing is possible with a screening curve analytical approach). The author has identified at least four different ways in which levelized costs in screening curve analyses are currently being calculated at the time the second edition of this book is written. Comparing results that have used different calculation methods is a meaningless exercise.

is $191,501 (in $000). The new assumption used is that the utility system would actually realize a net savings of 10% of the fuel cost used by the new CC unit due to the new CC unit reducing the operation of existing, more expensive generating units. Thus, there will be a savings in fuel costs for the remaining generating units on the utility system of $210,651 ($000) (= 1.10 × $191,501 ($000)).

Consequently, a net system fuel *savings* of $19,150 ($000) (= $210,651 ($000) in fuel savings from reduced operation of the utility's existing generating units minus $191,501 ($000) in fuel cost from operating the new generating unit) would occur. A similar calculation is made for all years for the fuel costs and for the environmental compliance costs. None of the values in the other columns in Table D.1 are changed as a result of these assumptions for system fuel and environmental compliance cost savings.

The results of these assumption changes are presented in Table D.2 that is presented on the next two pages. The resulting fuel cost impacts are presented in Column (12) and the resulting environmental compliance cost impacts are presented in Columns (9) (10), and (11). The negative values in those columns represent cost savings.

TABLE D.2
Typical Screening Curve Results for a CC Unit: With Two System Impacts

(1) Capacity Factor (%)	(2) Levelized Cost ($/kW)	(3) Levelized Cost ($/MWh)	(4) Year	(5) Capital $000	(6) Fixed O&M $000	(7) Capital Repl. $000	(8) Firm Gas Transport $000	(9) NO_x Emission $000	(10) SO_2 Emission $000	(11) CO_2 Emission $000	(12) Fuel Costs $000	(13) Variable O&M $000	(14) Total $000
0	223	0	Current Year (CY)	0	0	0	0	0	0	0	0	0	0
5	220	503	CY + 1	0	0	0	0	0	0	0	0	0	0
10	217	248	CY + 2	0	0	0	0	0	0	0	0	0	0
15	214	163	CY + 3	0	0	0	0	0	0	0	0	0	0
20	211	121	CY + 4	0	0	0	0	0	0	0	0	0	0
25	208	95	CY + 5	95,000	3,000	5,000	38,325	(3)	(0)	(3,382)	(19,150)	438	119,228
30	205	78	CY + 6	91,200	3,060	5,100	38,325	(3)	(0)	(3,890)	(19,533)	447	114,706
35	202	66	CY + 7	87,400	3,121	5,202	38,325	(3)	(0)	(4,397)	(19,924)	456	110,180
40	199	57	CY + 8	83,600	3,184	5,306	38,325	(3)	(0)	(4,904)	(20,322)	465	105,650
45	196	50	CY + 9	79,800	3,247	5,412	38,325	(3)	(0)	(5,412)	(20,729)	474	101,115
50	193	44	CY + 10	76,000	3,312	5,520	38,325	(4)	(0)	(5,919)	(21,143)	484	96,576
55	190	40	CY + 11	72,200	3,378	5,631	38,325	(4)	(0)	(6,426)	(21,566)	493	92,032
60	187	36	CY + 12	68,400	3,446	5,743	38,325	(4)	(0)	(6,934)	(21,997)	503	87,483
65	184	32	CY + 13	64,600	3,515	5,858	38,325	(4)	(0)	(7,441)	(22,437)	513	82,929
70	181	30	CY + 14	60,800	3,585	5,975	38,325	(4)	(0)	(7,948)	(22,886)	523	78,371
75	178	27	CY + 15	57,000	3,657	6,095	38,325	(4)	(0)	(8,456)	(23,344)	534	73,808
80	175	25	CY + 16	53,200	3,730	6,217	38,325	(4)	(0)	(8,963)	(23,811)	545	69,239

85	173	23
90	170	22
95	167	20
100	164	19

CY + 17	49,400	3,805	6,341	38,325	(4)	(9,470)	0	555	64,665
CY + 18	45,600	3,881	6,468	38,325	(4)	(9,978)	0	567	60,086
CY + 19	41,800	3,958	6,597	38,325	(4)	(10,485)	0	578	55,502
CY + 20	38,000	4,038	6,729	38,325	(4)	(10,992)	0	589	50,911
CY + 21	34,200	4,118	6,864	38,325	(4)	(11,500)	0	601	46,316
CY + 22	30,400	4,201	7,001	38,325	(4)	(12,007)	0	613	41,714
CY + 23	26,600	4,285	7,141	38,325	(5)	(12,514)	0	626	37,107
CY + 24	22,800	4,370	7,284	38,325	(5)	(13,022)	0	638	32,493
CY + 25	19,000	4,458	7,430	38,325	(5)	(13,529)	0	651	27,874
CY + 26	15,200	4,547	7,578	38,325	(5)	(14,036)	0	664	23,248
CY + 27	11,400	4,638	7,730	38,325	(5)	(14,544)	0	677	18,616
CY + 28	7,600	4,731	7,884	38,325	(5)	(15,051)	0	691	13,977
CY + 29	3,800	4,825	8,042	38,325	(5)	(15,558)	0	704	9,332
NPV total costs =	734,884	41,064	68,440	441,840	(44)	(87,103)	0	5,995	942,951
NPV $/kW at 100% Capacity Factor =	127	7	12	77	(0)	(15)	0	1	164

Even accounting for only two of the many system impacts that would occur from adding a new fuel-efficient CC unit to the utility system, the modified screening curve's levelized cost value for the CC unit at an 80% capacity factor in Table D.2 has now dropped dramatically from $101 per MWh (or 10.1 cents per kWh) to $25 per MWh (or 2.5 cents per kWh). The adjustment begins to provide a more complete picture of the actual net system cost impact when the CC unit is connected to the utility system. However, this more complete picture is still not a full picture because only two of the impacts the CC unit will have on the system have been accounted for.

Stated another way, the typical screening curve result presented in Table D.1 projected a CC unit cost value that is four times higher than what a projection of the net system costs from the new CC unit would be, as shown in Table D.2, if only two of many system cost impacts had been accounted for.

However, the more complete picture presented in Table D.2 is still not a full picture because only two of the impacts the CC unit will have on the system have been accounted for. Other system cost impacts still unaccounted for include, but are not limited to, system variable O&M costs and impacts on the utility's resource plans in subsequent years compared to what those plans might be if another resource option had been selected.

THE MORAL OF THE STORY

The moral of the story is that, by failing to account for a variety of system cost impacts, typical screening curve analyses are based on very incomplete information and can provide very misleading results. Just how misleading these results can be is demonstrated by the just concluded quantitative example.

As previously stated, a screening curve approach may be useful to screen out resource options (but only if the resource options are identical, or nearly so, in regard to each of four key characteristics). Therefore, this approach may be useful in certain preliminary economic screening analyses, but only in those limited cases in which the resource options being compared are identical, or virtually identical, in those four key characteristics.

However, the quantitative example just discussed clearly points out the fallacy of using a screening curve analysis result and jumping to conclusions about which resource option is best for a particular utility system. Resource option decisions should only be made based on a full economic analysis that accounts for all known system cost impacts.

If an individual or organization attempts to justify a resource option selection solely with the results of a screening curve analysis, the individual or organization attempting to use such an analysis as justification either: (i) does not understand how utility systems work and/or the significant limitations inherent in a screening curve approach or (ii) knows better, but is hoping to convince someone else who is not aware of the severe limitations of screening curve analyses. In either case, such an attempt does not speak highly of that individual or organization.

Appendix E: Mini-Lesson #3
Further Discussion of the RIM and TRC Preliminary Cost-Effectiveness Screening Tests for DSM

The rate impact measure (RIM) and total resource cost (TRC) preliminary cost-effectiveness screening tests for demand side management (DSM) options have probably been the subject of more prolonged discussion over the last several decades than any other single utility resource planning-related issue that focused on "how-to-calculate mechanics" that I am aware of. This has certainly been the case in the state of Florida, and I often ran across cases in other states where the issue of the two preliminary economic screening tests had resurfaced again and again. We discussed the basics of these two preliminary cost-effectiveness screening tests in Chapter 6. In this mini-lesson, we shall discuss the two tests a bit more.

As mentioned during the discussion of resource planning analyses in Part I of this book, neither of these preliminary cost-effectiveness screening tests should be used to make a final resource option decision for a utility. That is because no preliminary DSM screening test is able to fully account for all of the utility system impacts that will result if the utility selects a DSM option. Such tests should only be used to perform *preliminary* economic analyses of DSM options. The proper role of these tests is to allow a comparison of a number of DSM options quickly against a common standard (which is usually the Supply option that the utility would build absent any incremental DSM). In this way, the least economically attractive DSM options can be identified and screened out.

As discussed in some detail in Chapter 6, the two cost-effectiveness screening tests do not differ in regard to their calculations of the benefits (avoided costs) a DSM program is projected to provide. The RIM and TRC tests produce identical projections of system benefits for a given DSM program. However, these two screening tests differ significantly in regard to what DSM-related cost impacts, which will be passed on to customers in the form of electric rates, are accounted for. As we saw in Chapter 6, the RIM test does account for all of these DSM-related cost impacts, but the TRC test does not.

Consequently, the TRC test represents a much easier "test" for DSM options and, in general, fewer DSM options are screened out with the TRC test than with the RIM test. In other words, more DSM options will "pass" the TRC test than will pass the RIM test. For that reason, proponents of DSM options have tended to support the TRC test, even though it clearly is fundamentally flawed in its role as an economic tool because it does not account for all of the DSM-related cost impacts that are passed on to a utility's customers. An example should help crystalize these points.

AN EXAMPLE FROM A 2020 FPL RATE CASE

As the second edition of this book is being written, a recently concluded FPL base rate case offers an excellent example of DSM preliminary cost-effectiveness analysis using both the RIM and TRC tests. This example shows an excellent depiction of the correctness of using the RIM test, and the shortcomings of the TRC test.

One of the (many) issues in this base rate case was whether FPL should alter an existing DSM program because it was (at that time) currently projected to no longer be cost-effective. This was due in part to steadily declining costs for Supply options (thus resulting in decreased DSM program benefits), as well as to incentive payments that had increased in the preceding years (thus resulting in increased DSM program costs). The DSM program in question was FPL's Commercial/Industrial Demand Reduction program. This program offers participating customers monthly incentive payments, in the form of reductions on the monthly bills, in exchange for allowing FPL to reduce the amount of electric service FPL provides at times of very high load and/or during system emergencies.

Among the analyses FPL presented in the base rate case filing were RIM and TRC screening analyses of this DSM program. The results of those two preliminary economic screening analyses are presented in Table E.1 in which both the benefits and costs are presented in terms of thousands of dollars of cumulative present value of revenue requirements (CPVRR, $000).

The total benefits of the DSM program under both tests are shown near the top of the table. As expected, those total benefits[1] are identical under each of the two tests:

TABLE E.1

FPL Example of a DSM Preliminary Cost-Effectiveness Analysis (Commercial/Industrial Demand Reduction Program)

	RIM Test Results	TRC Test Results
Benefits (CPVRR, $000)		
Total Benefits	**44,526**	**44,526**
Costs (CPVRR, $000)		
Utility Administrative Costs	904	904
Participant Equipment Costs	N.A.	0
Utility Incentive Payments to Participants	44,476	N.A.
Unrecovered Revenue Requirements	372	N.A.
Total Costs	**45,752**	**904**
Benefit-to-Cost Ratio:	**0.97**	**49.3**

Note: "N.A." = Not Accounted for in this test

---

[1] Both the benefit and cost values are shown in terms of $000 CPVRR.

$44,526. Likewise, the first category of costs, Utility Administrative Costs, shows identical values under each test of $904. However, from that point on, the costs accounted for under the two tests differ significantly:

- Regarding participant equipment costs, this category of costs is intended to account for any costs that a participating customer incurs to participate in the program. Because these participating customer-only costs that are not passed on to all customers, including non-participating customers, through electric rates, the RIM test does not account for these costs. The TRC test does account for these costs. However, it is a moot point for this particular DSM program because there are no equipment costs for a customer who participates in the program.
- Because the incentive payments to participating customers are recovered from all customers through electric rates, the RIM test correctly accounts for these costs which total $44,476 (which is almost as much as the total benefits of the program: $44,526). The TRC test ignores these incentive payments.
- There will be some reduction in total sales from this program due to FPL using the program to reduce service to participating customers during very high loads and/or emergency conditions. These reduced GWh sales affect the denominator of electric rate calculations. Therefore, the RIM test correctly accounts for these unrecovered revenue requirements of $372. The TRC test ignores these cost impacts.

As a result, the RIM test shows a total cost of $45,752, while the TRC test shows a total cost of only $904. The benefit-to-cost ratio under the RIM test is 0.97 (= $44,526/$45,752) for the RIM test, thus slightly falling short of the program being cost-effective for all customers. Conversely, the benefit-to-cost ratio under the TRC test is 49.3 (= $44,526/$904). As a result, the TRC test projects the DSM program to be wildly cost-effective for all customers.

A simple comparison of two benefit-to-cost ratios that differ by a factor of approximately 49-to-1 is a clear indication that something is amiss with one of the two tests. Then, just looking at the fact that only one of the program costs (incentive payments) is $44,476, and total benefits are $44,526, indicates that it is the TRC test's benefit-to-cost ratio of 49.3 that is illogical.

Furthermore, the objective of the DSM program analysis in the rate case was to determine if FPL needed to lower the incentive payments to participating customers to return the DSM program to a point where the program was again cost-effective for all customers. However, if the TRC test is used, not only would the TRC test results show that a lowering of the incentive is not needed, the TRC test results would also tell one that the incentive could be increased by a factor of 10 (or more), resulting in incentive payments totaling $440,476—versus total benefits of $44,526—and still the program would be viewed as cost-effective for all customers under the TRC test because incentive payments to customers are not accounted for in the TRC test. Such a result makes no sense at all and only serves to show that the TRC test is fundamentally flawed.

ATTEMPTS TO DELEGITIMIZE THE RIM TEST

Over the years, some supporters of the TRC test have attempted to defend the TRC test by attempting to delegitimize the RIM test. In these attempts, these supporters have made a number of ill-informed statements about the RIM test. A few of these incorrect statements have been repeated so often over the years that the incorrect statements have, unfortunately, seemed to be taken for granted by some. As a result, these statements are seldom examined to see if they are accurate.

Therefore, in order to set the record straight, I shall address three of the most often repeated incorrect and misleading statements about the RIM test. I shall characterize those three statements as follows: (i) "the RIM test is the 'most restrictive' cost-effectiveness test"; (ii) "Supply options are not evaluated using the RIM test"; and (iii) "energy conservation or efficiency programs cannot pass the RIM test." Each of these misleading and incorrect comments will be discussed in turn.

Delegitimizing Statement #1: The RIM Test is the "Most Restrictive" Cost-Effectiveness Test

This misleading statement has actually been intended as a disparaging remark regarding the RIM test by individuals/organizations that seem to want the outcome of any utility resource planning analysis to end up with either more, or only, DSM options being selected. However, from a resource planning perspective that seeks to evaluate all types of resource options on a level playing field, this statement is actually a ringing endorsement for the RIM test.

As we discussed in Chapter 6, when comparing the RIM and TRC tests, the benefits calculations are identical for both tests. The two tests differ only in regard to DSM-related costs that are included in each calculation as just demonstrated by the FPL base rate case example. Because the RIM test accounts for all DSM-related costs and cost impacts that are passed on to a utility's customers through electric rates, and the TRC test omits two significant costs (incentive payments to participants and unrecovered revenue requirements), the RIM test will naturally present a tougher hurdle for DSM options to pass. In this sense, it is definitely "more restrictive" than the TRC test.

However, because the real objective in integrated resource planning is to evaluate resource options with a full accounting of all cost impacts associated with the resources options that will be passed on to the utility's customers, the phrase "more restrictive" in regard to the RIM test actually means: "most accountable," "most complete," or "most informative." After all, when comparing different Supply options such as new generating units, a utility ensures that it accounts for all of the costs of new generating units that will be passed on to its customers. It only makes sense to do the same for DSM options as well.

In other words, the meant-to-be-disparaging phrase "most restrictive" when applied to the RIM test is actually an unintended compliment that acknowledges that the RIM test is the only DSM cost-effectiveness test that fully accounts for all of the DSM-related costs and cost impacts that will be passed on to customers through electric rates.

Delegitimizing Statement #2: "Supply Options Are Not Evaluated Using the RIM Test"

This statement is intended to give the impression that somehow utilities are evaluating DSM options with a more restrictive/different economic evaluation approach than is used to evaluate Supply options, thus giving an unfair advantage to Supply options in economic comparisons with DSM options. However, the statement is both incorrect and misleading.

First, the statement is incorrect at face value. When using the RIM test, both the DSM option and the competing Supply option are being evaluated against each other. Therefore, when using the RIM test, one is evaluating a Supply option at the same time one is evaluating a DSM option. (Note that this is also the case when a TRC test is used to compare a Supply option to a DSM option.)

Second, the statement is also misleading. When making this statement, the point that the one making the statement is probably trying to convey is as follows: when only Supply options are evaluated against each other, the RIM test is not used. This better phrased statement is a correct statement, but an irrelevant one that makes the statement misleading. When evaluating only Supply options, neither the RIM nor the TRC test is used.

As previously discussed, Supply options are typically large resource options ranging from a few dozen MW up to perhaps 2,000 MW in size. There are typically only a relatively small number of types of Supply options that are suitable for a given utility to consider at a given time. Due to its size, each Supply option will generally have a noticeable impact on the utility system if it is chosen.

Therefore, Supply options readily lend themselves to analyses of resource plans in which one or more of the competing Supply options are incorporated into resource plans. The resource plans are then analyzed using sophisticated computer models. Stated another way, evaluations of Supply options often forego any preliminary economic screening analyses and proceed directly with the final (or system) economic analyses.

On the other hand, the impact of an individual DSM option per participating customer is much smaller than the impact of a single Supply option. The majority of DSM options, especially for residential and small commercial customers, may have demand reduction values close to 1 kW per installation. In addition, the utility may have hundreds, or even thousands, of DSM measures that are potentially applicable for it to consider. Because of the small nature of individual DSM measures, individual DSM measures do not lend themselves as well to direct analysis of resource plans (such as used in final or system economic analyses) because the small size of individual DSM measures would result in small system impacts that would be very difficult (if not impossible) to accurately judge on a resource plan scale. Nor would it be practical to even attempt to evaluate hundreds of DSM measures individually in resource plan analyses due to the time it takes to set up and perform these analyses.

Consequently, several DSM preliminary cost-effectiveness screening tests have been created so that preliminary economic screening analyses of individual DSM measures could be carried out quickly. In this way, large numbers of DSM measures can be "screened" using these tests. Then the best DSM measures (*i.e.*, those measures that pass the preliminary screening) can be combined into DSM portfolios for the much more meaningful analyses of resource plans with portfolios of these DSM options versus resource plans with either different DSM portfolios or with only Supply options.

But let's also take a closer look at how two Supply options are actually compared. There are two key characteristics of such an evaluation. The first key characteristic is that the evaluation is performed with a full accounting of all costs for the two Supply options including both costs for the individual Supply option itself and utility system cost impacts. The second key characteristic is that an evaluation between two Supply options is simultaneously an evaluation of both system costs and system average electric rates.

In regard to the second point, because the utility system will be serving the same amount of GWh regardless of which Supply option is selected, the typical analytical approach is to evaluate the CPVRR cost, *i.e.*, the total costs, for a resource plan featuring each Supply option. However, this simultaneously represents a comparative evaluation from an electric rate perspective. The system costs for each Supply option-based resource plan represent the numerator, and the identical number of GWh served represents the denominator, in an electric rate calculation. Due to the fact that the denominator does not change when analyzing competing Supply option resource plans, the Supply option resource plan with the lowest system cost will also result in the Supply option resource plan with the lowest system average electric rate. Consequently, an evaluation of only Supply options is simultaneously an evaluation of both costs and system average electric rates.

Now let's return to the RIM and TRC preliminary economic screening tests. By design, the TRC test does not address electric rate impacts, but the RIM test is focused on electric rates. Furthermore, the TRC test omits two significant DSM-related costs: incentive payments and unrecovered revenue requirements.

Therefore, the TRC test is definitely not evaluating a DSM option versus a Supply option in a manner that is consistent with how two competing Supply options are evaluated, *i.e.*, with a full accounting of costs related to the resource option that will be passed on to a utility's customers, and from an electric rate perspective.

However, the RIM test both accounts for all DSM-related costs and takes an electric rate perspective. Consequently, even though the RIM test itself is not used when two competing Supply options are evaluated, the RIM test does evaluate a DSM option versus a Supply option in a manner that is consistent with how two Supply options are evaluated.

DELEGITIMIZING STATEMENT #3: ENERGY CONSERVATION OR EFFICIENCY PROGRAMS "CANNOT PASS THE RIM TEST"

This statement is incorrect and misleading. The gist of the statement is that, although load management type DSM programs that have relatively low kWh-per-kW reduction ratios can pass the RIM test, energy conservation or energy efficiency type DSM programs that have much larger kWh-per-kW reduction ratios cannot pass the RIM test.

This statement is simply incorrect. The state of Florida has primarily used the RIM test (in combination with the Participant Test) for more than 30 years at the time the second edition of this book is being written and the electric utilities in that state have offered a large number of energy conservation programs that passed the RIM preliminary screening test. Using FPL's DSM efforts as an example, the utility has achieved approximately 5,000 MW of Summer MW reduction through 2022 from a combination of load management and energy conservation programs. Of that total, more than half of the total 5,000 MW has been achieved with energy conservation programs.[2]

Therefore, the statement that "use of the RIM test does not support the implementation of large quantities of energy conservation DSM programs" has been proven by actual results to be incorrect.

The unspoken premise of this statement is that many more energy conservation programs pass the TRC test than pass the RIM test. This premise is essentially meaningless for two reasons. The first reason is that large numbers of energy conservation DSM programs have passed the RIM test as evidenced by DSM efforts in Florida. Second, the premise is also meaningless because one *would expect* the use of a "test" such as TRC that does not account for all DSM-related costs to "pass" more DSM programs of any type. Having more DSM programs of all types "pass" an incomplete economic screening test is not only expected, but it is not much of an accomplishment.

A FINAL THOUGHT ON RIM VERSUS TRC

Although I believe that the RIM test is the correct DSM preliminary cost-effectiveness screening test to use, I also believe that the RIM versus TRC debate has diverted attention away from a more important issue.

Recall that both the RIM and TRC tests are designed as *preliminary* economic analysis screening tools. As such, the two tests are intended to be used to screen out the economically inferior DSM options. They are definitely *not* intended to be used to make final resource option decisions. Final resource option decisions should only be made after analyses have examined all of the important system impacts of resource plans that feature specific resource options.

Because RIM and TRC tests are merely preliminary economic analysis screening tools, these screening tests could, in theory, be scrapped. A utility *could* attempt to include all DSM options in various resource plans that would then compete in final (or system) economic analyses. However, that would not be practical due to the sheer volume of potential DSM measures, and the corresponding volume of resource plans, that would have to be evaluated. Therefore, for practical reasons, utilities will likely continue to use these preliminary economic screening tools.

So why mention the theoretical possibility of scrapping the use of the RIM and TRC tests? This was done to help point out that preliminary economic screening should *not* be the primary issue when discussing resource planning involving DSM options (although, unfortunately, it too often is). The most important issue is whether

---

[2] Some opinions regarding the future of utility DSM programs are presented in Chapter 14.

the utility, and the regulatory body that governs the utility, have a thorough set of analysis results of different resource plans (featuring different resource options) that account for both system economic and system non-economic impacts.

A far more productive use of time in regulatory hearings involving DSM options would be to focus on the electric rate, system fuel use, system emissions, etc., impacts of varying levels of DSM versus Supply options, rather than focus on heated debates regarding the two *preliminary* economic screening tools of RIM and TRC.

Another key issue that is often neglected in these discussions is how much DSM is *actually needed* to meet the utility's projected resource needs. As we have previously discussed in Part I of this book, a utility should not be adding an amount of DSM that exceeds its near-term resource needs. To do so will definitely increase the electric rate impact of the DSM portfolio due to the earlier-than-needed introduction of DSM program costs, combined with the greater kWh reduction values from the additional DSM further lowering the number of GWh over which the utility recovers its revenue requirements. In addition to raising near-term electric rates, the additional DSM may result in the entire DSM portfolio being non-cost-effective.

In my opinion, regulatory bodies, interveners, and utilities spend too much time discussing the results of these two *preliminary* economic screening tests and too little time discussing both the amount of new resources that are needed, and the variety of system impacts that occur as a result of the potential resource options being evaluated. Stated another way, it is almost irrelevant whether the RIM or TRC test is used solely to identify a number of DSM options that will then be evaluated in a final (or system) economic analysis.[3] (However, this assumes that one does *not* incorrectly stop the analytical process at the conclusion of this preliminary economic screening step in the IRP process.)

What is important is that the final (or system) economic analyses correctly takes into account the actual resource need(s) of the utility, correctly uses an electric rate perspective in regard to the economics of the various resource plans, and accounts for all of the system impacts, both economic and non-economic of the resource plans. Conversely, a utility or regulatory body that prematurely stops the process at the end of the preliminary economic screening of DSM options, and then decides that the utility should perform all of the DSM that passed either of the RIM or TRC preliminary economic screening tests without accounting for the utility's resource needs or all system impacts, is almost certainly doing that utility's customers a disservice. This is because such a decision would be based on preliminary, and therefore very incomplete, information regarding all of the electric rates and other impacts to the utility system that would be passed on to the utility's customers.

---

[3] Because the TRC test does not include all relevant DSM costs and, therefore, represents a much lower "test" to pass, more DSM options will be identified with TRC than with RIM. In a true final (or system) economic analysis, many/all of these "additional" TRC-based DSM options will likely be eliminated. Therefore, the use of the TRC test in the preliminary economic screening is not as efficient a process as would be the case if the RIM test were used in the preliminary economic screening.

Appendix F: Mini-Lesson #4
How Can a Resource Option Result in Lower Costs but Increase Electric Rates?

When a discussion of utility resource options reaches the point at which DSM options are compared to Supply options, one often encounters the statement that certain DSM options (*i.e.*, DSM options that do not pass the Rate Impact Measure (RIM) preliminary economic screening test) can result in lower utility system costs, but will also result in higher electric rates.

For some, the first reaction to hearing this statement is (and I quote): "*Huh?*" To them, it seems counterintuitive that something that lowers costs could also result in higher electric rates. (Their next thought in regard to an explanation of this statement may be something along the lines that "*Those pesky utility people are up to something!*")

However, the explanation of why this statement is true is a lot less exciting and is based on simple fractions that we all learned in grade school. We shall now walk through an explanation of how this statement can be true using another hypothetical utility system.

Let's assume that this new hypothetical utility system currently has sales of 50,000 GWh (or 50,000,000,000 kWh once we remember that 1 GWh = 1,000,000 kWh). Let's also assume that the utility has total costs (or revenue requirements) of $5 billion dollars for the current year.

Now, recalling that an electric rate is basically a simple fraction in which the utility's costs ($) are divided by the utility's sales (kWh), we have $5 billion in costs divided by 50,000,000,000 kWh of sales or $0.10/kWh. Because electric rates are typically expressed in terms of cents/kWh, we will state that the electric rate for the current year is 10.00 cents/kWh. This calculation is presented in Table F.1.

TABLE F.1

How Can a Resource Option Result in Lower Costs, but Increase Electric Rates? (Current Year)

I. Current Situation (for a Hypothetical Electric Utility):

a. Current total costs (revenue requirements) =	$5,000,000,000
b. Current total sales (GWh) =	50,000
c. Current electric rate (cents per kWh) =	**10.00**

Let's now assume that this particular hypothetical utility is growing both in load and costs at 2% per year. Therefore, because of the growth in load, the utility projects that it will have to add a resource option in the future. We assume this point will occur 5 years in the future. The utility is considering two resource options that will result in acceptable system reliability (*i.e.*, either of the two options will allow the utility to meet its reliability criteria). The first resource option (a Supply option) will have an annual net cost (after accounting for capital and operating costs of the new Supply option as well as system fuel and environmental compliance cost net savings) of $100 million dollars in Current Year + 5.

The second resource option (a DSM option) will actually lower the utility's total costs by $10 million due to reduced system fixed and operating cost savings being greater than the cost to implement the DSM option. However, the second option will also lower the utility's annual sales (GWh) by 3%.

The utility is, naturally, interested in determining the impact that each resource option will have on the electric rates that will be charged to all of its customers. The utility knows that its costs will have increased by 2% for each of 5 years to $5,520,404,016 (= $5,000,000,000 × 1.02 × 1.02 × 1.02 × 1.02 × 1.02) before accounting for the addition of one of the resource options. In addition, this utility's sales will have increased by 2% for each of 5 years to 55,204 GWh (= 50,000 GWh × 1.02 × 1.02 × 1.02 × 1.02 × 1.02) before accounting for the addition of one of the resource options.

Using those values as starting points, the utility first calculates the electric rate if Resource Option #1 (the Supply option) is selected. The result is presented in Table F.2 (which is an expansion of Table F.1).

TABLE F.2

How Can a Resource Option Result in Lower Costs but Increase Electric Rates? (With Future Scenario 1)

I. Current Situation (for a Hypothetical Electric Utility)

a. Current total costs (revenue requirements) =	$5,000,000,000
b. Current total sales (GWh) =	50,000
c. Current electric rate (cents per kWh) =	**10.00**

II. Future Scenario #1

a1. Projected costs w/o resource option =	$5,520,404,016
a2. Projected cost of potential resource option #1 =	$100,000,000
a3. Projected total future costs (revenue requirements) =	$5,620,404,016
b1. Projected sales w/o resource option (GWh) =	55,204
b2. Impact on sales from resource option #1 (GWh) =	0
b3. Projected total future sales (GWh) =	55,204
c. Projected future electric rate (cents per kWh) =	**10.18**

In this scenario, we see in the table's section II, Row (a1), that the utility's total costs for this future year of $5,520,404,016 that we just discussed. Rows (a2) and (a3) show that the utility's total revenue requirements are increased by $100 million if Resource Option #1 is selected.

Row (b1) presents the projected sales in the future year of 55,204 GWh that we also previously discussed. Rows (b2) and (b3) show that the number of GWh sold with which the utility's total revenue requirements are collected is not affected by the selection of Resource Option #1 because this option is a Supply option.

The projected electric rate is shown in Row (c) and that value is derived by dividing the value in Row (a3) by the value in Row (b3) and converting from $/GWh to cents/kWh. The projected electric rate for this year if Resource Option #1 is selected is 10.18 cents/kWh.

Table F.3 now expands the picture to add similar information if the utility were to select Resource Option #2 (the DSM option) that lowers the projected system costs by $10 million and decreases the number of GWh served by 3% or 1,656 GWh (= 55,204 GWh × 0.03).

TABLE F.3
How Can a Resource Option Result in Lower Costs but Increase Electric Rates? (with Future Scenarios 1 and 2)

I. Current Situation (for a Hypothetical Electric Utility)

a. Current total costs (revenue requirements) =	$5,000,000,000
b. Current total sales (GWh) =	50,000
(c. Current electric rate (cents per kWh) =	**10.00**

II. Future Scenario #1

a1. Projected costs w/o resource option =	$5,520,404,016
a2. Projected cost of potential resource option #1 =	$100,000,000
a3. Projected total future costs (revenue requirements) =	$5,620,404,016
b1. Projected sales w/o resource option (GWh) =	55,204
b2. Impact on sales from resource option #1 (GWh) =	0
b3. Projected total future sales (GWh) =	55,204
c. Projected future electric rate (cents per kWh) =	**10.18**

III. Future Scenario #2

d1. Projected costs w/o resource option =	$5,520,404,016
d2. Projected cost of potential resource option #2 =	($10,000,000)
d3. Projected total future costs (revenue requirements) =	$5,510,404,016
e1. Projected sales w/o resource option (GWh) =	55,204
e2. Impact on sales from resource option #2 (GWh) =	(1,656)
e3. Projected total future sales (GWh) =	53,548
f. Projected future electric rates (cents per kWh) =	**10.29**

In this scenario, we see in the table's section III, Rows (d2) and (d3), that the utility's total revenue requirements are indeed decreased by $10 million if Resource Option #2 is selected. We also see in Rows (e2) and (e3) that the number of GWh sold with which the utility's total revenue requirements are collected is decreased by 3% due to Resource Option #2. The end result, shown in Row (f), is that the utility's projected electric rate for this year if Resource Option #2 is selected is 10.29 cents/kWh, *i.e.*, a higher electric rate than if Resource Option #1 is selected.

By comparing the results of the two scenarios, it is clear that although Resource Option #2 results in significantly lower costs of $110 million (= approximately $5,620 million – approximately $5,510 million), it also results in a higher electric rate (10.29 cents/kWh versus 10.18 cents/kWh).

Therefore, assuming all else equal, the utility's customers would prefer that Resource Option #1 were selected (even though the utility's total costs would be lower if Resource Option #2 were selected). After all, a customer's monthly bill is directly driven by the electric rate that applies to the customer. A customer's monthly bill is only partially or indirectly driven by the utility's total costs. This is because the utility's total costs are only one part of the calculation that determines the actual electric rate as shown by this example. The number of kWh with which those costs are recovered is the other part.

Returning to our past work in grade school with simple fractions, we are reminded that each fraction has a numerator (the value that is the top part of the fraction) and a denominator (the value that is the bottom part of the fraction). Using these terms, we can explain in simple terms what is happening in the example just discussed. Using the scenario with Resource Option #1 as a starting point, we have a numerator (utility costs) of $5,620,404,016 and a denominator (GWh of sales) of 55,204 GWh that yields an electric rate of 10.18 cents/kWh.

The scenario with Resource Option #2 not only lowers the numerator to $5,510,404,016, but also lowers the denominator to 53,548 GWh. The reason that this yields a higher electric rate of 10.29 cents/kWh is that the denominator is lowered by a greater percentage than the numerator is lowered. In this example, the differences between Resource Option #1's and Resource Option #2's "fractions" (*i.e.*, electric rates) were that the numerator (cost) value was lowered by approximately 0.18% (= $10,000,000/$5,520,404,016), while the denominator (sales) value was lowered by 3% (= 1,656/55,204), if Resource Option #2 is selected. Consequently, the resulting fraction, or electric rate, (cents/kWh) increased.

Hopefully, the concept of how a resource option can lower costs but increase electric rates is no longer counterintuitive.[1]

---

[1] This should be especially true for those of you who were paying attention in grade school when fractions were taught (even if the only reason you concentrated on fractions was to avoid thinking about what would likely happen to you during the next recess period when the dodge ball game resumed).

Appendix G: Mini-Lesson #5
How Can a Resource Option That Produces Emissions Result in Low Utility Total System Emissions? ("The Taxi Cab Example")

In Chapter 7, as part of our non-economic analyses of our hypothetical utility system, we evaluated each of the five resource plans in terms of projected system air emissions for SO_2, NO_x, and CO_2. The results were provided in Table 7.7 in terms of millions of tons of each emission over the entire analysis period. The results can be summarized as follows:

- **SO_2:** all five resource plans were projected to have identical emissions of 1.011 million tons.
- **NO_x:** Two resource plans, Supply Only Resource Plan 1 (CC) and With DSM Resource Plan 2, had the lowest projected emissions of 0.252 million tons. The results for the remaining three resource plans were, respectively, as follows: Supply Only Resource Plan 3 (PV) with 0.253 million tons, With DSM Resource Plan 1 with 0.254 million tons, and Supply Only Resource Plan 2 (CT) with 0.255 million tons.
- **CO_2:** The With DSM Resource Plan 2 was projected to have the lowest emissions with 1,113 million tons. The remaining four resource plans, respectively, were as follows: Supply Only Resource Plan 3 (PV) with 1,116 million tons, Supply Only Resource Plan 1 (CC) and With DSM Resource Plan 1 (a tie) with 1,119 million tons, and Supply Only Resource Plan 2 (CT) with 1,122 million tons.

These results may have been surprising, and even counterintuitive, to some readers. The results for the Supply Only Resource Plan 1 (CC) may have been particularly surprising because this resource plan adds a Supply option (a CC unit) that burns fossil fuel and it operates 80% of the hours per year. This resource plan tied for lowest SO_2 and NO_x emissions, and its CO_2 emissions were only (approximately) one-fourth of one percent (0.25%) higher than the resource plan projected to have the lowest CO_2 emissions.

The question such a reader may have asked is framed as the main title of this appendix: *"How can a resource option that produces emissions result in low utility total system's emissions?"* After all, commonly held beliefs are that energy conservation efforts and renewable energy *have* to result in lower air emissions.[1] Because the new CC unit was compared not only to two other Supply options, including a PV option, but also to two DSM options, the system emission outcome may be even more surprising or counterintuitive.

A reader may have been tempted to ask one or both of the following questions:

- *"Is this some trick that utility companies utilize in their analyses?"* and/or
- *"Is this something that is peculiar to electric utility companies; i.e., could a similar result occur for another type of business?"*

The answer to the first question is "no" and the answer to the second question is "yes."

In order to demonstrate why these are the correct answers, we will examine the same air emission question using another type of business that is also reliant upon using fossil fuel. This business is one that most people are more familiar with than they are with electric utilities: a traditional taxi cab-type company.

THE TAXI CAB COMPANY ANALOGY

The analogy of using a taxi cab company as a proxy for an electric utility system for the purpose of examining air emissions, while not perfect, is useful. This is because both electric utilities and taxi companies share at least three fundamental operating characteristics. These three shared operating characteristics are as follows:

1. Both power plants and taxis use fossil fuel when operating.
2. Air emissions result from the use of this fuel.
3. All else equal, the greater amount of fuel that is used, the more air emissions which result.

We shall now take a look at a taxi company and how its total air emissions may change as it considers its "resource options." The taxi company analogy will be kept very simple in order to keep the calculations simple. However, this simplicity does not change the underlying principles that will emerge.

We shall start by listing certain assumptions that describe the operation of our (very small) taxi company.[2] These assumptions are as follows:

- The taxi company has one cab which gets 20 miles per gallon of gasoline.
- This cab operates 24 hours each day and drives an average of 20 miles each hour.
- For each gallon of fuel that the cab uses, 1 "unit" (e.g., 1 lb) of air emission is released.

---

[1] At this point, the reader should recall Fundamental Principle #4 of Electric Utility Resource Planning which states that one must always keep in mind the question of: "compared to what?"

[2] You read this correctly; this is now "our" taxi company. When you picked up this book, I bet you never thought that you'd own (at least on paper) an electric utility company and a taxi company. No need to worry; not a word of this will be mentioned to the U.S. Internal Revenue Service.

Therefore, our cab is driven 480 miles per day (480 miles = 24 hours × 20 miles each hour). Our cab also consumes 24 gallons of fuel per day (24 gallons = 480 miles/20 miles per gallon). From this information, we can now calculate the current daily emissions for our taxi company:

24 gallons of fuel consumed per day × 1 unit of air emission per gallon

= 24 emission units per day.

The current situation in regard to daily mileage, fuel usage, and emissions for our taxi company is summarized in Table G.1 which presents our taxi company's current situation.

TABLE G.1
Taxi Company Analogy: Current Situation

	Miles Driven	Gallons of Fuel Consumed	Emission Units
1. Current Situation	480	24	24

However, our taxi company is now forecasting that changes will need to be made because of growing customer demand for our taxi company's services. The fore-casted change is that the demand for service will increase for 5 "peak hours" of each day and remain unchanged for all other hours. For each of these 5 peak hours, an additional 20 miles of driving, or an additional 100 miles per day of driving, is forecast. Therefore, the forecast is that the taxi company's customers will soon be requesting 580 miles (= 480 + 100) of transportation per day.

The taxi company realizes that something needs to be done because its sole cab is already occupied during each hour of the day. The taxi company must somehow address this increased demand.

The conclusion that is reached is that the taxi company has two basic options:

1. The "Conservation" option: Make all of the expected increased demand vanish by persuading the new "peak hour" prospective riders, or an equivalent number of its existing riders during the peak hours, to participate in "conservation" activities that result in these folks not using the taxi company's services. (For example, the taxi company could lend bicycles for these people to use, could give them bus tokens, or could persuade these riders to use another private transportation service (such as Uber).
2. The "Supply" option: Purchase a second gasoline-fueled cab and use it. The new cab the taxi company is considering purchasing gets 30 miles per gallon.[3]

---

[3] The assumption that the new cab is more fuel-efficient than its current cab is consistent with the fact that new generating units are typically more fuel-efficient than existing generating units. Note also that, although new car models typically have lower emission rates than prior models, we will assume for simplicity's sake that the new cab still has the same emission rate of 1 emission unit per gallon of fuel consumed as the current cab.

Our taxi company now needs to make a decision as to which of these two "resource options" it will choose. The company decides to base its decision solely on one criterion: *"which option results in the lowest air emissions from the operation of the taxi company?"*

The taxi company now analyzes its system emissions for each of the two resource options. It starts its analysis with the Conservation option.

Because the Conservation option removes all of the projected new miles of travel, and there are no changes made to operation of the taxi company's sole cab, neither the daily miles driven nor the daily amount of fuel consumed change. Consequently, the resulting daily emission level remains unchanged as shown in Table G.2.

TABLE G.2

Taxi Company Analogy: Current Situation, and Forecasts with Conservation Option

	Miles Driven	Gallons of Fuel Consumed	Emission Units
1. Current Situation	480	24	24
2. Forecast w/ the Conservation Option	480	24	24

Our taxi company feels pretty good about the projected emission level outcome with the Conservation option. It will have satisfied all of its customers' needs for transportation (otherwise no customers would have voluntarily selected the Conservation option) and it will have done so without increasing its daily emission levels. However, in order to satisfy itself that it has evaluated all of its options, the taxi company now evaluates its Supply option.

The taxi company decides that, because the Supply option (a new cab) gets 30 miles per gallon while its existing cab gets only 20 miles per gallon, it will change the way it operates. The new 30 miles per gallon cab will now be operated (or dispatched) as often as possible (24 hours per day) and the existing 20 miles per gallon cab will be operated only during the 5 "peak" hours of the day. In other words, the taxi company will operate the cab with the lower fuel use (and lower fuel cost) as often as possible (as a "baseload" cab) and will operate the cab with the higher fuel cost only when needed to meet peak load (as a "peaking" cab).

The daily emission profile for the taxi company if the Supply option is chosen is now calculated as follows:

- The new cab will be driven 24 hours per day and will cover 20 miles each hour. This results in 480 miles driven per day (= 24 hours × 20 miles/hour). The new cab gets 30 miles per gallon, so the new cab consumes 16 gallons of fuel per day (= 480 miles/30 miles per gallon). Because cabs emit one emission unit for each gallon of fuel consumed, the new cab operation results in 16 emission units per day.
- The existing cab will now be operated 5 hours per day and will also cover 20 miles each hour. This results in 100 miles driven per day (= 5 hours ×

20 miles/hour). The existing cab gets 20 miles per gallon, so the new cab consumes 5 gallons of fuel per day (= 100 miles/20 miles/gallon). Because cabs emit one emission unit for each gallon of fuel consumed, the existing cab operation results in 5 emission units per day.

Consequently, the total daily emission level for the taxi company if the Supply option is chosen is 21 emission units (= 16 + 5).

Our taxi company has now evaluated each of its available options and the results of its analysis are summarized in Table G.3.

TABLE G.3

Taxi Company Analogy: Current Situation, and Forecasts with Conservation and Supply Options

	Miles Driven	Gallons of Fuel Consumed	Emission Units
1. Current Situation	480	24	24
2. Forecast w/ the Conservation Option	480	24	24
3. Forecast w/ the Supply Option	580	21	21

The results show that, with its Supply option, although the taxi company will have doubled the number of its cabs and will drive significantly more miles, its daily fuel consumption and emissions will be lower if it selects the new Supply option (a second cab) than if it had selected its Conservation option.

This result was probably counterintuitive to the taxi company prior to performing the analysis. Nevertheless, the Supply option (adding another, more fuel-efficient cab) emerges as a better choice for reducing air emissions than the Conservation option.

Now let's return to why we examined the taxi company in the first place. The taxi company was used as a proxy for an electric utility company because both types of companies have certain characteristics in common as previously mentioned. In other words, the taxi company example provided an analogy to an electric utility company. But what did we learn from the analogy?

This analogy tells us several things. First, we have learned that it is clearly possible for the addition of a specific "Supply" option that uses fossil fuels to result in lower system air emissions than a specific "Conservation" option for a specific system. Second, because it is possible for a Supply option to be a better choice in regard to system emissions than a Conservation option for one such system (*i.e.*, the taxi company), it is logical to conclude that it may be true for other similar systems (such as an electric utility). And, as we were reminded at the beginning of this Appendix, the addition of a new CC unit with Supply Only Resource Plan 1 (CC) resulted in a tie for lowest SO_2 and NO_x emissions and was within approximately 0.25% of the lowest CO_2 emissions, when comparing the five competing resource plans.

Third, the analogy reminds us that what is important is the view of the entire system, not the view of a resource option itself. If the taxi company had simply looked at the two resource options, noted that the Conservation option by itself has no emissions while the Supply option (another taxi cab) itself has emissions, it may have leapt to the conclusion that the taxi company's system emissions have to be lower if it selects the Conservation option. As we have seen, a look solely at the resource options themselves, without accounting for all of the system impacts, would have led it to the wrong decision.

Fourth (and finally), the analogy tells us that a number of factors must be accounted for in analyzing the system in question, including:

- The current characteristics of the system in question (such as the number of cabs, the fuel efficiency of those cabs, and the emission rates of those cabs in our taxi company example);
- These same characteristics for each resource option being considered (the characteristics for the Conservation and Supply options); and
- How each option will be utilized on the system and how the operation of the system's existing components will be affected (how the existing cab's operation will be affected by each of the two resource options being considered).[4]

With these factors in mind, one can go back to the taxi company example and see how the outcome might have been different if even one of the key assumptions had been changed. One example would be if the new cab had only gotten 25 miles per gallon rather than 30 miles per gallon, use of the new cab for 24 hours a day would result in 19.2 emission units (= 480/25). As a result, the Supply option would have 24.2 emission units per day (= 19.2 + 5). The Conservation option's 24 emission units would now just edge out the Supply option's 24.2 emission units, resulting in the Conservation option emerging as the resource option which resulted in the lowest system air emissions.

Another example would be if the new Supply option is now an electric vehicle (EV) instead of a conventional (gasoline-fueled) vehicle. Let's conservatively assume that the new Supply option EV will be used 200 miles each day (before it is returned to the taxi company location for recharging to get ready for operation the next day). For those 200 miles, there will be no fuel consumed by the EV and, therefore, no emissions.[5] The remaining 380 miles (= 580 − 200) will now be covered by the original 20 miles-per-gallon taxi. This taxi will produce 19 air emission units (= 380 miles/20 miles per gallon × 1 air emission unit per gallon). The total emissions for our taxi cab company are now 19 air emission units (= 0 + 19). This is now the lowest air emission projection our cab company has considered as shown in Table G.4.

---

[4] The reader should immediately be reminded of Fundamental Principles of Electric Utility Planning #1 and #4.

[5] Our taxi company does not attempt to account for any emissions that may have occurred when the EV cab's batteries were charged.

TABLE G.4

Taxi Company Analogy: Current Situation, and Forecasts with Conservation and Two Supply Options

	Miles Driven	Gallons of Fuel Consumed	Emission Units
1. Current Situation	480	24	24
2. Forecast w/ the Conservation Option	480	24	24
3. Forecast w/ the Conventional Supply Option	580	21	21
4. Forecast w/ the EV Supply Option	580	19	19

If one is truly interested in determining which resource option lowers system air emissions the most for a fossil fuel-based system, one *cannot* immediately select the resource option that may seem intuitively correct. All of the resource options must be carefully analyzed to account for all impacts to the specific utility system.

CONCLUSIONS

This discussion started by questioning the seemingly counterintuitive result we saw in Chapter 7 in which the selection of a new generating unit that would operate most of the hours in the year resulted in a tie for lowest system air emissions for two of the three types of air emissions, even when compared to resource options whose individual operations produced no air emissions (*i.e.*, the DSM and PV options). In order to explain how such a result could occur, we used the analogy of a taxi company.

This analogy allowed us to see how it was possible for a system that consumes fossil fuels to add a new resource that also consumes fossil fuel and actually end up with lower (or at least low) system emissions. This simple analogy example, through Table G.3, established the fact that such an outcome, which we first saw in Chapter 7, was definitely possible for a system that consumes fossil fuels.

In turn, this leads us back to the conclusion that one cannot rely upon one's intuition regarding which types of resource options will result in the lowest system air emissions for a specific utility system. As pointed out in Fundamental Principles of Electric Utility Planning #1 and #4, such a leap to an intuitive conclusion is very likely to be proven incorrect once a complete analysis of a specific utility system is performed. We have now seen this in two different ways: the multi-year utility system analysis in Chapter 7, and the taxi company analogy.

There should now be no doubt that one needs to do a complete analysis of a utility system in order to accurately determine how different types of resource options will impact the air emission profile of a specific utility system.

Index

Printed in the United States
by Baker & Taylor Publisher Services